T0140729

# On Linear-Quadratic Optimal Control and Robustness of Differential-Algebraic Systems

**Dissertation**

zur Erlangung des akademischen Grades

**doctor rerum naturalium**
**(Dr. rer. nat.)**

von Dipl.-Math. Matthias Voigt

geb. am 16. April 1986 in Erlabrunn

genehmigt durch die Fakultät für Mathematik der
Otto-von-Guericke-Universität Magdeburg

Gutachter: Prof. Dr. rer. nat. Peter Benner

Prof. Dr. Paul Van Dooren

eingereicht am: 27. Januar 2015

Verteidigung am: 04. Mai 2015

Bibliographic information published by the Deutsche Nationalbibliothek

The Deutsche Nationalbibliothek lists this publication in the Deutsche Nationalbibliografie; detailed bibliographic data are available on the Internet at http://dnb.d-nb.de .



ISBN 978-3-8325-4118-7

Logos Verlag Berlin GmbH
Comeniushof, Gubener Str. 47,
10243 Berlin
Tel.: +49 (0)30 42 85 10 90
Fax: +49 (0)30 42 85 10 92
INTERNET: http://www.logos-verlag.de

Acknowledgement

Many people contributed to the success of this thesis and it is hard not to forget any of them. First of all I thank my advisor and teacher Peter Benner for his patience and constant support. During the last years I could always count on his advice and his vision. Our meetings were always very inspiring and stimulating to me and helped me with my personal and professional development. Most importantly, he always encouraged me to pursue my own ideas. I further thank Paul Van Dooren who instantly agreed to be the second reviewer of this thesis. I highly appreciate the assistance and support of my colleague and friend Timo Reis who introduced me into the fascinating field of linear-quadratic optimal control theory of DAEs and who supported me in my career. Without him, this thesis would not look like it does now. I further thank Volker Mehrmann, Achim Ilchmann, and Michael Overton for the fruitful discussions and for supporting me with all the job and grant applications in the past years. In particular, I am grateful to Michael for his great hospitality during my research stay at New York University. I really enjoyed my time there. Moreover, I am indebted to Mert Gürbüzbalaban for the fruitful discussions on the $\mathcal{H}_\infty$-norm and for proofreading a draft of Chapter 5 as well as Daniel Bankmann for carefully reading parts of Chapter 3.

Many results of this thesis would have not be achieved in this form without the hard work of my former students Maximilian Bremer, Peihong Jiang, and Ryan Lowe. I am also grateful to the DAAD for financially supporting my students and my research stay as well as the Max Planck Institute for Dynamics of Complex Technical Systems and the IMPRS Magdeburg for the general funding of this research.

Special thanks go to my colleagues from the MPI Magdeburg for the very nice atmosphere during work. In particular, I want to thank my good friends Jens Saak, Norman Lang, Heiko Weichelt, Patrick Kürschner, Martin Hess, Tobias Breiten, Sara Grundel, Steffen Klamt, Philipp Erdrich, Alfredo Remón, and Zvonimir Bujanović. I will miss all the movie and pub evenings as well as the regular coffee breaks with them. They were very important to find my balance again, especially in periods of high pressure and stress. Most importantly, I want to express my gratitude to my family for the love and support.

## Zusammenfassung

Diese Arbeit betrachtet das linear-quadratische Optimalsteuerungsproblem für differentiell-algebraische Gleichungen. Im ersten Teil präsentieren wir eine vollständige theoretische Analyse dieses Problems. Die Grundlage bildet eine neue differentiell-algebraische Version des Kalman-Yakubovich-Popov-Lemmas. Wir leiten äquivalente Kriterien für die Zulässigkeit des Optimalsteuerungsproblems und der Lösbarkeit einer Deskriptor-Lur'e-Gleichung her. Des weiteren analysieren wir den Zusammenhang zwischen dieser Matrixgleichung und den Spektraleigenschaften eines zugehörigen geraden Matrixbüschels. Insbesondere zeigen wir, wie man die Lösung mithilfe bestimmter invarianter Unterräume des geraden Matrixbüschels konstruieren kann. Ein weiterer Schwerpunkt ist die Untersuchung der Lösungsstruktur der Deskriptor-Lur'e-Gleichung. Wir formulieren und beweisen Bedingungen für die Existenz extremaler und nichtpositiver Lösungen. Abschließend diskutieren wir Anwendungen der neuen Theorie, beispielsweise bei der Untersuchung dissipativer und zyklodissipativer Systeme oder der Faktorisierung rationaler Funktionen.

Eng verwandt zu dissipativen Systemen sind Systeme mit gegen den Uhrzeigersinn laufender Eingangs-/Ausgangsdynamik. Ähnlich wie bei dissipativen Systemen geben wir eine spektrale Charakterisierung mittels eines geraden Matrixbüschels an. Weiterhin studieren wir das Problem der Erzwingung dieser Struktur für den Fall, daß diese während des Modellierungsprozesses verloren wurde.

Ein weiterer Schwerpunkt dieser Arbeit sind Robustheitsfragen, d. h. wir betrachten den Einfluß von Störungen auf Systemeigenschaften wie Zyklodissipativität und Stabilität. Wir leiten einen Algorithmus zur Berechnung des Abstands eines zyklodissipativen Systems zur Menge der nichtzyklodissipativen Systeme her. Dafür betrachten wir die Störungstheorie eines zugehörigen geraden Matrixbüschels. Eine besondere Schwierigkeit in diesem Zusammenhang stellen mögliche Störungen des Büschels im singulären Teil oder der defektiven unendlichen Eigenwerte dar.

Wir betrachten außerdem das Problem der Berechnung der $\mathcal{H}_\infty$-Norm für große, dünnbesetzte Deskriptorsysteme. Wir stellen zwei Verfahren vor, um dieses Problem zu lösen. Der erste Zugang basiert auf einem Zusammenhang zum komplexen $\mathcal{H}_\infty$-Radius einer Übertragungsfunktion. Dies erlaubt die Anwendung eines Ansatzes mit Pseudopolmengen. Das Verfahren beruht auf der schnellen Berechung des extremalen Punktes einers gegebenen Pseudopolmenge. Dieser Punkt kann effizient über eine Folge von Störungen vom Rang eins approximiert werden. Die zweite Methode geht auf den klassischen Zugang zur Berechung der $\mathcal{H}_\infty$-Norm mittels Niveaumengen zurück. Wir verwenden einen iterativen strukturerhaltenden Algorithmus zur Berechnung der gesuchten Eigenwerte der auftretenden geraden Matrixbüschel und präsentieren Heuristiken zur Bestimmung geeigneter Shifts.

# Contents

# List of Figures

# List of Tables

# List of Algorithms

# List of Acronyms

| | |
|---|---|
| ARE | algebraic Riccati equation |
| BLAS | Basic Linear Algebra Subprograms |
| ccw | counterclockwise |
| DAE | differential-algebraic equation |
| EKCF | even Kronecker canonical form |
| GUPTRI | generalized upper triangular |
| ODE | ordinary differential equation |
| I/O | input/output |
| KCF | Kronecker canonical form |
| KYP | Kalman-Yakubovich-Popov |
| LAPACK | Linear Algebra Package |
| LMI | linear matrix inequality |
| MIMO | multi-input multi-output |
| QKF | quasi-Kronecker form |
| QWF | quasi-Weierstraß form |
| SAMDP | subspace accelerated MIMO dominant pole algorithm |
| SISO | single-input single-output |
| SLICOT | Subroutine Library in Control Theory |

# List of Symbols

## Basic Sets

| | |
|---|---|
| $\mathbb{N}$, $\mathbb{N}_0$ | set of natural numbers, $\mathbb{N}_0 = \mathbb{N} \cup \{0\}$, respectively |
| $\mathbb{R}$, $\mathbb{C}$ | the fields of real and complex numbers, respectively |
| $\mathbb{C}^+$, $\mathbb{C}^-$ | the open sets of complex numbers with positive and negative real part, respectively |
| $\overline{S}$ | closure of the set $S$ |
| $\mathbb{R}[s]$, $\mathbb{C}[s]$ | the rings of polynomials with coefficients in $\mathbb{R}$ and $\mathbb{C}$, respectively |
| $\mathbb{R}(s)$, $\mathbb{C}(s)$ | the fields of rational functions with coefficients in $\mathbb{R}$ and $\mathbb{C}$, respectively |
| $\mathcal{R}^{m\times n}$ | the set of $m \times n$ matrices with entries in a ring $\mathcal{R}$ |
| $\mathrm{Gl}_n(\mathbb{R})$, $\mathrm{Gl}_n(\mathbb{C})$ | the groups of invertible $n \times n$ matrices with entries in $\mathbb{R}$ and $\mathbb{C}$, respectively |

## Matrices and Matrix Pencils

| | |
|---|---|
| $I_n$ | the $n \times n$ identity matrix |
| $0_{m\times n}$ | the $m\times n$ zero matrix (subscripts can be omitted, if clear from context) |
| $A^{\mathsf{T}}$ | transpose of the matrix $A$ |
| $A^{\mathsf{H}}$ | conjugate transpose of a matrix $A$, $A^{\mathsf{H}} = \overline{A}^{\mathsf{T}}$ |
| $A^{-1}$ | inverse of a matrix $A \in \mathrm{Gl}_n(\mathbb{C})$ |
| $A^+$ | Moore-Penrose pseudoinverse of a matrix $A \in \mathbb{C}^{m\times n}$ |

| Symbol | | Meaning |
|---|---|---|
| $\operatorname{im} A$, $\ker A$ | | image and kernel of a matrix $A \in \mathbb{C}^{m \times n}$, respectively |
| $\operatorname{rank} A$ | | rank of a matrix $A \in \mathbb{C}^{m \times n}$, respectively |
| $\sigma_{\max}(A)$ | | maximum singular value of the matrix $A \in \mathbb{C}^{m \times n}$ |
| $\operatorname{tr}(A)$ | $:=$ | $\sum_{i=1}^{n} a_{ii}$ (trace of the matrix $A \in \mathbb{C}^{n \times n}$) |
| $\Lambda(A)$ | | spectrum of $A \in \mathbb{C}^{n \times n}$ |
| $\Lambda(E, A)$ | | set of finite eigenvalues of the matrix pencil $sE - A \in \mathbb{C}[s]^{m \times n}$ |
| $\operatorname{In}(A)$ | | inertia of a Hermitian matrix $A \in \mathbb{C}^{n \times n}$, i. e., $\operatorname{In}(A) = (\pi_+, \pi_0, \pi_-) \in \mathbb{N}_0^3$ consisting of the numbers of positive, zero, and negative eigenvalues of $A$, respectively |
| $A > (\geq, >, \leq)\, B$ | | for two Hermitian matrices $A, B \in \mathbb{C}^{n \times n}$, the matrix $A - B$ is positive definite (positive semidefinite, negative definite, negative semidefinite) |
| $A \neq_{\mathcal{V}} B$ | | for two Hermitian matrices $A, B \in \mathbb{C}^{n \times n}$ and a subspace $\mathcal{V} \in \mathbb{C}^n$, there exists an $v \in \mathcal{V}$ such that $v^{\mathsf{H}} A v \neq v^{\mathsf{H}} B v$ |
| $A =_{\mathcal{V}} (>_{\mathcal{V}}, \geq_{\mathcal{V}}, >_{\mathcal{V}}, \leq_{\mathcal{V}})\, B$ | | for two Hermitian matrices $A, B \in \mathbb{C}^{n \times n}$ and a subspace $\mathcal{V} \in \mathbb{C}^n$, we have $v^{\mathsf{H}}(A - B)v = (>, \geq, >, \leq)\, 0$ for all $v \in \mathcal{V} \setminus \{0\}$ |
| $\operatorname{diag}(A_1, \ldots, A_k)$ | | blockdiagonal matrix with $A_i \in \mathbb{C}^{m_i \times n_i}$ with $m_i, n_i \in \mathbb{N}_0$ for $i = 1, \ldots, k$ (i. e., $A \in \mathbb{C}^{m \times n}$ with $m = m_1 + \ldots + m_k$, $n = n_1 + \ldots + n_k$) |
| $A \otimes B$ | | Kronecker product of $A \in \mathbb{C}^{m \times n}$ with $B \in \mathbb{C}^{p \times q}$, i. e., $\begin{bmatrix} a_{11}B & \ldots & a_{1n}B \\ \vdots & & \vdots \\ a_{m1}B & \ldots & a_{mn}B \end{bmatrix}$ |
| $\operatorname{vec}(A)$ | | vectorization of $A = \begin{bmatrix} a_1 & \ldots & a_n \end{bmatrix} \in \mathbb{C}^{m \times n}$, i. e., $\begin{pmatrix} a_1^{\mathsf{H}} & \ldots & a_n^{\mathsf{H}} \end{pmatrix}^{\mathsf{H}}$ |
| $\lVert A \rVert_2$ | $:=$ | $\max_{\lVert x \rVert_2 = 1} \lVert Ax \rVert_2 = \sqrt{\max_{\lambda \in \Lambda(A^{\mathsf{H}} A)} \lambda}$ for $A \in \mathbb{C}^{m \times n}$ (spectral norm) |

| | | |
|---|---|---|
| $\|A\|_{\mathrm{F}}$ | $:=$ | $\sqrt{\sum_{i=1}^{m}\sum_{j=1}^{n}\|a_{ij}\|^2}$ for $A \in \mathbb{C}^{m\times n}$ (Frobenius norm) |

## Dynamical Systems and Behaviors

For more detailed explanations see Subsections 2.2.1 and 2.2.4.

| | |
|---|---|
| $(E,A,B)$, $(E,A,B,C,D)$, $(E,A)$, $(E,A,B,C)$ | short notation for a dynamical system of the form $E\dot{x}(t) = Ax(t) + Bu(t)$ without and with output equation $y(t) = Cx(t) + Du(t)$, respectively; $(E,A) := (E,A,0_{n\times 0})$, $(E,A,B,C) := (E,A,B,C,0)$ |
| $\Sigma_n, \Sigma_{n,m}, \Sigma_{n,m,p}$ | sets of dynamical systems $(E,A)$, $(E,A,B)$, and $(E,A,B,C,D)$ with regular $sE - A \in \mathbb{R}[s]^{n\times n}$, $B \in \mathbb{R}^{n\times m}$, $C \in \mathbb{R}^{p\times n}$, and $D \in \mathbb{R}^{p\times m}$, respectively |
| $\mathfrak{B}_{(E,A)}$, $\mathfrak{B}_{(E,A,B)}$, $\mathfrak{B}_{(E,A,B,C,D)}$ | behavior of $(E,A) \in \Sigma_n$, $(E,A,B) \in \Sigma_{n,m}$, and $(E,A,B,C,D) \in \Sigma_{n,m,p}$, respectively |
| $\mathfrak{B}_{(E,A)}(x_0)$, $\mathfrak{B}_{(E,A,B)}(x_0)$, $\mathfrak{B}_{(E,A,B,C,D)}(x_0)$ | behaviors $\mathfrak{B}_{(E,A)}$, $\mathfrak{B}_{(E,A,B)}$, and $\mathfrak{B}_{(E,A,B,C,D)}$ with initial differential variable $Ex(0) = Ex_0$, respectively |
| $\mathcal{ZD}_{(E,A,B,C,D)}$ | zero dynamics of $(E,A,B,C,D) \in \Sigma_{n,m,p}$ |
| $\mathcal{ZD}_{(E,A,B,C,D)}(x_0)$ | zero dynamics of $(E,A,B,C,D) \in \Sigma_{n,m,p}$ with consistent initial differential variable $Ex(0) = Ex_0$ |

## Rational Functions

As a convention we consistently use the variable $s$ as the indeterminate of a rational function $G(s)$, whereas, e. g., $G(\lambda)$ is the value of $G(s)$ at $\lambda$. For details on the spaces and norms see Definition 2.2.20.

| | | |
|---|---|---|
| $G^{\sim}(s)$ | $:=$ | $G^{\mathsf{H}}(-\overline{s})$ (conjugate of $G(s) \in \mathbb{R}(s)^{p\times m}$) |
| $\mathfrak{Z}(G)$, $\mathfrak{P}(G)$ | | sets of zeros and poles of $G(s) \in \mathbb{R}(s)^{p\times m}$ |
| $\operatorname{im}_{\mathbb{R}(s)} G(s)$, $\ker_{\mathbb{R}(s)} G(s)$ | | image and kernel of $G(s) \in \mathbb{R}(s)^{p\times m}$ over the field $\mathbb{R}(s)$ |

| | | |
|---|---|---|
| $\operatorname{rank}_{\mathbb{R}(s)} G(s)$, $\operatorname{rank}_{\mathbb{C}(s)} G(s)$ | | rank of $G(s) \in \mathbb{R}(s)^{p\times m}$ over the field $\mathbb{R}(s)$ or $\mathbb{C}(s)$, respectively |
| $\mathrm{Gl}_n(\mathbb{R}(s))$, $\mathrm{Gl}_n(\mathbb{C}(s))$ | | the groups of invertible functions in $\mathbb{R}(s)^{n\times n}$ and $\mathbb{C}(s)^{n\times n}$, respectively |
| $\mathcal{RH}_2^{p\times m}$ | | space of real-rational $p \times m$ matrix-valued functions that are analytic in the closed right half-plane and square-integrable on the imaginary axis |
| $\mathcal{RHL}_2^{p\times m}$ | | space of real-rational $p \times m$ matrix-valued functions whose strictly proper part is in $\mathcal{RH}_2^{p\times m}$ |
| $\mathcal{RH}_\infty^{p\times m}$ | | space of real-rational $p \times m$ matrix-valued functions that are analytic and bounded in the closed right half-plane |
| $\mathcal{RL}_\infty^{p\times m}$ | | space of real-rational $p \times m$ matrix-valued functions that are bounded on the imaginary axis |
| $\\|G\\|_{\mathcal{H}_2}$ | $:=$ | $\left(\frac{1}{2\pi}\int_{-\infty}^{\infty} \\|G(\mathrm{i}\omega)\\|_{\mathrm{F}}^2 \,\mathrm{d}\omega\right)^{1/2}$ for $G \in \mathcal{RH}_2^{p\times m}$ |
| $\\|G\\|_{\mathcal{HL}_2}$ | $:=$ | $\left(\\|G_{\mathrm{sp}}\\|_{\mathcal{H}_2} + \frac{1}{2\pi}\int_0^{2\pi} \\|G_{\mathrm{poly}}(e^{\mathrm{i}\omega})\\|_{\mathrm{F}}^2 \,\mathrm{d}\omega\right)^{1/2}$ for $G = G_{\mathrm{sp}} + G_{\mathrm{poly}} \in \mathcal{RHL}_2^{p\times m}$ with strictly proper part $G_{\mathrm{sp}}(s)$ and polynomial part $G_{\mathrm{poly}}(s)$ |
| $\\|G\\|_{\mathcal{H}_\infty}$ | $:=$ | $\sup_{\lambda\in\mathbb{C}^+} \\|G(\lambda)\\|_2$ for $G \in \mathcal{RH}_\infty^{p\times m}$ |
| $\\|G\\|_{\mathcal{L}_\infty}$ | $:=$ | $\sup_{\omega\in\mathbb{R}} \\|G(\mathrm{i}\omega)\\|_2$ for $G \in \mathcal{RL}_\infty^{p\times m}$ |

## Functions and Functions Spaces

| | | |
|---|---|---|
| $\mathcal{C}(\mathcal{I},\mathbb{R}^n)$ | $:=$ | $\{f : \mathcal{I} \to \mathbb{R}^n : f$ is continuous$\}$ with $\mathcal{I} \subseteq \mathbb{R}$ |
| $\mathcal{C}^k(\mathcal{I},\mathbb{R}^n)$ | $:=$ | $\{f : \mathcal{I} \to \mathbb{R}^n : f$ is $k$-times continuously differentiable$\}$ with $\mathcal{I} \subseteq \mathbb{R}$ |
| $\mathcal{C}_{\mathrm{pw}}(\mathcal{I},\mathbb{R}^n)$ | $:=$ | $\{f : \mathcal{I} \to \mathbb{R}^n : f$ is piecewise continuous$\}$ with $\mathcal{I} \subseteq \mathbb{R}$ |
| $\mathcal{C}_{\mathrm{pw}}^k(\mathcal{I},\mathbb{R}^n)$ | $:=$ | $\{f : \mathcal{I} \to \mathbb{R}^n : f$ is $k$-times piecewise continuously differentiable$\}$ with $\mathcal{I} \subseteq \mathbb{R}$ |

| | | |
|---|---|---|
| $\dot{f}$ | | distributional derivative of $f : \mathcal{I} \to \mathbb{R}^n$ with $\mathcal{I} \subseteq \mathbb{R}$ |
| $\operatorname{ess\,sup}_{t\in\mathcal{I}} f(t)$ | | essential supremum of $f : \mathcal{I} \to \mathbb{R}$ on the set $\mathcal{I} \subseteq \mathbb{R}$ |
| $w_1 \underset{t}{\Diamond} w_2$ | | the $t$-concatenation of $w_1, w_2 : \mathbb{R} \to \mathbb{R}^n$. That is, $(w_1 \underset{t}{\Diamond} w_2) : \mathbb{R} \to \mathbb{R}^n$ with $(w_1 \underset{t}{\Diamond} w_2)(\tau) = w_1(\tau)$ for $\tau < t$, and $(w_1 \underset{t}{\Diamond} w_2)(\tau) = w_2(\tau - t)$ for $\tau \geq t$ |
| $\mathcal{L}^2(\mathcal{I}, \mathbb{R}^n)$ | | the set of measurable and square integrable functions $f : \mathcal{I} \to \mathbb{R}^n$ on the set $\mathcal{I} \subseteq \mathbb{R}$ |
| $\mathcal{L}^2_{\mathrm{loc}}(\mathcal{I}, \mathbb{R}^n)$ | | the set of measurable and locally square integrable functions $f : \mathcal{I} \to \mathbb{R}^n$ on the set $\mathcal{I} \subseteq \mathbb{R}$ |
| $\lVert f \rVert_{\mathcal{L}^2(\mathcal{I},\mathbb{R}^n)}$ | $:=$ | $\left(\int_{\mathcal{I}} \lVert f(\tau) \rVert_2^2 \,\mathrm{d}\tau\right)^{1/2}$ for $f \in \mathcal{L}^2(\mathcal{I}, \mathbb{R}^n)$ |

If the restriction $f|_{\widetilde{\mathcal{I}}}$ of a function $f : \mathcal{I} \to \mathbb{R}^n$ with $\widetilde{\mathcal{I}} \subset \mathcal{I}$ belongs to a space of functions defined on $\widetilde{\mathcal{I}}$, we slightly abuse the notation and say that $f$ belongs to this function space (instead of its restriction to $\widetilde{\mathcal{I}}$). For instance, if we have $f|_{\widetilde{\mathcal{I}}} \in \mathcal{L}^2\big(\widetilde{\mathcal{I}}, \mathbb{R}^n\big)$ we simply write $f \in \mathcal{L}^2\big(\widetilde{\mathcal{I}}, \mathbb{R}^n\big)$.

# 1 Introduction

## 1.1 Motivation

The modeling of dynamical processes plays an increasingly important role in science and engineering. Often, the resulting mathematical models are given by ordinary differential equations. However, in many applications, the differential equations underlie additional hidden algebraic constraints that restrict the dynamics of the system. Equations of this kind are called *differential-algebraic equations (DAEs)* . Typically, these arise in the modeling of network structures such as electrical circuits [Ria08, Rei10] or gas networks [GJH+13], where the algebraic constraints are induced by the network topology. However, there are also applications in the treatment of semidiscretized partial differential equations such as the Navier-Stokes equation [Wei96, BBSW13], or holonomically constrained mechanical systems [RS88, ESF98, MS05]. The analysis and numerical treatment of DAEs as well as their optimization and control has obtained a lot of attention in the past, see e. g., [GM86, KM06, LMT13, BCM12].

In this thesis we consider we consider DAEs in the control theoretic framework which are also often called differential-algebraic systems, descriptor systems, or singular systems. In general, these are given by systems of the nonlinear equations

$$F(t, x(t), \dot{x}(t), u(t)) = 0, \tag{1.1a}$$

$$y(t) - H(t, x(t), u(t)) = 0, \tag{1.1b}$$

where $F : \mathcal{I} \times \mathcal{D}_x \times \mathcal{D}_{\dot{x}} \times \mathcal{D}_u \to \mathbb{R}^n$ and $H : \mathcal{I} \times \mathcal{D}_x \times \mathcal{D}_u \to \mathbb{R}^p$ with an open interval $\mathcal{I} \subseteq \mathbb{R}$, and open sets $\mathcal{D}_x, \mathcal{D}_{\dot{x}} \subseteq \mathbb{R}^n$, and $\mathcal{D}_u \subseteq \mathbb{R}^m$. Here, the solution trajectory $x : \mathcal{I} \to \mathcal{D}_x$ of (1.1a) is called the state of the system. Furthermore, $u : \mathcal{I} \to \mathcal{D}_u$ is an input control signal that can be designed under various aspects. In practice, the state of the system is not completely known, but only a part of it is available from measurements. Therefore, the output equation (1.1b) is introduced, where the output signal $y : \mathcal{I} \to \mathcal{D}_y \subseteq \mathbb{R}^p$ contains part of the state information.

The special feature of DAEs is that the local Jacobian of $F$ with respect to $\dot{x}(t)$ may be *singular*. This means that $F$ can in general *not* be equivalently rewritten as

$$\dot{x}(t) = f(t, x(t), u(t)),$$

but additional algebraic equations may restrict the solution $x(\cdot)$ to a submanifold. In general, this manifold is difficult to determine, in particular when the system dimension is large.

To analyze the behavior of nonlinear DAEs, one typically linearizes along a solution trajectory [Cam95]. This generally leads to a linear time-varying DAE of the form

$$\begin{aligned} E(t)\dot{x}(t) &= A(t)x(t) + B(t)u(t), \\ y(t) &= C(t)x(t) + D(t)u(t), \end{aligned}$$

with matrix-valued functions $E$, $A : \mathcal{I} \to \mathbb{R}^{n\times n}$, $B : \mathcal{I} \to \mathbb{R}^{n\times m}$, $C : \mathcal{I} \to \mathbb{R}^{p\times n}$, and $D : \mathcal{I} \to \mathbb{R}^{p\times m}$. If the problem is autonomous in the sense that $F$ and $H$ do not explicitly depend on $t$, and the linearization is around a stationary solution, then one usually obtains a linear time-invariant descriptor system

$$E\dot{x}(t) = Ax(t) + Bu(t), \tag{1.2a}$$

$$y(t) = Cx(t) + Du(t), \tag{1.2b}$$

with $E$, $A \in \mathbb{R}^{n\times n}$, $B \in \mathbb{R}^{n\times m}$, $C \in \mathbb{R}^{p\times n}$, and $D \in \mathbb{R}^{p\times m}$. Moreover, many processes can be directly modeled as linear time-invariant systems of the form (1.2).

To illustrate this, we consider the holonomically constrained mass-spring-damper system taken from [MS05] and illustrated in Figure 1.1. In this system, the $i$-th mass $m_i$ is connected to the $(i+1)$-st mass by a spring with stiffness $k_i$ and a damper with viscosity $d_i$, and also to the ground by a spring with stiffness $\kappa_i$ and a damper with viscosity $\delta_i$, respectively. Additionally, the first mass is connected to the last one by a rigid bar and it can be externally excited by the control $u(\cdot)$. The vibration of this system can be described by a second order descriptor system

$$\begin{aligned} M\ddot{p}(t) &= D\dot{p}(t) + Kp(t) - G^{\mathsf{T}}\lambda(t) + B_2u(t), \\ 0 &= Gp(t), \end{aligned} \tag{1.3}$$

where $p : \mathcal{I} \to \mathbb{R}^g$ is the position vector and $\lambda : \mathcal{I} \to \mathbb{R}$ is the Lagrange multiplier.

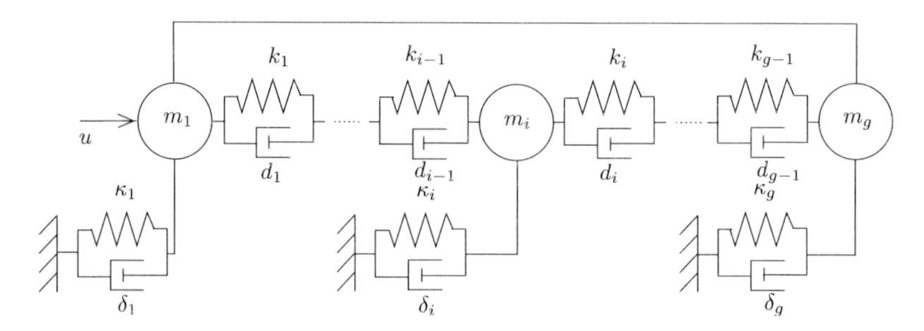


Figure 1.1: Constrained mass-spring-damper system with $g$ masses

Moreover, $M = \operatorname{diag}(m_1, \ldots, m_g) \in \mathbb{R}^{g \times g}$ is the mass matrix,

$$D = \begin{bmatrix} \delta_1 + d_1 & -d_1 & & & \\ -d_1 & d_1 + \delta_2 + d_2 & -d_2 & & \\ & \ddots & \ddots & \ddots & \\ & & -d_{g-2} & d_{g-2} + \delta_{g-1} + d_{g-1} & -d_{g-1} \\ & & & -d_{g-1} & d_{g-1} + \delta_g \end{bmatrix} \in \mathbb{R}^{g \times g}$$

is the damping matrix,

$$K = \begin{bmatrix} \kappa_1 + k_1 & -k_1 & & & \\ -k_1 & k_1 + \kappa_2 + k_2 & -k_2 & & \\ & \ddots & \ddots & \ddots & \\ & & -k_{g-2} & k_{g-2} + \kappa_{g-1} + k_{g-1} & -k_{g-1} \\ & & & -k_{g-1} & k_{g-1} + \kappa_g \end{bmatrix} \in \mathbb{R}^{g \times g}$$

is the stiffness matrix, and $G = \begin{bmatrix} 1 & 0 & \ldots & 0 & -1 \end{bmatrix} \in \mathbb{R}^{1 \times g}$ is the constraint matrix. Since we excite the first mass only, we have $B_2 = \begin{bmatrix} 1 & 0 & \ldots & 0 \end{bmatrix}^\mathsf{T} \in \mathbb{R}^{g \times 1}$. Now, we introduce the velocity vector $v : \mathcal{I} \to \mathbb{R}^g$ which is given by $v(t) = \dot{p}(t)$. By measuring the position of the first mass, we finally obtain a linearized system of the form (1.2), namely

$$\begin{aligned} \begin{bmatrix} I_g & 0 & 0 \\ 0 & M & 0 \\ 0 & 0 & 0 \end{bmatrix} \begin{pmatrix} \dot{p}(t) \\ \dot{v}(t) \\ \dot{\lambda}(t) \end{pmatrix} &= \begin{bmatrix} 0 & I_g & 0 \\ K & D & -G^\mathsf{T} \\ G & 0 & 0 \end{bmatrix} \begin{pmatrix} p(t) \\ v(t) \\ \lambda(t) \end{pmatrix} + \begin{bmatrix} 0 \\ B_2 \\ 0 \end{bmatrix} u(t), \\ y(t) &= \begin{bmatrix} C_1 & 0 & 0 \end{bmatrix} \begin{pmatrix} p(t) \\ v(t) \\ \lambda(t) \end{pmatrix}, \end{aligned} \tag{1.4}$$

with $C_1 = \begin{bmatrix} 1 & 0 & \ldots & 0 \end{bmatrix} \in \mathbb{R}^{1 \times g}$. Later, we will often use this system for illustrational purposes.

In this thesis we consider questions that are related to the infinite time-horizon linear-quadratic optimal control problem of minimizing

$$\int_{t_0}^{t_1} \begin{pmatrix} x(\tau) \\ u(\tau) \end{pmatrix}^\mathsf{T} \begin{bmatrix} Q & S \\ S^\mathsf{T} & R \end{bmatrix} \begin{pmatrix} x(\tau) \\ u(\tau) \end{pmatrix} \mathrm{d}\tau \tag{1.5}$$

with $Q = Q^\mathsf{T} \in \mathbb{R}^{n \times n}$, $S \in \mathbb{R}^{n \times m}$, and $R = R^\mathsf{T} \in \mathbb{R}^{m \times m}$ subject to the DAE (1.2a) with some initial and terminal conditions. Alternatively, if an output equation (1.2b) is given, then we consider the optimal control problem of minimizing

$$\int_{t_0}^{t_1} \begin{pmatrix} y(\tau) \\ u(\tau) \end{pmatrix}^\mathsf{T} \begin{bmatrix} \widetilde{Q} & \widetilde{S} \\ \widetilde{S}^\mathsf{T} & \widetilde{R} \end{bmatrix} \begin{pmatrix} y(\tau) \\ u(\tau) \end{pmatrix} \mathrm{d}\tau \tag{1.6}$$

with $\widetilde{Q} = \widetilde{Q}^\mathsf{T} \in \mathbb{R}^{p \times p}$, $\widetilde{S} \in \mathbb{R}^{p \times m}$, and $\widetilde{R} = \widetilde{R}^\mathsf{T} \in \mathbb{R}^{m \times m}$ subject to the system (1.2) with some initial and terminal conditions.

Our goal is to determine an optimal control signal $\widehat{u}(\cdot)$ that for the infinite time-horizon minimizes one of the two integrals above and satisfies the system dynamics and the boundary values. In this way we steer the system from a given state to a desired state with least possible costs. For systems with $E = I_n$, the reader might immediately think of the algebraic Riccati equation [LR95]. However, this equation has various shortcomings. For instance, it is not an appropriate tool for so-called *singular* optimal control problems. Then one has to turn to Lur'e equations as discussed in [Rei11]. Moreover, there exist generalizations of the algebraic Riccati equation for DAEs but they pose certain unnatural conditions on the system. In this work we close these gaps and present a complete linear-quadratic control theory for DAEs.

In this work we also consider another interpretation of the integrals (1.5) and (1.6). Namely, they might also be understood as the energy that is supplied to the system in the time interval $[t_0, t_1]$. Loosely speaking, if one of the integrals above is non-negative for all time intervals and all possible solution trajectories of the system, then the system can only dissipate as much energy to its environment as energy that has been supplied to the system before. For the physical process this means that a part of the energy is transformed into, e. g., heat, electromagnetic radiation, or an increase in entropy. This leads to the notions of dissipative and cyclo-dissipative systems and naturally arises in manifold applications. The theory of dissipative and cyclo-dissipative systems is closely related to the linear-quadratic optimal control problem and will be studied in detail in this thesis.

Moreover, mechanical systems or electrical circuits where controls and measurements are collocated, i. e., taken at the same position, then the system might have a so-called counterclockwise input/output dynamics [PL10]. This property has some relations to passive systems (a special case of dissipative systems) and will therefore also be shortly studied in this work.

Another point of interest is concerned with the robustness of dynamical systems under the influence of perturbations. For instance, if a physical process is cyclo-dissipative, it is important to reflect this property in the structure of the mathematical model. Otherwise, simulations could produce physically meaningless results. However, due to modeling errors, introduced, e. g., by model order reduction, linearizations, or uncertainties in the parameters of the system, it could easily happen that the cyclo-dissipativity in the model structure is lost. In other words, a physically cyclo-dissipative process is modeled by a non-cyclo-dissipative mathematical model. Then, it is necessary to restore this structure by a post-processing procedure, typically known as dissipativity enforcement [BS13] (or passivity enforcement in certain special cases, e. g., [GT04, GTU06]). On the other hand, even if the model is cyclo-dissipative, it might be close to a non-cyclo-dissipative model. Then it is desirable to assess the robustness of cyclo-dissipativity with respect to perturbations of the model. Therefore, we derive an algorithm that computes the cyclo-dissipativity radius, i. e., the distance of a cyclo-dissipative model to the set of non-cyclo-dissipative models.

Besides cyclo-dissipativity, an important property of descriptor systems is stability. When designing feedback laws one is in general not only interested in minimizing the cost functionals (1.5) or (1.6), but also to make the system as stable as possible. More precisely, the goal is to make stability of the closed-loop system as robust as possible with respect to perturbations in the controller. In order to assess this robustness, unstructured and structured stability radii were introduced [HP86a, HP86b, HP90]. It turns out that the structured stability radius is furthermore reciprocal to the $\mathcal{H}_\infty$-norm of a certain transfer function. The goal of robust control is now to determine a controller that maximizes the structured stability radius or minimizes the corresponding $\mathcal{H}_\infty$-norm. In order to do this optimization it is necessary to be able to compute this norm. However, up to now, there are only algorithms that are capable of dealing with small and dense problems due to computational complexity and memory requirements. Therefore, in this thesis we develop two methods to overcome this issue, so that we can also deal with large and sparse problems.

## 1.2 Structure of this Thesis

This thesis is structured as follows. In Chapter 2 we will introduce the basic theory this work is based on. In particular, this includes an overview over the theory of matrix pencils and an introduction of certain condensed forms. A particular focus will be on even matrix pencils, their *structured* normal forms and structure-preserving methods to compute eigenvalues and deflating subspaces. Furthermore, we give a short introduction into system theoretic aspects of DAEs. We review different controllability, stabilizability, observability, and detectability notions and discuss properties of the zero dynamics. Moreover, we summarize the most important concepts from frequency domain analysis and the theory of rational functions.

In Chapter 3 we study the linear-quadratic optimal control problem for DAEs. We relate the feasibility of this problem to the solvability of a certain linear matrix inequality, the so-called descriptor KYP inequality, and obtain a new differential-algebraic version of the well-known Kalman-Yakubovich-Popov lemma. From this we derive a generalized version of the Lure' matrix equation for DAEs (the descriptor Lur'e equation) and analyze its solution structure in detail. In particular, we analyze conditions for the existence of stabilizing, anti-stabilizing, and extremal solutions. Moreover, we show relations to even matrix pencils and derive conditions for solvability of the descriptor Lur'e equation in terms of the spectral structure of even matrix pencils. We also show how to construct a solution by means of the associated deflating subspaces. This result can be further employed to solve the spectral factorization problem for the associated Popov function. A special emphasis of our theory is the characterization of the existence of a nonpositive solution of the descriptor KYP inequality with applications to the characterization of dissipativity of dynamical systems. A further result is the application of our theory to construct inner-outer factorizations and normalized coprime factorizations of general transfer functions. These results have been obtained in a joint work with Timo Reis and will be published in a sequence of forthcoming papers. The results of the first part of this chapter have appeared as a preprint in

> T. Reis, O. Rendel, and M. Voigt. The Kalman-Yakubovich-Popov inequality for differential-algebraic systems, Hamburger Beiträge zur angewandten Mathematik 2014-27, Universität Hamburg, Fachbereich Mathematik, 2014.

In Chapter 4 we consider systems with counterclockwise input/output dynamics. In frequency domain this leads to the concept of negative imaginary transfer functions. Similarly as in the case of dissipative systems, we show that there is a characterization of this property in terms of an even matrix pencil. Moreover, we discuss the problem of enforcing negative imaginariness of a system which is necessary if this property is theoretically satisfied but has been destroyed during the modeling process. The results of this chapter are published in

> P. Benner and M. Voigt. Spectral characterization and enforcement of negative imaginariness for descriptor systems, *Linear Algebra Appl.*, 439(4):1104–1129, 2013.

Moreover, in the enforcement process we need to compute the eigenvectors corresponding to the purely imaginary eigenvalues of a skew-Hamiltonian/Hamiltonian matrix pencil. We introduce a new, structure-exploiting way to compute these. This procedure is available in the technical report

> P. Jiang and M. Voigt. MB04BV – A FORTRAN 77 subroutine to compute the eigenvectors associated to the purely imaginary eigenvalues of skew-Hamiltonian/Hamiltonian matrix pencils, SLICOT Working Note 2013-03, NICONET e.V., 2013. Available from `http://slicot.org/objects/software/reports/SLWN2013_3.pdf`.

In Chapter 5 we turn to the problem of computing the complex cyclo-dissipativity radius. For this purpose we make use of the characterization of cyclo-dissipativity in terms of the spectrum of an even matrix pencil. To compute the cyclo-dissipativity radius we consider the perturbation theory of this pencil. A particular issue in this context are possible perturbations of the singular part or the defective infinite eigenvalues which have to be treated separately. In the end the problem reduces to the solution of an eigenvalue optimization problem. These results are subject of a forthcoming article.

The problem of computing the $\mathcal{H}_\infty$-norm for large-scale descriptor systems is discussed in Chapter 6. We propose two methods to achieve this. The first method makes use of a relation of the $\mathcal{H}_\infty$-norm to the complex $\mathcal{H}_\infty$-radius. To compute it, we consider so-called $\varepsilon$-pseudopole sets for a transfer function. Namely, we have to find the value of $\varepsilon$ for which the $\varepsilon$-pseudopole set touches the imaginary axis. The method is based on a two-level iteration. In the inner iteration we construct a sequence of rank-1 perturbations that drives one of the poles of the original transfer function to the rightmost pseudopole. In the outer iteration we vary $\varepsilon$ by utilizing Newton's method. The results of this chapter are already published in

> P. Benner and M. Voigt. A structured pseudospectral method for $\mathcal{H}_\infty$-norm computation of large-scale descriptor systems, *Math. Control Signals Systems*, 26(2):303–338 (2014).

The second method goes back to the original approach for computing the $\mathcal{H}_\infty$-norm. We discuss a way to modify it in a way that it is also suitable for large-scale computations. In particular, we replace the dense eigensolvers by sparse methods and discuss a heuristic to generate suitable shifts for it. This method is described and compared with the first approach in

> R. Lowe and M. Voigt. $\mathcal{L}_\infty$-Norm computation for large-scale descriptor systems using structured iterative eigensolvers, Preprint MPIMD/13-20, Max Planck Institute Magdeburg, 2013. Available from `http://www2.mpi-magdeburg.mpg.de/preprints/2013/20/`.

Finally, in Chapter 7 we summarize the results of this thesis and discuss possible future research directions.

## 1.3 System Setup

In this thesis we perform a number of numerical experiments. These tests have been performed on a 2.6.32-23-generic-pae Ubuntu machine with Intel® Core™2 Duo CPU with 3.00GHz and 2GB RAM. The algorithms have been implemented and tested in MATLAB® or FORTRAN. FORTRAN software has been compiled using the `gfortran` compiler with option `-O2`. For testing purposes, mex files have been written for calling FORTRAN codes from MATLAB. The following software libraries and programs have been used:

- MATLAB version 7.14.0.739 (R2012a);
- BLAS and LAPACK version 3.4.1;
- SLICOT version 4.5.

# 2 Mathematical Preliminaries

In this chapter we introduce the main concepts this thesis is based on. First, we discuss general matrix pencils, their properties and condensed forms. Then we turn to the analysis of matrix pencils with particular structure, namely even and skew-Hamiltonian/Hamiltonian pencils. Furthermore, we introduce differential-algebraic control systems and give details about different concepts for controllability, stabilizability, observability, and detectability. Moreover, we discuss transfer functions of such systems which arise when turning to the frequency domain. Finally, we give details on the zero dynamics and its relations to the concept of outer systems.

## 2.1 Matrix Theoretic Preliminaries

### 2.1.1 General Matrix Pencils

Here we introduce some fundamentals and condensed forms for matrix pencils, i. e., first order matrix polynomials $sE - A \in \mathbb{C}[s]^{m \times n}$ with $E, A \in \mathbb{C}^{m \times n}$. To analyze their properties, we introduce the Kronecker canonical form which is a canonical representative of the class of pencils equivalent to $sE - A$ as follows.

**Definition 2.1.1** (Equivalence of pencils). [KM06, Chap. 2] Two pencils $sE_1 - A_1 \in \mathbb{C}[s]^{m \times n}$ and $sE_2 - A_2 \in \mathbb{C}[s]^{m \times n}$ are said to be *equivalent* if there exist $U_{\mathrm{l}} \in \mathrm{Gl}_m(\mathbb{C})$ and $U_{\mathrm{r}} \in \mathrm{Gl}_n(\mathbb{C})$ such that

$$sE_2 - A_2 = U_{\mathrm{l}}(sE_1 - A_1)U_{\mathrm{r}}.$$

**Theorem 2.1.2** (Kronecker canonical form (KCF)). *[KM06, Thm. 2.3] Let $sE - A \in \mathbb{C}[s]^{m \times n}$ be given. Then there exist $U_{\mathrm{l}} \in \mathrm{Gl}_m(\mathbb{C})$ and $U_{\mathrm{r}} \in \mathrm{Gl}_n(\mathbb{C})$ such that $U_{\mathrm{l}}(sE - A)U_{\mathrm{r}} = \mathrm{diag}\,(\mathcal{C}_1(s), \ldots, \mathcal{C}_\ell(s))$, where each $\mathcal{C}_j(s)$, $j = 1, \ldots, \ell$, is of one of the following structures:*

*Type K1:* $$\begin{bmatrix} s - \lambda_j & -1 & & \\ & \ddots & \ddots & \\ & & \ddots & -1 \\ & & & s - \lambda_j \end{bmatrix} \in \mathbb{C}[s]^{k_j \times k_j} \text{ with } \lambda_j \in \mathbb{C};$$

*Type K2:* $\begin{bmatrix} -1 & s & & \\ & \ddots & \ddots & \\ & & \ddots & s \\ & & & -1 \end{bmatrix} \in \mathbb{C}[s]^{k_j \times k_j}$;

*Type K3:* $\begin{bmatrix} -1 & s & & \\ & \ddots & \ddots & \\ & & -1 & s \end{bmatrix} \in \mathbb{C}[s]^{k_j - 1 \times k_j}$;

*Type K4:* $\begin{bmatrix} -1 & & \\ s & \ddots & \\ & \ddots & -1 \\ & & s \end{bmatrix} \in \mathbb{C}[s]^{k_j \times k_j - 1}$.

*The Kronecker canonical form is unique up to permutation of the blocks, i. e., kind, size, and number of the blocks are invariants of the pencil $sE - A$.*

Using the KCF we can define the following terms.

**Definition 2.1.3** (Regularity, rank, eigenvalues, multiplicity, index)**.** Consider a pencil $sE - A \in \mathbb{C}[s]^{m \times n}$.

a) The pencil $sE - A$ is called *regular* if $m = n$ and the characteristic polynomial $\det(sE - A)$ is not the zero polynomial, otherwise it is called *singular*.

b) The *rank* of $sE - A$ (over the field of rational functions) is given by

$$\operatorname{rank}_{\mathbb{C}(s)}(sE - A) = \max_{\lambda \in \mathbb{C}} \operatorname{rank}(\lambda E - A).$$

c) The numbers $\lambda_j$ appearing in the blocks of type K1 in the KCF are called *(generalized) finite eigenvalues* of the pencil $sE - A$. Blocks of type K2 are said to be *corresponding to infinite eigenvalues* .

d) Let $sE - A$ have an eigenvalue $\lambda \in \mathbb{C}$ or an infinite eigenvalue represented by blocks $\mathcal{C}_1(s), \ldots, \mathcal{C}_p(s)$ of type K1 or K2, respectively with $\mathcal{C}_j(s) \in \mathbb{C}[s]^{k_j \times k_j}$ for $j = 1, \ldots, p$. Then we call

   i) the number $\sum_{j=1}^{p} k_j$ the *algebraic multiplicity*;

   ii) the number $p$ the *geometric multiplicity*;

   iii) the numbers $k_1, \ldots, k_p$ the *partial multiplicities*

   of the respective eigenvalue. The eigenvalue is called *defective*, if $p \neq \sum_{j=1}^{p} k_j$, i. e., if there exists a $k_j > 1$.

e) The *(Kronecker) index* of $sE - A$ is the size of the largest block of type K2 or K4 in the KCF [BR13, Def. 3.2]. If no such block exists, then the index is zero.

*Remark* 2.1.4.

a) The finite eigenvalues of $sE - A$ are exactly the values $\lambda \in \mathbb{C}$ satisfying $\operatorname{rank}(\lambda E - A) < \operatorname{rank}_{\mathbb{C}(s)}(sE - A)$.

b) The pencil $sE - A$ is regular if and only if there do not occur any blocks of type K3 or K4 in its KCF.

c) If $sE - A$ is regular, then the finite eigenvalues are the roots of the characteristic polynomial $\det(sE - A)$.

d) If there are no blocks of types K3 and K4, then the KCF is also often called *Weierstraß canonical form*.

In the following we will introduce a generalization of the concept of *principal vectors* and *Jordan chains* to matrix pencils, see [Gan59, BT12].

**Definition 2.1.5** (Kronecker chain)**.** Assume that $sE - A \in \mathbb{C}[s]^{m \times n}$ is given.

a) A tuple $(y_1, \dots, y_k) \in (\mathbb{C}^n \setminus \{0\})^k$ such that

$$\lambda E y_1 = A y_1, \quad E(\lambda y_2 + y_1) = A y_2, \ \dots, \ E(\lambda y_k + y_{k-1}) = A y_k$$

is called a *Kronecker chain* associated to the eigenvalue $\lambda \in \mathbb{C}$.

b) A tuple $(y_1, \dots, y_k) \in (\mathbb{C}^n \setminus \{0\})^k$ such that

$$E y_1 = 0, \quad E y_2 = A y_1, \ \dots, \ E y_k = A y_{k-1}$$

is called a *Kronecker chain* associated to an infinite eigenvalue.

c) A tuple $(y_1, \dots, y_k) \in (\mathbb{C}^n \setminus \{0\})^k$ such that

$$E y_1 = 0, \quad E y_2 = A y_1, \ \dots, \ E y_k = A y_{k-1}, \quad A y_k = 0$$

is called a *Kronecker chain* associated to the singular part.

*Remark* 2.1.6.

a) In the above definition, the vector $y_1$ in a) and b) is called an *eigenvector* to the eigenvalue $\lambda \in \mathbb{C}$ or an infinite eigenvalue, respectively.

b) Definition 2.1.5 only introduces the concept of *right Kronecker chains*. Analogously, one can introduce *left Kronecker chains* by looking for tuples of vectors that fulfill analogous properties when being multiplied from the left.

c) Kronecker chains associated to the singular part are called *singular chains* in [BT12]. For ease of terminology, we also call these Kronecker chains. Note that such Kronecker chains might be void, for instance for the singular pencil

$$s\begin{bmatrix}0\\0\end{bmatrix}-\begin{bmatrix}1\\0\end{bmatrix}\in\mathbb{R}[s]^{2\times 1}.$$

Note, that even if $sE-A\in\mathbb{R}[s]^{m\times n}$, the transformation matrices $U_\mathrm{l}$ and $U_\mathrm{r}$ that lead to KCF will in general be complex. However, there exists an alternative where also the transformation matrices are real.

**Theorem 2.1.7** (Quasi-Kronecker form (QKF)). *[BIT12, BT12, BT13] Let $sE-A\in\mathbb{R}[s]^{m\times n}$ be given. Then there exist $U_\mathrm{l}\in\mathrm{Gl}_m(\mathbb{R})$ and $U_\mathrm{r}\in\mathrm{Gl}_n(\mathbb{R})$ such that $U_\mathrm{l}(sE-A)U_\mathrm{r}=\operatorname{diag}(\mathcal{K}_1(s),\mathcal{K}_2(s),\mathcal{K}_3(s),\mathcal{K}_4(s))\in\mathbb{R}[s]^{m\times n}$ where the KCFs of $\mathcal{K}_1(s),\ldots,\mathcal{K}_4(s)$ have only blocks of types K1, …, K4, respectively.*

Similarly as above, if the subpencils $\mathcal{K}_3(s)$ and $\mathcal{K}_4(s)$ are empty, the QKF is also called *quasi-Weierstraß form*.

Even if the KCF displays the complete spectral information of the pencil, it cannot generally be computed numerically, since the transformation matrices $U_\mathrm{l}$ and $U_\mathrm{r}$ might be arbitrarily ill-conditioned. On the other hand, one can compute less condensed normal forms, that still reveal the Kronecker structure of a general matrix pencil, for instance the GUPTRI form [DK87, DK93a, DK93b]. Its computation is based on sequences of orthogonal or unitary matrix multiplications, singular value decompositions and rank decisions. This method might also suffer from numerical problems, for instance rank decisions are troublesome if there is no clear gap in the small singular values close to the desired truncation tolerance. However, the transformation matrices can be chosen to be orthogonal or unitary and therefore, one can at least compute the Kronecker structure of a nearby "least generic" pencil within a desired tolerance [EEK97, EEK99]. If $sE-A$ is square and one is not interested in the full Kronecker structure, but only in the eigenvalues, the generalized Schur decomposition is an attractive alternative. This decomposition is summarized in the next theorem, where we directly state the result for pencils with real coefficients.

**Theorem 2.1.8** (Generalized real Schur decomposition). *[GL96, Thm. 7.7.2] Let $sE-A\in\mathbb{R}[s]^{n\times n}$. Then there exist orthogonal matrices $Q\in\mathbb{R}^{n\times n}$ and $Z\in\mathbb{R}^{n\times n}$ such that*

$$Q^\mathsf{T}(sE-A)Z=\begin{bmatrix}sE_{11}-A_{11} & \ldots & sE_{1k}-A_{1k}\\ & \ddots & \vdots\\ & & sE_{kk}-A_{kk}\end{bmatrix},$$

*where $sE_{ii}-A_{ii}\in\mathbb{R}[s]^{l_i\times l_i}$ with $l_i\in\{1,2\}$ for $i=1,\ldots,k$. Moreover, if $l_i=1$, then $sE_{ii}-A_{ii}$ corresponds to a real eigenvalue, an infinite eigenvalue or a singular block, if $l_i=2$, then it corresponds to a pair of complex conjugate eigenvalues.*

To compute the generalized Schur decompostion, one can use the well-known QZ algorithm [GL96, Subsect. 7.7]. However, it should be noted, that in the case of a singular pencil, the eigenvalues cannot be reliably computed, since arbitrarily small round-off errors might change the Kronecker structure and can introduce and move eigenvalues to arbitrary positions in the complex plane. In this case it is necessary to extract the regular part of the pencil before, e. g., by transformation to GUPTRI form.

The following concept generalizes the notion of invariant subspaces to matrix pencils.

**Definition 2.1.9** (Basis matrix, deflating subspace)**.**

a) A matrix $Y$ is called a *basis matrix* for a subspace $\mathcal{Y} \subseteq \mathbb{C}^n$ if it has full column rank and $\operatorname{im} Y = \mathcal{Y}$.

b) A subspace $\mathcal{Y} \subseteq \mathbb{C}^n$ is called *(right) deflating subspace* for the pencil $sE - A \in \mathbb{C}[s]^{m \times n}$ if, for a basis matrix $Y \in \mathbb{C}^{n \times k}$ of $\mathcal{Y}$, there exists some $l \in \mathbb{N}_0$, a matrix $Z \in \mathbb{C}^{m \times l}$ and a pencil $s\widetilde{E} - \widetilde{A} \in \mathbb{C}[s]^{l \times k}$ with $\operatorname{rank}_{\mathbb{C}(s)} \big(s\widetilde{E} - \widetilde{A}\big) = l$, such that

$$(sE - A)Y = Z\big(s\widetilde{E} - \widetilde{A}\big).$$

*Remark* 2.1.10.

a) Similarly as for invariant subspaces and Jordan chains in the matrix case, the deflating subspaces can be spanned by the vectors of Kronecker chains.

b) In the literature, deflating subspaces corresponding to singular blocks of $sE - A$ are also often called *reducing subspaces*, see, e. g., [Doo83, DK87]. For ease of terminology, we will also call them deflating subspaces.

An important property a deflating subspace might possess is $E$-neutrality. This property is of particular importance in the context of even pencils, discussed in the next subsection, but it can also be defined in more general terms.

**Definition 2.1.11** ($E$-neutrality)**.** [Rei11] Let a matrix $E \in \mathbb{C}^{n \times n}$ be given. Then a subspace $\mathcal{Y} \subseteq \mathbb{C}^n$ is called

a) *$E$-neutral* if $y_1^{\mathsf{H}} E y_2 = 0$ for all $y_1, y_2 \in \mathcal{Y}$;

b) *maximally $E$-neutral* if $\mathcal{Y}$ is $E$-neutral and every proper superspace $\mathcal{Y}_L \supset \mathcal{Y}$ is not $E$-neutral.

### 2.1.2 Even Matrix Pencils and their Condensed Forms

A particularly important role in this thesis is played by even matrix pencils. These are pencils whose coefficients have special structure, and therefore, they allow for structured condensed forms that we briefly introduce in this subsection.

**Definition 2.1.12** (Even matrix pencil). A matrix pencil $P(s) = s\mathcal{E} - \mathcal{A} \in \mathbb{C}[s]^{n \times n}$ is called *even* if $P^{\sim}(s) = P(s)$, i. e., $\mathcal{E} = -\mathcal{E}^{\mathsf{H}}$ and $\mathcal{A} = \mathcal{A}^{\mathsf{H}}$.

In general, a transformation to KCF does not preserve the evenness of an even pencil. However, when restricting the class of allowable transformation matrices, it is possible to find an even Kronecker-like form given in the next theorem. Note that the following result was initially formulated for *Hermitian matrix pencils* $s\mathcal{E} - \mathcal{A}$ (i. e., with $\mathcal{E} = \mathcal{E}^{\mathsf{H}}$ and $\mathcal{A} = \mathcal{A}^{\mathsf{H}}$). In [Rei11] it has been reformulated for even pencils by replacing $\mathcal{E}$ by $\mathrm{i}\mathcal{E}$ and applying some permutations.

**Theorem 2.1.13** (Even Kronecker canonical form (EKCF)). *[Tho76, Rei11] Let $s\mathcal{E} - \mathcal{A} \in \mathbb{C}[s]^{n \times n}$ be an even pencil. Then there exists a $U \in \mathrm{Gl}_n(\mathbb{C})$ such that $U^{\mathsf{H}}(s\mathcal{E} - \mathcal{A})U = \mathrm{diag}\,(\mathcal{D}_1(s), \ldots, \mathcal{D}_\ell(s))$, where each $\mathcal{D}_j(s)$, $j = 1, \ldots, \ell$, is of one of the following structures:*

$$\text{Type E1:} \quad \left[\begin{array}{cccc|cccc} & & & & -s+\mu_j & -1 & & \\ & & & & & \ddots & \ddots & \\ & & & & & & \ddots & -1 \\ & & & & & & & -s+\mu_j \\ \hline s+\overline{\mu}_j & & & & & & & \\ -1 & \ddots & & & & & & \\ & \ddots & \ddots & & & & & \\ & & -1 & s+\overline{\mu}_j & & & & \end{array}\right] \in \mathbb{C}[s]^{2k_j \times 2k_j}$$

*with $\mu_j \in \mathbb{C}^+$;*

$$\text{Type E2:} \quad \varepsilon_j \begin{bmatrix} & & & 1 & -s\mathrm{i}-\mu_j \\ & & \iddots & \iddots & \\ & \iddots & \iddots & & \\ 1 & \iddots & & & \\ -s\mathrm{i}-\mu_j & & & & \end{bmatrix} \in \mathbb{C}[s]^{k_j \times k_j} \text{ with } \mu_j \in \mathbb{R},\ \varepsilon_j \in \{-1, 1\};$$

$$\text{Type E3:} \quad \varepsilon_j \begin{bmatrix} & & s\mathrm{i} & 1 \\ & \iddots & \iddots & \\ s\mathrm{i} & \iddots & & \\ 1 & & & \end{bmatrix} \in \mathbb{C}[s]^{k_j \times k_j} \text{ with } \varepsilon_j \in \{-1, 1\};$$

*Type E4:*
$$\left[\begin{array}{ccc|ccc} & & & 1 & -s & & \\ & & & & \ddots & \ddots & \\ & & & & & 1 & -s \\ \hline 1 & & & & & & \\ s & \ddots & & & & & \\ & \ddots & 1 & & & & \\ & & s & & & & \end{array}\right] \in \mathbb{C}[s]^{(2k_j-1)\times(2k_j-1)}.$$

*The numbers $\varepsilon_j$ in the blocks of type E2 and E3 are called the* sign-characteristics. *The even Kronecker canonical form is unique up to permutation of the blocks, i. e., kind, size, sign-characteristic, and number of the blocks are invariants of the pencil $s\mathcal{E} - \mathcal{A}$.*

*Remark* 2.1.14.

a) The structure of a block of type E1 shows that for even matrix pencils, finite eigenvalues $\lambda \notin \mathrm{i}\mathbb{R}$ occur in pairs $(\lambda, -\overline{\lambda})$.

b) Blocks of type E2 and E3 correspond to the purely imaginary and infinite eigenvalues, respectively. The additional sign parameter appears basically due to the fact that for a fixed $\lambda_0 \in \mathrm{i}\mathbb{R}$ the congruence transformation with $U$ preserves the inertia of the Hermitian matrix $\lambda_0\mathcal{E} - \mathcal{A}$.

c) Blocks of type E4 consist of a combination of blocks that are equivalent to those of type K3 and K4.

d) Even if $s\mathcal{E} - \mathcal{A} \in \mathbb{R}[s]^{n\times n}$, the transformation matrix $U$ will generally be complex. However, there exists an even Kronecker canonical form that also preserves the realness of the coefficient matrices, see [Tho91].

Now we classify the inertia of the matrices $\mathcal{D}_j(\mathrm{i}\omega)$ in dependence of the corresponding parameters and $\omega \in \mathbb{R}$. This result will be of importance in the proofs of Theorems 3.4.2 and 4.3.3.

**Lemma 2.1.15** (Inertia of blocks in the EKCF). *[CG89, Cle00]*

*a) If $\mathcal{D}_j(s)$ is of type E1, then for all $\omega \in \mathbb{R}$ it holds that*

$$\mathrm{In}\left(\mathcal{D}_j(\mathrm{i}\omega)\right) = \left(k_j, 0, k_j\right).$$

*b) If $\mathcal{D}_j(s)$ is of type E2 and $k_j$ is even, then it holds that*

$$\mathrm{In}\left(\mathcal{D}_j(\mathrm{i}\omega)\right) = \begin{cases} \left(k_j/2, 0, k_j/2\right), & \text{if } \mu_j \neq \omega, \\ \left(k_j/2 - 1, 1, k_j/2 - 1\right) + \mathrm{In}\left(\varepsilon_j\right), & \text{if } \mu_j = \omega. \end{cases}$$

*c) If $\mathcal{D}_j(s)$ is of type E2 and $k_j$ is odd, then it holds that*

$$\operatorname{In}(\mathcal{D}_j(\mathrm{i}\omega)) = \begin{cases} ((k_j-1)/2, 0, (k_j-1)/2) + \operatorname{In}(\varepsilon_j(\omega-\mu_j)), & \text{if } \mu_j \neq \omega, \\ ((k_j-1)/2, 1, (k_j-1)/2), & \text{if } \mu_j = \omega. \end{cases}$$

*d) If $\mathcal{D}_j(s)$ is of type E3 and $k_j$ is even, then for all $\omega \in \mathbb{R}$ it holds that*

$$\operatorname{In}(\mathcal{D}_j(\mathrm{i}\omega)) = (k_j/2, 0, k_j/2).$$

*e) If $\mathcal{D}_j(s)$ is of type E3 and $k_j$ is odd, then for all $\omega \in \mathbb{R}$ it holds that*

$$\operatorname{In}(\mathcal{D}_j(\mathrm{i}\omega)) = ((k_j-1)/2, 0, (k_j-1)/2) + \operatorname{In}(\varepsilon_j).$$

*f) If $\mathcal{D}_j(s)$ is of type E4, then for all $\omega \in \mathbb{R}$ it holds that*

$$\operatorname{In}(\mathcal{D}_j(\mathrm{i}\omega)) = (k_j-1, 1, k_j-1).$$

Even if the EKCF reveals the full information of the structure of the pencil, it cannot be computed numerically, because arbitrary small perturbations may change the structural information and the transformation matrices may be unbounded, i. e., arbitrarily ill-conditioned.

For even pencils $s\mathcal{E} - \mathcal{A} \in \mathbb{R}[s]^{n \times n}$, there exists a computationally attractive alternative, namely the even staircase form under orthogonal transformations. It allows to check regularity and to determine the index within the usual limitations of rank computations in finite precision arithmetic as described above for the general case.

**Theorem 2.1.16.** *[BMX07] For every even pencil $s\mathcal{E} - \mathcal{A} \in \mathbb{R}[s]^{n \times n}$, there exists a real orthogonal matrix $U \in \mathbb{R}^{n \times n}$, such that*

$$U^{\mathsf{T}}\mathcal{E}U =$$

$$\begin{array}{c} s_1 \\ \vdots \\ \vdots \\ s_w \\ \hline l \\ \hline q_w \\ \vdots \\ \vdots \\ q_1 \end{array}
\left[\begin{array}{cccc|c|cccc}
\mathcal{E}_{1,1} & \cdots & \cdots & \mathcal{E}_{1,w} & \mathcal{E}_{1,w+1} & \mathcal{E}_{1,w+2} & \cdots & \mathcal{E}_{1,2w} & 0 \\
\vdots & \ddots & & \vdots & \vdots & \vdots & \iddots & \iddots & \\
\vdots & & \ddots & \vdots & \vdots & \mathcal{E}_{w-1,w+2} & \iddots & & \\
-\mathcal{E}_{1,w}^{\mathsf{T}} & \cdots & \cdots & \mathcal{E}_{w,w} & \mathcal{E}_{w,w+1} & 0 & & & \\
\hline
-\mathcal{E}_{1,w+1}^{\mathsf{T}} & \cdots & \cdots & -\mathcal{E}_{w,w+1}^{\mathsf{T}} & \mathcal{E}_{w+1,w+1} & & & & \\
\hline
-\mathcal{E}_{1,w+2}^{\mathsf{T}} & \cdots & -\mathcal{E}_{w-1,w+2}^{\mathsf{T}} & 0 & & & & & \\
\vdots & \iddots & \iddots & & & & & & \\
-\mathcal{E}_{1,2w}^{\mathsf{T}} & \iddots & & & & & & & \\
0 & & & & & & & &
\end{array}\right],$$

$$
U^{\mathsf{T}}\mathcal{A}U =
\begin{array}{c}
s_1 \\ \vdots \\ \vdots \\ s_w \\ \hline l \\ \hline q_w \\ \vdots \\ \vdots \\ q_1
\end{array}
\left[\begin{array}{cccc|c|cccc}
\mathcal{A}_{1,1} & \cdots & \cdots & \mathcal{A}_{1,w} & \mathcal{A}_{1,w+1} & \mathcal{A}_{1,w+2} & \cdots & \cdots & \mathcal{A}_{1,2w+1} \\
\vdots & \ddots & & \vdots & \vdots & \vdots & & \iddots & \\
\vdots & & \ddots & \vdots & \vdots & \vdots & \iddots & & \\
\mathcal{A}_{1,w}^{\mathsf{T}} & \cdots & \cdots & \mathcal{A}_{w,w} & \mathcal{A}_{w,w+1} & \mathcal{A}_{w,w+2} & & & \\ \hline
\mathcal{A}_{1,w+1}^{\mathsf{T}} & \cdots & \cdots & \mathcal{A}_{w,w+1}^{\mathsf{T}} & \mathcal{A}_{w+1,w+1} & & & & \\ \hline
\mathcal{A}_{1,w+2}^{\mathsf{T}} & \cdots & \cdots & \mathcal{A}_{w,w+2}^{\mathsf{T}} & & & & & \\
\vdots & & \iddots & & & & & & \\
\vdots & \iddots & & & & & & & \\
\mathcal{A}_{1,2w+1}^{\mathsf{T}} & & & & & & & &
\end{array}\right],
\tag{2.1}
$$

*where* $q_1 \geq s_1 \geq q_2 \geq s_2 \geq \ldots \geq q_w \geq s_w$, $l = d_{w+1} + r_{w+1}$, *and for* $i = 1, \ldots, w$ *we have* $\mathcal{E}_{i,i} = -\mathcal{E}_{i,i}^{\mathsf{T}}$, $\mathcal{A}_{i,i} = \mathcal{A}_{i,i}^{\mathsf{T}}$. *Furthermore,*

$$
\begin{aligned}
\mathcal{E}_{j,2w+1-j} &\in \mathbb{R}^{s_j \times q_{j+1}}, \quad 1 \leq j \leq w-1, \\
\mathcal{E}_{w+1,w+1} &= \begin{bmatrix} \Delta & 0 \\ 0 & 0 \end{bmatrix}, \quad \Delta = -\Delta^{\mathsf{T}} \in \mathbb{R}^{d_{w+1} \times d_{w+1}}, \\
\mathcal{A}_{j,2w+2-j} &= \begin{bmatrix} \Gamma_j & 0 \end{bmatrix} \in \mathbb{R}^{s_j \times q_j}, \quad \Gamma_j \in \mathbb{R}^{s_j \times s_j}, \quad 1 \leq j \leq w, \\
\mathcal{A}_{w+1,w+1} &= \begin{bmatrix} \Sigma_{11} & \Sigma_{12} \\ \Sigma_{21} & \Sigma_{22} \end{bmatrix}, \quad \Sigma_{11} \in \mathbb{R}^{d_{w+1} \times d_{w+1}}, \quad \Sigma_{22} \in \mathbb{R}^{r_{w+1} \times r_{w+1}}, \\
\mathcal{A}_{w+1,w+1} &= \mathcal{A}_{w+1,w+1}^{\mathsf{T}},
\end{aligned}
$$

*and the blocks* $\Sigma_{22}$ *and* $\Delta$ *and* $\Gamma_j$, $j = 1, \ldots, w$ *(if they occur) are nonsingular.*

FORTRAN 77 implementations for the computation of these and other related structured staircase forms via a sequence of singular value decompositions have been presented in [BM07]. Since the staircase form uses congruence transformations, all the invariants of the EKCF are preserved. The recursive construction of the even staircase form also generates a sequence of inertias $\{(\pi_{+,j}, 0, \pi_{-,j})\}_{j=1}^{w+1}$ of certain ephemeral symmetric submatrices that appear during the construction and will be returned by the even staircase algorithm [BMX07]. The following theorem shows that the characteristic quantities describing the singular part and the eigenvalue infinity of $s\mathcal{E} - \mathcal{A}$ can be extracted from the even staircase form.

**Theorem 2.1.17.** *Suppose that an even pencil* $s\mathcal{E} - \mathcal{A}$ *has been reduced to the condensed form in Theorem 2.1.16 by the algorithm in [BMX07] with the inertia sequence* $\{(\pi_{+,j}, 0, \pi_{-,j})\}_{j=1}^{w+1}$, *and* $r_j = \pi_{+,j} + \pi_{-,j}$. *Then* $s\mathcal{E} - \mathcal{A}$ *has the following block structures associated to the singular part and the eigenvalue* $\infty$ *in the EKCF of Theorem 2.1.13.*

*a) For every $j = 1, \ldots, w$, the pencil has $s_j - q_{j+1} - (r_{j+1} - r_j)$ blocks of type E3 and size $2j \times 2j$, among which $\frac{1}{2}(s_j - q_{j+1} - (r_{j+1} - r_j))$ blocks have positive sign-characteristic and $\frac{1}{2}(s_j - q_{j+1} - (r_{j+1} - r_j))$ blocks have negative sign-characteristic. (Here we set $q_{w+1} = 0$.)*

*b) For every $j = 1, \ldots, w+1$, the pencil has $r_j - r_{j-1}$ blocks of type E3 and size $(2j-1) \times (2j-1)$, among which $\pi_{-,j} - \pi_{-,j-1}$ blocks have positive sign-characteristic, and $\pi_{+,j} - \pi_{+,j-1}$ blocks have negative sign-characteristic. (Here we set $\pi_{+,0} = \pi_{-,0} = r_0 = 0$.)*

*c) For every $j = 1, \ldots, w$, the pencil has $q_j - s_j$ blocks of type E4 and size $(2j-1) \times (2j-1)$.*

*Proof.* The theorem is an adaption of [BMX07, Thm. 3.3] to the *complex* EKCF. The proof can be carried out by identifying blocks from the real EKCF with those of the complex one. Note that in contrast to [BMX07, Thm. 3.3], the sign-characteristics of blocks corresponding to the infinite eigenvalues is changed. □

From the above theorem, the following corollary is a direct consequence.

**Corollary 2.1.18.** *[BMX07] Consider an even pencil and its staircase form* (2.1). *Then the following statements hold.*

*a) The pencil is regular if and only if $s_i = q_i$ for $i = 1, \ldots, w$.*

*b) The pencil is regular and of index at most one if and only if $w = 0$.*

*c) The subpencil $s\mathcal{E}_{w+1,w+1} - \mathcal{A}_{w+1,w+1} \in \mathbb{R}[s]^{l \times l}$ contains the regular part associated to finite eigenvalues and blocks associated to the infinite eigenvalues of index at most one.*

*d) The finite eigenvalues of the pencil are the eigenvalues of*

$$s\Delta - \left(\Sigma_{11} - \Sigma_{12}\Sigma_{22}^{-1}\Sigma_{21}\right).$$

*e) For every eigenvalue $\lambda_0 \in \mathrm{i}\mathbb{R}$, satisfying*

$$\left(\lambda_0\Delta - \left(\Sigma_{11} - \Sigma_{12}\Sigma_{22}^{-1}\Sigma_{21}\right)\right) x_0 = 0$$

*for $x_0 \in \mathbb{C}^{d_{w+1}} \setminus \{0\}$, the sign-characteristic of $\lambda_0$ is given by the sign of the real number $\mathrm{i}x_0^{\mathsf{H}}\Delta x_0$.*

Thus, once the staircase form has been computed, for the computation of eigenvalues and deflating subspaces one can restrict the methods to the middle regular index one block of the staircase form. The appropriate numerical methods rely on a transformation to a closely related skew-Hamiltonian/Hamiltonian pencil. At the moment, numerical algorithms for such pencils are better developed than those for even pencils and therefore, we always turn an even eigenvalue problem to a related skew-Hamiltonian/Hamiltonian one. The respective factorizations and numerical methods are summarized in the next subsection.

### 2.1.3 Skew-Hamiltonian/Hamiltonian Matrix Pencils

In this subsection we analyze skew-Hamiltonian/Hamiltonian matrix pencils. Here, we only consider the case of real coefficients, since these are of particular importance for the results of this thesis. Pencils with complex coefficients can be treated as well, but there are non-trivial differences in their numerical treatment compared to the real case. In fact, the real case is much more involved. We consider the following matrix structures.

**Definition 2.1.19.** [BBMX02] Let $\mathcal{J}_n := \begin{bmatrix} 0 & I_n \\ -I_n & 0 \end{bmatrix}$ be given.

a) A matrix $\mathcal{H} \in \mathbb{R}^{2n\times 2n}$ is *Hamiltonian* if $(\mathcal{H}\mathcal{J}_n)^\mathsf{T} = \mathcal{H}\mathcal{J}_n$.

b) A matrix $\mathcal{S} \in \mathbb{R}^{2n\times 2n}$ is *skew-Hamiltonian* if $(\mathcal{S}\mathcal{J}_n)^\mathsf{T} = -\mathcal{S}\mathcal{J}_n$.

c) A matrix pencil $s\mathcal{S} - \mathcal{H} \in \mathbb{R}[s]^{2n\times 2n}$ is *skew-Hamiltonian/Hamiltonian* if $\mathcal{S}$ is skew-Hamiltonian and $\mathcal{H}$ is Hamiltonian.

Even and skew-Hamiltonian/Hamiltonian pencils are closely related. Clearly, if $s\mathcal{S} - \mathcal{H} \in \mathbb{R}[s]^{2n\times 2n}$ is skew-Hamiltonian/Hamiltonian, then $\mathcal{J}_n(s\mathcal{S} - \mathcal{H})$ is even. However, the converse is not true, since the dimension of an even pencil can be odd. In this case one must inflate or deflate the pencil by one dimension to introduce or remove one of the infinite eigenvalues. We usually inflate the pencil and incorporate one additional infinite eigenvalue. If $s\mathcal{E} - \mathcal{A} \in \mathbb{R}[s]^{n\times n}$ with odd $n$ is an even pencil, then

$$\mathcal{J}_{(n+1)/2} \begin{bmatrix} s\mathcal{E} - \mathcal{A} & 0 \\ 0 & 1 \end{bmatrix}$$

is skew-Hamiltonian/Hamiltonian with an additional infinite eigenvalue. Due to these relations it is clear that the finite eigenvalues of skew-Hamiltonian/Hamiltonian pencils have the same spectral symmetry as even pencils. This means that in the real case they appear in quadruples $(\lambda, -\lambda, \overline{\lambda}, -\overline{\lambda})$ if $\lambda \notin \mathbb{R} \cup i\mathbb{R}$, and otherwise in pairs $(\lambda, -\lambda)$. Moreover, the concept of sign-characteristics for purely imaginary and infinite eigenvalues can be transfered into the skew-Hamiltonian/Hamiltonian context.

For the computation of the eigenvalues and deflating subspaces of skew-Hamiltonian/Hamiltonian pencils $s\mathcal{S} - \mathcal{H} \in \mathbb{R}[s]^{2n \times 2n}$ we could in general use the standard QZ algorithm. However, this algorithm does not exploit the structural properties of the pencil. Furthermore, the spectral symmetry is destroyed by unstructured round-off errors. Luckily, there exist structured equivalence transformations that preserve the skew-Hamiltonian/Hamiltonian structure and the spectral symmetry. These allow the development of algorithms which are faster, more accurate, and more robust than the QZ algorithm. The basic ideas are present in the following paragraphs.

We make use of *$\mathcal{J}_n$-congruence transformations* of the form

$$s\widetilde{\mathcal{S}} - \widetilde{\mathcal{H}} := \mathcal{J}_n \mathcal{Q}^\mathsf{T} \mathcal{J}_n^\mathsf{T} (s\mathcal{S} - \mathcal{H})\mathcal{Q}$$

with $\mathcal{Q} \in \mathrm{Gl}_{2n}(\mathbb{R})$ which preserve the skew-Hamiltonian/Hamiltonian structure. In general we would hope that we can compute an *orthogonal* matrix $\mathcal{Q} \in \mathbb{R}^{2n \times 2n}$ such that

$$\mathcal{J}_n \mathcal{Q}^\mathsf{T} \mathcal{J}_n^\mathsf{T} (s\mathcal{S} - \mathcal{H})\mathcal{Q} = \begin{bmatrix} s\mathcal{S}_{11} - \mathcal{H}_{11} & s\mathcal{S}_{12} - \mathcal{H}_{12} \\ 0 & s\mathcal{S}_{11}^\mathsf{T} + \mathcal{H}_{11}^\mathsf{T} \end{bmatrix}$$

is in skew-Hamiltonian/Hamiltonian Schur form, i. e., the subpencil $s\mathcal{S}_{11} - \mathcal{H}_{11} \in \mathbb{R}[s]^{n \times n}$ is in generalized real Schur form. Unfortunately, not every skew-Hamiltonian/Hamiltonian pencil admits such a Schur form, since certain simple purely imaginary eigenvalues, or multiple purely imaginary eigenvalues with even algebraic multiplicity, but uniform sign-characteristic, cannot be represented in this structure. An embedding into a pencil of the double size solves this issue as follows, see [BBMX99].

We introduce the orthogonal matrices

$$\mathcal{Y}_n := \frac{\sqrt{2}}{2} \begin{bmatrix} I_{2n} & I_{2n} \\ -I_{2n} & I_{2n} \end{bmatrix}, \quad \mathcal{P}_n = \begin{bmatrix} I_n & 0 & 0 & 0 \\ 0 & 0 & I_n & 0 \\ 0 & I_n & 0 & 0 \\ 0 & 0 & 0 & I_n \end{bmatrix}, \quad \mathcal{X}_n = \mathcal{Y}_n \mathcal{P}_n, \tag{2.2}$$

and define the $4n \times 4n$ pencil

$$s\mathcal{B}_\mathcal{S} - \mathcal{B}_\mathcal{H} := \mathcal{X}_n^\mathsf{T} \begin{bmatrix} s\mathcal{S} - \mathcal{H} & 0 \\ 0 & s\mathcal{S} + \mathcal{H} \end{bmatrix} \mathcal{X}_n. \tag{2.3}$$

It can be easily observed that $s\mathcal{B}_\mathcal{S} - \mathcal{B}_\mathcal{H}$ is again real skew-Hamiltonian/Hamiltonian with the same eigenvalues (now with double algebraic, geometric, and partial multiplicities, but with appropriate mixed sign-characteristic) as the pencil $s\mathcal{S} - \mathcal{H}$. To compute the eigenvalues of $s\mathcal{B}_\mathcal{S} - \mathcal{B}_\mathcal{H}$ one uses the generalized symplectic URV decomposition of $s\mathcal{S} - \mathcal{H}$, formulated in the next theorem.

**Theorem 2.1.20** (Generalized symplectic URV decomposition). *[BBMX99] Let $s\mathcal{S}-\mathcal{H} \in \mathbb{R}[s]^{2n\times 2n}$ be a skew-Hamiltonian/Hamiltonian pencil. Then there exist orthogonal matrices $\mathcal{Q}_1$, $\mathcal{Q}_2 \in \mathbb{R}^{2n\times 2n}$ such that*

$$\begin{aligned} \mathcal{Q}_1^\mathsf{T} \mathcal{S} \mathcal{J}_n \mathcal{Q}_1 \mathcal{J}_n^\mathsf{T} &= \begin{bmatrix} S_{11} & S_{12} \\ 0 & S_{11}^\mathsf{T} \end{bmatrix}, \\ \mathcal{J}_n \mathcal{Q}_2^\mathsf{T} \mathcal{J}_n^\mathsf{T} \mathcal{S} \mathcal{Q}_2 &= \begin{bmatrix} T_{11} & T_{12} \\ 0 & T_{11}^\mathsf{T} \end{bmatrix}, \\ \mathcal{Q}_1^\mathsf{T} \mathcal{H} \mathcal{Q}_2 &= \begin{bmatrix} H_{11} & H_{12} \\ 0 & H_{22} \end{bmatrix}, \end{aligned} \tag{2.4}$$

*where $S_{12}$ and $T_{12}$ are skew-symmetric and the formal matrix product $S_{11}^{-1} H_{11} T_{11}^{-1} H_{22}^\mathsf{T}$ is in real periodic Schur form [BGD92, HL94, Kre01].*

Efficient implementations of algorithms for computing this and related factorizations are available in the FORTRAN 77 library SLICOT[1], see also [BSV13a, BSV13b] for more technical details. Applying this result to the specially structured pencil $s\mathcal{B}_\mathcal{S} - \mathcal{B}_\mathcal{H}$, we can compute an orthogonal matrix $\mathcal{Q} \in \mathbb{R}^{4n\times 4n}$ such that

$$\mathcal{J}_{2n} \mathcal{Q}^\mathsf{T} \mathcal{J}_{2n}^\mathsf{T} (s\mathcal{B}_\mathcal{S} - \mathcal{B}_\mathcal{H}) \mathcal{Q} = s \left[\begin{array}{cc|cc} S_{11} & 0 & S_{12} & 0 \\ 0 & T_{11} & 0 & T_{12} \\ \hline 0 & 0 & S_{11}^\mathsf{T} & 0 \\ 0 & 0 & 0 & T_{11}^\mathsf{T} \end{array}\right] - \left[\begin{array}{cc|cc} 0 & H_{11} & 0 & H_{12} \\ -H_{22}^\mathsf{T} & 0 & H_{12}^\mathsf{T} & 0 \\ \hline 0 & 0 & 0 & H_{22} \\ 0 & 0 & -H_{11}^\mathsf{T} & 0 \end{array}\right]$$

with $\mathcal{Q} = \mathcal{P}_n^\mathsf{T} \begin{bmatrix} \mathcal{J}_n \mathcal{Q}_1 \mathcal{J}_n^\mathsf{T} & 0 \\ 0 & \mathcal{Q}_2 \end{bmatrix} \mathcal{P}_n$.

Note that for these computations we never explicitly construct the embedded pencils. It is sufficient to compute the necessary parts of the matrices in (2.4).

The eigenvalues of $s\mathcal{S} - \mathcal{H}$ can then be computed as $\pm\mathrm{i}\sqrt{\lambda_j}$ where the $\lambda_j$, $j = 1, \ldots, n$, are the eigenvalues of $S_{11}^{-1} H_{11} T_{11}^{-1} H_{22}^\mathsf{T}$ which can be determined by evaluating the entries on the $1 \times 1$ and $2 \times 2$ diagonal blocks of the matrices only. In particular, the finite, purely imaginary eigenvalues correspond to the $1 \times 1$ diagonal blocks of this formal matrix product. Provided that the pairwise distance of the simple, finite, purely imaginary eigenvalues with mixed sign-characteristics is sufficiently large, they can be computed in a robust way without any error in the real part. This property of the algorithm plays an essential role for many of the applications that we will consider in subsequent sections. For an illustration see Figure 2.1 where the eigenvalues computed by the standard QZ algorithm as well as those computed by the structure-preserving method are depicted.

[1] `http://slicot.org`

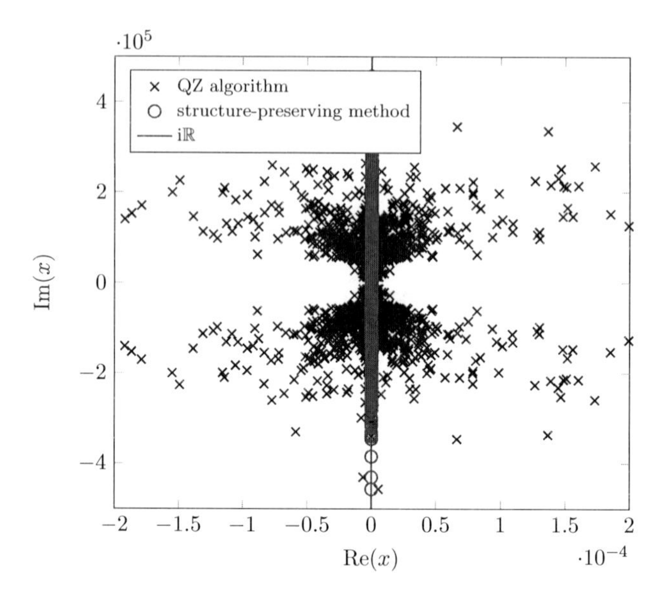


Figure 2.1: Computed eigenvalues of a skew-Hamiltonian/Hamiltonian pencil with only purely imaginary eigenvalues resulting from a linearized gyroscopic system

## 2.2 System Theoretic Basics

In this subsection we discuss the system theoretic preliminaries of this work. Here we mainly discuss systems given by real matrices. However, all of the following considerations can be extended to systems given by complex matrices in a straightforward way.

### 2.2.1 Descriptor Systems, Behaviors, and Stability

In this thesis we consider differential-algebraic equations of the form

$$E\dot{x}(t) = Ax(t) \tag{2.5}$$

where $sE - A \in \mathbb{R}[s]^{n \times n}$ is regular. The set of these systems is denoted by $\Sigma_n$ and we write $(E, A) \in \Sigma_n$. Moreover, a trajectory $x : \mathbb{R} \to \mathbb{R}^n$ is said to be a *solution* of

(2.5) if and only if it belongs to the *behavior* of (2.5), defined by

$$\mathfrak{B}_{(E,A)} := \big\{x \in \mathcal{L}^2_{\mathrm{loc}}(\mathbb{R},\mathbb{R}^n) : E\dot{x} \in \mathcal{L}^2_{\mathrm{loc}}(\mathbb{R},\mathbb{R}^n) \text{ and } x \text{ solves (2.5)} \\ \text{for almost all } t \in \mathbb{R}\big\}.$$

Note that $\mathfrak{B}_{(E,A)}$ contains all continuously differentiable trajectories, but furthermore includes solutions with jumps in the algebraic variables.

The first property that we consider is stability. Unlike in the ODE case, there are various definitions. We only focus on the following concept which is called *stability in the behavioral sense* in [BR13]. We use a modified definition of [BR13].

**Definition 2.2.1** (Stability). The system $(E,A) \in \Sigma_n$ is called

a) *polynomially bounded* if for all $x \in \mathfrak{B}_{(E,A)}$, there exists some $p(s) \in \mathbb{R}[s]$ and $M \geq 0$ such that

$$\|x(\tau)\|_2 \leq M\,|p(\tau)| \text{ for almost all } \tau \in \mathbb{R};$$

b) *asymptotically stable* if for all $x \in \mathfrak{B}_{(E,A)}$ holds

$$\lim_{t\to\infty} \operatorname*{ess\,sup}_{\tau>t} \|x(\tau)\|_2 = 0.$$

**Proposition 2.2.2** (Algebraic characterizations for asymptotic stability). *[BR13]. The system $(E,A) \in \Sigma_n$ is polynomially bounded if and only if $\Lambda(E,A) \subset \overline{\mathbb{C}^-}$. It is asymptotically stable if and only if $\Lambda(E,A) \subset \mathbb{C}^-$*

The main focus of this thesis are differential-algebraic equations that can be controlled by an input signal, i. e., we consider equations of the form

$$E\dot{x}(t) = Ax(t) + Bu(t), \tag{2.6}$$

where $(E,A) \in \Sigma_n$ and $B \in \mathbb{R}^{n\times m}$. We will also often use the term descriptor system. The set of these systems is denoted by $\Sigma_{n,m}$ and we write $(E,A,B) \in \Sigma_{n,m}$. The function $u : \mathbb{R} \to \mathbb{R}^m$ is called *input* of the system; we call $x(t) \in \mathbb{R}^n$ the *(generalized) state* of $(E,A,B)$ at time $t \in \mathbb{R}$. The set of solution trajectories $(x,u) : \mathbb{R} \to \mathbb{R}^n \times \mathbb{R}^m$ induces the *behavior* of (2.6):

$$\mathfrak{B}_{(E,A,B)} := \big\{(x,u) \in \mathcal{L}^2_{\mathrm{loc}}(\mathbb{R},\mathbb{R}^n) \times \mathcal{L}^2_{\mathrm{loc}}(\mathbb{R},\mathbb{R}^m) : E\dot{x} \in \mathcal{L}^2_{\mathrm{loc}}(\mathbb{R},\mathbb{R}^n) \\ \text{and } (x,u) \text{ solves (2.6) for almost all } t \in \mathbb{R}\big\}.$$

Moreover, we define the *behavior with initial differential variable*

$$\mathfrak{B}_{(E,A,B)}(x_0) := \big\{(x,u) \in \mathfrak{B}_{(E,A,B)} : Ex(0) = Ex_0\big\}. \tag{2.7}$$

Often, the complete state information is not available. Then, besides the system dynamics (2.6), an additional output equation is introduced which reflects the information that can be retrieved via measurements. This leads to a differential-algebraic control system of the form

$$\begin{aligned} E\dot{x}(t) &= Ax(t) + Bu(t), \\ y(t) &= Cx(t) + Du(t) \end{aligned} \tag{2.8}$$

with $(E, A, B) \in \Sigma_{n,m}$ and given $C \in \mathbb{R}^{p\times n}$, $D \in \mathbb{R}^{p\times m}$. The class of such systems is denoted by $\Sigma_{n,m,p}$ and we write $(E, A, B, C, D) \in \Sigma_{n,m,p}$. The function $y : \mathbb{R} \to \mathbb{R}^n$ is then called *output* of the system. Finally, the set of solution trajectories $(x, u, y) : \mathbb{R} \to \mathbb{R}^n \times \mathbb{R}^m \times \mathbb{R}^p$ is given by the *behavior*

$$\begin{aligned} \mathfrak{B}_{(E,A,B,C,D)} := \big\{ &(x, u, y) \in \mathcal{L}^2_{\text{loc}}(\mathbb{R}, \mathbb{R}^n) \times \mathcal{L}^2_{\text{loc}}(\mathbb{R}, \mathbb{R}^m) \times \mathcal{L}^2_{\text{loc}}(\mathbb{R}, \mathbb{R}^p) : \\ &E\dot{x} \in \mathcal{L}^2_{\text{loc}}(\mathbb{R}, \mathbb{R}^n) \text{ and } (x, u, y) \text{ solves (2.8) for almost all } t \in \mathbb{R} \big\} . \end{aligned}$$

Again, we define the *behavior with initial differential variable* by

$$\mathfrak{B}_{(E,A,B,C,D)}(x_0) := \big\{ (x, u, y) \in \mathfrak{B}_{(E,A,B,C,D)} : Ex(0) = Ex_0 \big\} .$$

Note that when $D = 0$, we sometimes omit the last argument in $(E, A, B, C, D) \in \Sigma_{n,m,p}$ and write $(E, A, B, C) \in \Sigma_{n,m,p}$ instead. It is important to note that the latter should not be misunderstood as a standard state-space system $(I_n, A, B, C, D) \in \Sigma_{n,m,p}$.

### 2.2.2 Controllability, Stabilizability, Observability, and Detectability

In this subsection we present some concepts for controllability, stabilizability, observability, and detectability of descriptor systems (2.8). Hereby, the main focus will be on controllability and stabilizability concepts. For brevity, the corresponding observability and detectability notions will be introduced by duality. The main difficulty in this context is the existence of several different ways to define controllability at infinity which are also not uniformly treated in the literature. In [BR13, Ber13], a well-structured overview and comparison of these concepts is given which we will stick to.

First we consider controllability and observability concepts. Even if an output equation may be given, these are only properties of the state equation given by $(E, A, B) \in \Sigma_{n,m}$. Since the matrices $E$, $A$, $B$ do not depend on time, we can assume w. l. o. g. that the initial time is zero. If this is not the case we can apply a shift-operator to move the initial value to time $t = 0$, see [BR13].

We need the following two spaces, defined in [BR13]:

a) the space of *consistent initial values*

$$\mathcal{V} := \left\{ x_0 \in \mathbb{R}^n : \exists (x,u) \in \mathfrak{B}_{(E,A,B)} \text{ with } \dot{x} \in \mathcal{L}^2_{\text{loc}}(\mathbb{R},\mathbb{R}^n) \text{ and } x(0) = x_0 \right\}; \tag{2.9}$$

b) the space of *consistent initial differential variables*

$$\mathcal{V}_{\text{diff}} := \left\{ x_0 \in \mathbb{R}^n : \mathfrak{B}_{(E,A,B)}(x_0) \neq \emptyset \right\}. \tag{2.10}$$

**Definition 2.2.3** (Controllability and stabilizability concepts)**.** [Ber13, Def. 3.1.5] Let the system $(E, A, B) \in \Sigma_{n,m}$ with the spaces $\mathcal{V}$ as in (2.9) and $\mathcal{V}_{\text{diff}}$ as in (2.10) be given. Then the system $(E, A, B)$ is called

a) *behaviorally stabilizable*

$$:\Leftrightarrow \forall (x,u) \in \mathfrak{B}_{(E,A,B)} \; \exists (\widetilde{x},\widetilde{u}) \in \mathfrak{B}_{(E,A,B)} \text{ with } (x,u)|_{(-\infty,0)} = (\widetilde{x},\widetilde{u})|_{(-\infty,0)}$$
$$\text{and } \lim_{t\to\infty} \operatorname*{ess\,sup}_{\tau>t} \|(x(\tau),u(\tau))\|_2 = 0;$$

b) *behaviorally anti-stabilizable*

$$:\Leftrightarrow \forall (x,u) \in \mathfrak{B}_{(E,A,B)} \; \exists (\widetilde{x},\widetilde{u}) \in \mathfrak{B}_{(E,A,B)} \text{ with } (x,u)|_{(0,\infty)} = (\widetilde{x},\widetilde{u})|_{(0,\infty)}$$
$$\text{and } \lim_{t\to-\infty} \operatorname*{ess\,sup}_{\tau<t} \|(x(\tau),u(\tau))\|_2 = 0;$$

c) *behaviorally controllable*

$$:\Leftrightarrow \forall (x_1,u_1),\, (x_2,u_2) \in \mathfrak{B}_{(E,A,B)} \; \exists T > 0,\, (x,u) \in \mathfrak{B}_{(E,A,B)} \text{ with}$$
$$(x(t),u(t)) = \begin{cases} (x_1(t),u_1(t)), & \text{if } t < 0, \\ (x_2(t),u_2(t)), & \text{if } t > T; \end{cases}$$

d) *impulse controllable*

$$:\Leftrightarrow \forall x_0 \in \mathbb{R}^n : \mathfrak{B}_{(E,A,B)}(x_0) \neq \emptyset \;\Leftrightarrow\; \mathcal{V}_{\text{diff}} = \mathbb{R}^n;$$

e) *controllable at infinity*

$$:\Leftrightarrow \forall x_0 \in \mathbb{R}^n \; \exists (x,u) \in \mathfrak{B}_{(E,A,B)} \text{ with } \dot{x} \in \mathcal{L}^2_{\text{loc}}(\mathbb{R},\mathbb{R}^n) \text{ and } x(0) = x_0$$
$$\Leftrightarrow \mathcal{V} = \mathbb{R}^n;$$

f) *strongly stabilizable*

$$:\Leftrightarrow \forall x_0 \in \mathbb{R}^n \; \exists (x,u) \in \mathfrak{B}_{(E,A,B)}(x_0) \text{ with } \lim_{t\to\infty} Ex(t) = 0;$$

g) *strongly anti-stabilizable*

$$:\Leftrightarrow \forall x_0 \in \mathbb{R}^n \ \exists (x,u) \in \mathfrak{B}_{(E,A,B)}(x_0) \text{ with } \lim_{t\to-\infty} Ex(t) = 0;$$

h) *strongly controllable*

$$:\Leftrightarrow \forall x_0,\ x_1 \in \mathbb{R}^n \ \exists T > 0,\ (x,u) \in \mathfrak{B}_{(E,A,B)}(x_0) \text{ with } Ex(T) = Ex_1;$$

i) *completely stabilizable*

$$:\Leftrightarrow \forall x_0 \in \mathbb{R}^n \ \exists (x,u) \in \mathfrak{B}_{(E,A,B)} \text{ with } \dot{x} \in \mathcal{L}^2_{\text{loc}}(\mathbb{R},\mathbb{R}^n),\ x(0) = x_0 \text{ and } \lim_{t\to\infty} x(t) = 0;$$

j) *completely anti-stabilizable*

$$:\Leftrightarrow \forall x_0 \in \mathbb{R}^n \ \exists (x,u) \in \mathfrak{B}_{(E,A,B)} \text{ with } \dot{x} \in \mathcal{L}^2_{\text{loc}}(\mathbb{R},\mathbb{R}^n),\ x(0) = x_0 \text{ and } \lim_{t\to-\infty} x(t) = 0;$$

k) *completely controllable*

$$:\Leftrightarrow \forall x_0,\ x_1 \in \mathbb{R}^n \ \exists T > 0,\ (x,u) \in \mathfrak{B}_{(E,A,B)} \text{ with } \dot{x} \in \mathcal{L}^2_{\text{loc}}(\mathbb{R},\mathbb{R}^n),\ x(0) = x_0 \text{ and } x(T) = x_1.$$

*Remark* 2.2.4. Note that in some of the definitions we need the additional requirement that $\dot{x} \in \mathcal{L}^2_{\text{loc}}(\mathbb{R},\mathbb{R}^n)$. This is necessary in order to make $x(\cdot)$ continuous and to allow function evaluations as discussed in [Ber13].

*Remark* 2.2.5 (Stabilizability and anti-stabilizability). From [BR13, p. 8], we have $(x(\cdot),u(\cdot)) \in \mathfrak{B}_{(E,A,B)}$ if and only if the reflected trajectory $(x(-\cdot),u(-\cdot))$ is in the behavior of the *backward system* $(-E,A,B)$. As a consequence, $(E,A,B)$ is behaviorally stabilizable (strongly stabilizable, completely stabilizable) if and only if $(-E,A,B)$ is behaviorally anti-stabilizable (strongly anti-stabilizable, completely anti-stabilizable).

In order to check the above mentioned properties, there exist algebraic characterizations in terms of the matrices $E$, $A$, and $B$. These are summarized in the following proposition.

**Proposition 2.2.6** (Equivalent algebraic conditions). *Let $(E,A,B) \in \Sigma_{n,m}$ be given with $r = \operatorname{rank} E$. Let $S_\infty \in \mathbb{R}^{n\times n-r}$ be a matrix with $\operatorname{im} S_\infty = \ker E$. Then the system $(E,A,B)$ is*

a) *behaviorally stabilizable if and only if* $\operatorname{rank}\begin{bmatrix}\lambda E - A & B\end{bmatrix} = n$ *for all* $\lambda \in \overline{\mathbb{C}^+}$;

b) *behaviorally anti-stabilizable if and only if* $\operatorname{rank}\begin{bmatrix}\lambda E - A & B\end{bmatrix} = n$ *for all* $\lambda \in \overline{\mathbb{C}^-}$;

c) *behaviorally controllable if and only if* $\operatorname{rank}\begin{bmatrix}\lambda E - A & B\end{bmatrix} = n$ *for all* $\lambda \in \mathbb{C}$;

d) *impulse controllable if and only if* $\operatorname{rank}\begin{bmatrix}E & AS_\infty & B\end{bmatrix} = n$;

e) *controllable at infinity if and only if* $\operatorname{rank}\begin{bmatrix}E & B\end{bmatrix} = n$;

f) *strongly stabilizable (strongly anti-stabilizable, strongly controllable) if and only if it is behaviorally stabilizable (behaviorally anti-stabilizable, behaviorally controllable) and impulse controllable;*

g) *completely stabilizable (completely anti-stabilizable, completely controllable) if and only if it is behaviorally stabilizable (behaviorally anti-stabilizable, behaviorally controllable) and controllable at infinity.*

*The above properties are further invariant under input, state, and feedback equivalence. That is,* $(E, A, B)$ *has one of the properties in a)–g) if and only if for* $W, T \in \mathrm{Gl}_n(\mathbb{R})$, $V \in \mathrm{Gl}_m(\mathbb{R})$, *and* $F \in \mathbb{R}^{m \times n}$, *the control system*

$$(E_F, A_F, B_F) := (WET, W(A + BF)T, WBV)$$

*has the respective property [BR13].*

*Proof.* The statements about stabilizability and controllability have been proven in a more general fashion in [BR13]. The results for the anti-stabilizability concepts follow from the relation to stabilizability, see Remark 2.2.5. □

We further define a concept that is defined by a purely linear algebraic condition and does not have an interpretation in terms of the behavior $\mathfrak{B}_{(E,A,B)}$. It generalizes the concept of sign-controllability for systems governed by ordinary differential equations [Sch91b, Sch91a, CALM97]. This concept will be later used as an assumption for the differential-algebraic KYP lemma, see 3.3.1.

**Definition 2.2.7** (Sign-controllability)**.** The system $(E, A, B) \in \Sigma_{n,m}$ is called

a) *behaviorally sign-controllable* if for all $\lambda \in \mathbb{C}$ it holds that $\operatorname{rank}\begin{bmatrix}\lambda E - A & B\end{bmatrix} = n$ or $\operatorname{rank}\begin{bmatrix}-\overline{\lambda} E - A & B\end{bmatrix} = n$;

b) *strongly sign-controllable* if it is behaviorally sign-controllable and impulse controllable;

c) *completely sign-controllable* if it is behaviorally sign-controllable and controllable at infinity.

Now we briefly introduce the associated notions of observability and detectability of a system $(E, A, B, C, D) \in \Sigma_{n,m,p}$. As they do not play a very prominent role in this thesis we only introduce them as controllability and stabilizability concepts for the *dual system* $\left(E^\top, A^\top, C^\top, B^\top, D^\top\right) \in \Sigma_{n,p,m}$, in other words we consider controllability and stabilizability of $\left(E^\top, A^\top, C^\top\right) \in \Sigma_{n,p}$.

**Definition 2.2.8** (Observability and detectability concepts)**.** The descriptor system $(E, A, B, C, D) \in \Sigma_{n,m,p}$ is called

a) behaviorally detectable $:\Leftrightarrow$ $\left(E^\top, A^\top, C^\top\right)$ is behaviorally stabilizable;

b) behaviorally anti-detectable $:\Leftrightarrow$ $\left(E^\top, A^\top, C^\top\right)$ is behaviorally anti-stabilizable;

c) behaviorally observable $:\Leftrightarrow$ $\left(E^\top, A^\top, C^\top\right)$ is behaviorally controllable;

d) impulse observable $:\Leftrightarrow$ $\left(E^\top, A^\top, C^\top\right)$ is impulse controllable;

e) observable at infinity $:\Leftrightarrow$ $\left(E^\top, A^\top, C^\top\right)$ is controllable at infinity;

f) strongly detectable (strongly anti-detectable, strongly observable) $:\Leftrightarrow$ $\left(E^\top, A^\top, C^\top\right)$ is strongly stabilizable (strongly anti-stabilizable, strongly controllable);

g) completely detectable (completely anti-detectable, completely observable) $:\Leftrightarrow$ $\left(E^\top, A^\top, C^\top\right)$ is completely stabilizable (completely anti-stabilizable, completely controllable).

**Definition 2.2.9** (Uncontrollable/unobservable mode)**.** The number $\lambda \in \mathbb{C}$ is called an *uncontrollable mode* of the system $(E, A, B, C, D) \in \Sigma_{n,m,p}$ if $\operatorname{rank}\begin{bmatrix}\lambda E - A & B\end{bmatrix} < n$. Moreover, it is called an *unobservable mode* if $\operatorname{rank}\begin{bmatrix}\lambda E^\top - A^\top & C^\top\end{bmatrix} < n$.

### 2.2.3 Frequency Domain Analysis

In applications it is often useful to consider a dynamical system in the frequency domain, in particular if one is interested in the influence of the inputs on the outputs of the system.

#### Laplace Transformation and Transfer Functions

A function $f : [0, \infty) \to \mathbb{R}^n$ is called *exponentially bounded* if there exist numbers $M$ and $\alpha$ such that $\|f(t)\|_2 \leq M \mathrm{e}^{\alpha t}$ for all $t \geq 0$. The value $\alpha$ is called a *bounding exponent* [TSH01, p. 32].

**Definition 2.2.10** (Laplace transformation)**.** [TSH01, p. 32] Let $f : [0, \infty) \to \mathbb{R}^n$ be exponentially bounded with bounding exponent $\alpha$. Then

$$\mathcal{L}\{f\}(s) := \int_0^\infty f(\tau) \mathrm{e}^{-s\tau} \mathrm{d}\tau$$

for $\mathrm{Re}(s) > \alpha$ is called the *Laplace transform* of $f$. The process of forming the Laplace transform is called *Laplace transformation.*

It can be shown that the integral converges uniformly in a domain of the form $\mathrm{Re}(s) \geq \beta$ for all $\beta > \alpha$.

Moreover, the following two fundamental properties hold.

**Theorem 2.2.11.** *[TSH01, Thm. 2.19] Let $f$, $g$, $h : [0, \infty) \to \mathbb{R}^n$ be given. Then the following two statements hold true:*

*a) The Laplace transformation is linear, i. e., if $f$ and $g$ are exponentially bounded, then $h := \gamma f + \delta g$ is also exponentially bounded and*

$$\mathcal{L}\{h\} = \gamma \mathcal{L}\{f\} + \delta \mathcal{L}\{g\}$$

*holds for all $\gamma$, $\delta \in \mathbb{C}$.*

*b) If $f \in \mathcal{C}^1_{\mathrm{pw}}([0, \infty), \mathbb{R}^n)$ and $\dot{f}$ is exponentially bounded, then $f$ is exponentially bounded and*

$$\mathcal{L}\{\dot{f}\}(s) = s\mathcal{L}\{f\}(s) - f(0).$$

Now we apply the Laplace transformation to a system $(E, A, B, C, D) \in \Sigma_{n,m,p}$. Assume that each of the Laplace transforms $X(s) := \mathcal{L}\{x\}(s)$, $U(s) := \mathcal{L}\{u\}(s)$, and $Y(s) := \mathcal{L}\{y\}(s)$ exists. By using Theorem 2.2.11, we obtain the Laplace transformed system

$$\begin{aligned} sEX(s) - Ex(0) &= AX(s) + BU(s), \\ Y(s) &= CX(s) + DU(s). \end{aligned}$$

Under the assumption that $sE - A \in \mathbb{R}[s]^{n \times n}$ is regular and $Ex(0) = 0$, we obtain the relation

$$Y(s) = \big(C(sE - A)^{-1}B + D\big)\, U(s).$$

This leads to the following definitions.

**Definition 2.2.12** (Transfer function)**.** The function

$$G(s) := C(sE - A)^{-1}B + D \in \mathbb{R}(s)^{p \times m}$$

is called the *transfer function* of the system $(E, A, B, C, D) \in \Sigma_{n,m,p}$.

The transfer function is a direct relation between the inputs and the outputs in the frequency domain and therefore, it plays an important role for the analysis of the behavior of the system. When evaluating for $s = i\omega$, then $\omega$ can be interpreted as a frequency (scaled by $1/2\pi$). Note that $G(s)$ can be formally defined without the requirement on $X$, $U$, $Y$ to exist.

It is also possible to relate a dynamical system of the form (2.8) to a given transfer function $G(s) \in \mathbb{R}(s)^{p\times m}$ which is, however, not unique. This leads to the following definitions.

**Definition 2.2.13** ((Minimal) realization)**.** [OV00, VLK81] Assume that the system $(E, A, B, C, D) \in \Sigma_{n,m,p}$ has the transfer function $G(s) \in \mathbb{R}(s)^{p\times m}$. Then we say that $(E, A, B, C, D)$ is a *realization* of $G(s)$. A realization $(E, A, B, C, D) \in \Sigma_{n,m,p}$ is called

a) *weakly minimal*, if it is completely controllable and completely observable;

b) *minimal*, if it is weakly minimal and does not contain nondynamic modes, i. e.,

$$A \cdot \ker E \subseteq \operatorname{im} A.$$

*Remark* 2.2.14.

a) If a realization $(E, A, B, C, D) \in \Sigma_{n,m,p}$ of a transfer function $G(s) \in \mathbb{R}(s)^{p\times m}$ is minimal, then its state space dimension $n$ is minimal among all realizations of $G(s)$.

b) Realizations are not unique. If $(E, A, B, C, D) \in \Sigma_{n,m,p}$ is a realization of $G(s)$, then for any two matrices $W, T \in \mathrm{Gl}_n(\mathbb{R})$, the system

$$(WET, WAT, WB, CT, D) \in \Sigma_{n,m,p}$$

is also a realization of $G(s)$. Transformations of the above kind are also called *(generalized) state-space transformations*.

#### Polynomial and Rational Matrices

Since transfer functions are rational, we present some fundamental theory on this class of functions.

**Definition 2.2.15** (Unimodular matrix, monic/coprime polynomials)**.**

a) A polynomial matrix $U(s) \in \mathbb{R}[s]^{n\times n}$ is called *unimodular*, if it is a unit in the ring $\mathbb{R}[s]^{n\times n}$.

b) A polynomial $p(s) \in \mathbb{R}[s]$ is called *monic* if its leading coefficient is one.

c) Two polynomials $p(s)$, $q(s) \in \mathbb{R}[s]$ are called *coprime* if their greatest common divisor is 1.

Matrices with rational entries can, via multiplication with suitable unimodular matrices, be transformed to Smith-McMillan form, described in the next theorem.

**Theorem 2.2.16** (Smith-McMillan form). *[Kai80, Chap. 6] For $G(s) \in \mathbb{R}(s)^{p\times m}$ there exist unimodular matrices $U(s) \in \mathbb{R}[s]^{p\times p}$ and $V(s) \in \mathbb{R}[s]^{m\times m}$, such that*

$$U^{-1}(s)G(s)V^{-1}(s) = \operatorname{diag}\left(\frac{\varepsilon_1(s)}{\psi_1(s)}, \ldots, \frac{\varepsilon_r(s)}{\psi_r(s)}, 0, \ldots, 0\right) \tag{2.11}$$

*for some monic and coprime polynomials $\varepsilon_j(s)$, $\psi_j(s) \in \mathbb{R}[s]$ such that $\varepsilon_j(s)$ divides $\varepsilon_{j+1}(s)$ and $\psi_{j+1}(s)$ divides $\psi_j(s)$ for $j = 1, \ldots, r-1$.*

The Smith-McMillan form can be utilized to define poles and zeros of rational matrices, as well as outer, inner, and co-inner rational functions.

**Definition 2.2.17** (Poles, zeros, inner function, co-inner function, outer function). Let $G(s) \in \mathbb{R}(s)^{p\times m}$ with Smith-McMillan form (2.11) be given. Then $\lambda \in \mathbb{C}$ is called

a) a *zero* of $G(s)$ if $\varepsilon_r(\lambda) = 0$;

b) a *pole* of $G(s)$ if $\psi_1(\lambda) = 0$.

The sets of zeros and poles of $G(s)$ are denoted by $\mathfrak{Z}(G)$ and $\mathfrak{P}(G)$, respectively. Moreover, by [Gre88], the rational matrix $G(s)$ is called

a) *outer* if $r = p$ and $\mathfrak{Z}(G) \subset \overline{\mathbb{C}^-}$;

b) *inner* if $\mathfrak{P}(G) \subset \mathbb{C}^-$ and $G^\sim(s)G(s) = I_m$;

c) *co-inner* if $\mathfrak{P}(G) \subset \mathbb{C}^-$ and $G(s)G^\sim(s) = I_p$.

*Remark* 2.2.18.

a) The zeros of $G(s)$ are also called the *transmission zeros* of an associated realization $(E, A, B, C, D) \in \Sigma_{n,m,p}$, see, e. g., [BIR12].

b) Rational functions with no zeros in $\mathbb{C}^+$ are called *minimum phase* in [Ilc93]. Minimum phase functions with full row rank over the field $\mathbb{R}(s)$ are outer.

c) Rational functions with $G^\sim(s)G(s) = I_m$ are called *all-pass* [TSH01]. All-pass functions are unitary-valued on the imaginary axis. Any inner function is all-pass.

d) A rational function $G(s)$ is co-inner if and only if $G^{\mathsf{H}}(\overline{s})$ is inner.

The following properties of rational functions will play an important role in the characterization of transfer functions.

**Definition 2.2.19** (Stability, properness)**.** Let $G(s) \in \mathbb{R}(s)^{p\times m}$ be given. We say that $G(s)$ is stable if all its poles are located in $\mathbb{C}^-$. Moreover, we call $G(s)$

a) *strictly proper* if $\lim_{\omega\to\infty} \|G(\mathrm{i}\omega)\|_2 = 0$;

b) *proper* if $\lim_{\omega\to\infty} \|G(\mathrm{i}\omega)\|_2 < \infty$;

c) *improper* if $\lim_{\omega\to\infty} \|G(\mathrm{i}\omega)\|_2 = \infty$.

Rational function can be additively decomposed into a strictly proper and polynomial part as follows [Sty06]. Assume that $G(s) \in \mathbb{R}(s)^{p\times m}$ has a realization $(E, A, B, C, D) \in \Sigma_{n,m,p}$. Let $W, T \in \mathrm{Gl}_n(\mathbb{R})$ be transformation matrices leading to QWF, i. e., let

$$(WET, WAT, WB, CT, D) = \left( \begin{bmatrix} I_r & 0 \\ 0 & E_{22} \end{bmatrix}, \begin{bmatrix} A_{11} & 0 \\ 0 & I_{n-r} \end{bmatrix}, \begin{bmatrix} B_1 \\ B_2 \end{bmatrix}, \begin{bmatrix} C_1 & C_2 \end{bmatrix}, D \right), \tag{2.12}$$

where $A_{11} \in \mathbb{R}^{r\times r}$, and $E_{22} \in \mathbb{R}^{n-r\times n-r}$ is nilpotent with index of nilpotency $\nu$. Moreover, we have $B_1 \in \mathbb{R}^{r\times m}$, $B_2 \in \mathbb{R}^{n-r\times m}$, $C_1 \in \mathbb{R}^{p\times r}$, and $C_2 \in \mathbb{R}^{p\times n-r}$. Then we can write

$$G(s) = G_{\mathrm{sp}}(s) + G_{\mathrm{poly}}(s) \tag{2.13}$$

with the *strictly proper part*

$$G_{\mathrm{sp}}(s) = C_1(sI_r - A_{11})^{-1}B_1,$$

and the *polynomial part*

$$G_{\mathrm{poly}}(s) = C_2(sE_{22} - I_{n-r})^{-1}B_2 + D = \sum_{j=0}^{\nu-1} M_j s^j \in \mathbb{R}[s]^{p\times m},$$

where

$$\begin{aligned} M_0 &:= D - C_2B_2, \\ M_j &:= -C_2E_{22}^jB_2, \quad j = 1, \ldots, \nu - 1, \end{aligned}$$

are the *Markov parameters*.

### Some Rational Function Spaces

In many applications of systems and control such as model order reduction or robust control, it is necessary to quantify the size of a dynamical system or the distance between two of them [Sty06]. This can be done by turning to the frequency domain and considering certain spaces of transfer functions and the induced norms. In the following we introduce the most important rational function spaces that we need in this thesis. Note that many of the concepts can also be generalized to non-rational functions, however this will not play a role in this work.

**Definition 2.2.20** (Some rational function spaces)**.** [Sty06] Let $G(s) = G_{\mathrm{sp}}(s) + G_{\mathrm{poly}}(s) \in \mathbb{R}(s)^{p\times m}$ be a decomposition as in (2.13). We define the following spaces and the associated norms:

a) the inner product space

$$\mathcal{RH}_2^{p\times m} := \left\{ G(s) \in \mathbb{R}(s)^{p\times m} : \mathfrak{P}(G) \subset \mathbb{C}^- \text{ and } \int_{-\infty}^{\infty} \|G(\mathrm{i}\omega)\|_{\mathrm{F}}^2 \,\mathrm{d}\omega < \infty \right\}$$

with $\mathcal{H}_2$-norm

$$\|G\|_{\mathcal{H}_2} := \left( \frac{1}{2\pi} \int_{-\infty}^{\infty} \|G(\mathrm{i}\omega)\|_{\mathrm{F}}^2 \,\mathrm{d}\omega \right)^{1/2};$$

b) the normed space

$$\mathcal{RHL}_2^{p\times m} := \left\{ G(s) \in \mathbb{R}(s)^{p\times m} : G_{\mathrm{sp}} \in \mathcal{RH}_2^{p\times m} \right\}$$

with $\mathcal{HL}_2$-norm

$$\|G\|_{\mathcal{HL}_2} := \left( \|G_{\mathrm{sp}}\|_{\mathcal{H}_2} + \frac{1}{2\pi} \int_0^{2\pi} \left\|G_{\mathrm{poly}}(e^{\mathrm{i}\omega})\right\|_{\mathrm{F}}^2 \,\mathrm{d}\omega \right)^{1/2};$$

c) the normed space

$$\mathcal{RH}_\infty^{p\times m} := \left\{ G(s) \in \mathbb{R}(s)^{p\times m} : \mathfrak{P}(G) \subset \mathbb{C}^- \text{ and } \sup_{\lambda\in\mathbb{C}^+} \|G(\lambda)\|_2 < \infty \right\}$$

with $\mathcal{H}_\infty$-norm

$$\|G\|_{\mathcal{H}_\infty} := \sup_{\lambda\in\mathbb{C}^+} \|G(\lambda)\|_2;$$

d) the normed space

$$\mathcal{RL}_\infty^{p\times m} := \left\{ G(s) \in \mathbb{R}(s)^{p\times m} : \mathfrak{P}(G) \cap \mathrm{i}\mathbb{R} = \emptyset \text{ and } \sup_{\omega\in\mathbb{R}} \|G(\mathrm{i}\omega)\|_2 < \infty \right\}$$

with $\mathcal{L}_\infty$-norm

$$\|G\|_{\mathcal{L}_\infty} := \sup_{\omega\in\mathbb{R}} \|G(\mathrm{i}\omega)\|_2 .$$

### 2.2.4 Zero Dynamics and Outer Systems

An important concept for later sections are the *zero dynamics* of differential-algebraic systems.

**Definition 2.2.21** (Zero dynamics)**.** [BI84, Ber13]

a) The *zero dynamics* of $(E, A, B, C, D) \in \Sigma_{n,m,p}$ is defined as the set of trajectories resulting in a trivial output, i. e.,

$$\mathcal{ZD}_{(E,A,B,C,D)} := \left\{(x, u) \in \mathcal{L}^2_{\text{loc}}(\mathbb{R}, \mathbb{R}^n) \times \mathcal{L}^2_{\text{loc}}(\mathbb{R}, \mathbb{R}^m) : (x, u, 0) \in \mathfrak{B}_{(E,A,B,C,D)}\right\}.$$

b) The set of *zero dynamics with initial differential variable* $Ex_0$ is

$$\mathcal{ZD}_{(E,A,B,C,D)}(x_0) := \mathcal{ZD}_{(E,A,B,C,D)} \cap \mathfrak{B}_{(E,A,B)}(x_0).$$

c) The set of *consistent initial differential variables for the zero dynamics* is

$$\mathcal{W}_{\text{diff}} := \left\{x_0 \in \mathbb{R}^n : \mathcal{ZD}_{(E,A,B,C,D)}(x_0) \neq \emptyset\right\}. \tag{2.14}$$

The following are particularly important properties of the zero dynamics.

**Definition 2.2.22** (Stability, stabilizability, autonomy of zero dynamics)**.** [Ber13, IR14] Let the system $(E, A, B, C, D) \in \Sigma_{n,m,p}$ with the space $\mathcal{W}_{\text{diff}}$ as in (2.14) be given. Then the zero dynamics of $\mathcal{ZD}_{(E,A,B,C,D)}$ are called

a) *polynomially bounded* if for all $(x, u) \in \mathcal{ZD}_{(E,A,B,C,D)}$, there exists some $p(s) \in \mathbb{R}[s]$ and $M \geq 0$ such that

$$\|(x(\tau), u(\tau))\|_2 \leq M \cdot |p(\tau)| \text{ for almost all } \tau \in \mathbb{R};$$

b) *asymptotically stable* if for all $\mathcal{ZD}_{(E,A,B,C,D)}$ it holds that

$$\lim_{t\to\infty} \operatorname*{ess\,sup}_{\tau>t} \|(x(\tau), u(\tau))\|_2 = 0;$$

c) *strongly asymptotically stable* if it is asymptotically stable and $\mathcal{W}_{\text{diff}} = \mathbb{R}^n$;

d) *polynomially stabilizable* if for all $x_0 \in \mathcal{W}_{\text{diff}}$, there exist some $p(s) \in \mathbb{R}[s]$, $M \geq 0$ and some $(x, u) \in \mathcal{ZD}_{(E,A,B,C,D)}(x_0)$ such that

$$\|(x(\tau), u(\tau))\|_2 \leq M \cdot |p(\tau)| \text{ for almost all } \tau \in \mathbb{R};$$

e) *stabilizable* if for all $x_0 \in \mathcal{W}_{\text{diff}}$, there exists a $(x, u) \in \mathcal{ZD}_{(E,A,B,C,D)}(x_0)$ with

$$\lim_{t\to\infty} \operatorname*{ess\,sup}_{\tau>t} \|(x(\tau), u(\tau))\|_2 = 0;$$

f) *strongly stabilizable* if it is stabilizable and $\mathcal{W}_{\text{diff}} = \mathbb{R}^n$;

g) *autonomous* if for all $x_0 \in \mathbb{R}^n$, the set $\mathcal{ZD}_{(E,A,B,C,D)}(x_0)$ has at most one element.

Many properties of the zero dynamics can be characterized by the location of the invariant zeros which will be introduced in the next definition.

**Definition 2.2.23** (Rosenbrock pencil, invariant zero). [IR14] The *Rosenbrock pencil* of the system $(E, A, B, C, D) \in \Sigma_{n,m,p}$ is given by

$$R(s) = \begin{bmatrix} -sE + A & B \\ C & D \end{bmatrix} \in \mathbb{R}[s]^{n+p \times n+m}.$$

The finite eigenvalues of the Rosenbrock pencil are called the *invariant zeros* of the system $(E, A, B, C, D)$.

**Proposition 2.2.24** (Zero dynamics and invariant zeros). *[IR14] Let the system $(E, A, B, C, D) \in \Sigma_{n,m,p}$ with Rosenbrock pencil $R(s)$ be given. Then $\mathcal{ZD}_{(E,A,B,C,D)}$ is*

*a) polynomially bounded if and only if* $\operatorname{rank} R(\lambda) = n + m \quad \forall \lambda \in \mathbb{C}^+$;

*b) asymptotically stable if and only if* $\operatorname{rank} R(\lambda) = n + m \quad \forall \lambda \in \overline{\mathbb{C}^+}$;

*c) strongly asymptotically stable if and only if it is asymptotically stable and the index of $R(s)$ is at most one;*

*d) polynomially stabilizable if and only if $(E, A, B, C, D)$ has no invariant zeros in $\mathbb{C}^+$;*

*e) stabilizable if and only if $(E, A, B, C, D)$ has no invariant zeros in $\overline{\mathbb{C}^+}$;*

*f) strongly stabilizable if and only if it is stabilizable and the index of $R(s)$ is at most one;*

*g) autonomous if and only if* $\operatorname{rank}_{\mathbb{R}(s)} R(s) = n + m$.

*There are also further equivalences with respect to stability, namely $\mathcal{ZD}_{(E,A,B,C,D)}$ is*

*h) polynomially bounded if and only if it is autonomous and polynomially stabilizable;*

*i) asymptotically stable if and only if it is autonomous and stabilizable.*

*Furthermore, note that a square system $(E, A, B, C, D) \in \Sigma_{n,m,m}$ is strongly asymptotically stable if and only if $R(\lambda) \in \mathrm{Gl}_{n+m}(\mathbb{C})$ for all $\lambda \in \overline{\mathbb{C}^+}$.*

The following result states that invariant zeros are uncontrollable or unobservable modes or transmission zeros.

**Proposition 2.2.25.** *Let a system $(E, A, B, C, D) \in \Sigma_{n,m,p}$ with transfer function $G(s) \in \mathbb{R}(s)^{p\times m}$ be given. Then the following holds true:*

*a) If $\lambda \in \mathbb{C}$ is a transmission zero of $(E, A, B, C, D)$, then it is an invariant zero of $(E, A, B, C, D)$.*

*b) If $\lambda \in \mathbb{C}$ is an invariant zero of $(E, A, B, C, D)$, then at least one of the following statements holds true:*

*i) $\lambda$ is a transmission zero of $(E, A, B, C, D)$;*

*ii) $\lambda$ is an uncontrollable mode of $(E, A, B, C, D)$;*

*iii) $\lambda$ is an unobservable mode of $(E, A, B, C, D)$.*

*Proof.* This statement is a special case of a theorem for systems described by polynomial matrices in [Ros73]. □

Moreover, the following results connects the rank of the transfer function with the rank of the associated Rosenbrock pencil.

**Lemma 2.2.26.** *Let a system $(E, A, B, C, D) \in \Sigma_{n,m,p}$ with transfer function $G(s) = C(sE - A)^{-1}B + D \in \mathbb{R}(s)^{p\times m}$ and Rosenbrock pencil $R(s) \in \mathbb{R}[s]^{n+p\times n+m}$ be given. Then*

$$\operatorname{rank}_{\mathbb{R}(s)} R(s) = n + \operatorname{rank}_{\mathbb{R}(s)} G(s).$$

*In particular, $R(s)$ is a regular pencil if and only if $G(s) \in \mathrm{Gl}_m(\mathbb{R}(s))$.*

*Proof.* This statement follows from

$$\begin{bmatrix} I_n & 0 \\ C(sE - A)^{-1} & I_p \end{bmatrix} \begin{bmatrix} -sE + A & B \\ C & D \end{bmatrix} = \begin{bmatrix} -sE + A & B \\ 0 & G(s) \end{bmatrix}.$$

□

The zero dynamics is closely related to the concept of outer systems, defined below.

**Definition 2.2.27** (Outer system). [IR14] The system $(E, A, B, C, D) \in \Sigma_{n,m,p}$ is called *outer* if it does not contain invariant zeros in $\mathbb{C}^+$ and the Rosenbrock pencil has full row rank over the field $\mathbb{R}(s)$. Equivalently, it holds that

$$\operatorname{rank} R(\lambda) = n + p \quad \forall \lambda \in \mathbb{C}^+.$$

The following theorem summarizes the relation between outer systems and outer rational functions.

**Theorem 2.2.28** (Outer systems and outer transfer functions)**.** *[IR14, Thm. 3.3] Let $(E, A, B, C, D) \in \Sigma_{n,m,p}$ with transfer function $G(s) \in \mathbb{R}(s)^{p\times m}$ be given. Then the following statements hold true:*

*a) If $(E, A, B, C, D)$ is an outer system, then $G(s)$ is an outer transfer function.*

*b) If $G(s)$ is an outer transfer function and $(E, A, B, C, D)$ is strongly stabilizable and strongly detectable, then $(E, A, B, C, D)$ is an outer system.*

We finally discuss a crucial property of outer systems. Namely, for any outer and strongly stabilizable system and initial value, there exists a sequence of square integrable inputs such that the corresponding output sequence tends to zero in $\mathcal{L}^2$. In Chapter 3, we will show that an infimizing sequence in a linear-quadratic optimal control problem can be expressed as a sequence of controls for an outer system such that the corresponding output sequence tends to zero. The following two propositions make this precise. Moreover, the results of Proposition 2.2.24 will lead to existence and uniqueness results for the optimal control.

**Proposition 2.2.29.** *[IR14] Let $(E, A, B, C, D) \in \Sigma_{n,m,p}$ be outer and strongly stabilizable and assume that $x_0 \in \mathbb{R}^n$. Then there exists a sequence $(u_k(\cdot))_{k\in\mathbb{N}}$ in $\mathcal{L}^2([0,\infty), \mathbb{R}^m)$ with the following properties:*

*a) For all $k \in \mathbb{N}$, there exists a $(x_k, u_k, y_k) \in \mathfrak{B}_{(E,A,B,C,D)}(x_0)$ with $\lim_{t\to\infty} Ex_k(t) = 0$.*

*b) The sequence $(y_k(\cdot))_{k\in\mathbb{N}}$ tends to zero in $\mathcal{L}^2([0,\infty), \mathbb{R}^p)$.*

**Proposition 2.2.30.** *[IR14] Let $(E, A, B, C, D) \in \Sigma_{n,m,p}$ be strongly stabilizable. Assume that for all $x_0 \in \mathbb{R}^n$ there exists a sequence $(u_k(\cdot))_{k\in\mathbb{N}}$ in $\mathcal{L}^2([0,\infty), \mathbb{R}^m)$ with the following properties:*

*a) For all $k \in \mathbb{N}$, there exists a $(x_k, u_k, y_k) \in \mathfrak{B}_{(E,A,B,C,D)}(x_0)$ with $\lim_{t\to\infty} Ex_k(t) = 0$.*

*b) The sequence $(y_k(\cdot))_{k\in\mathbb{N}}$ tends to zero in $\mathcal{L}^2([0,\infty), \mathbb{R}^p)$.*

*Then there exists some matrix $U \in \mathbb{K}^{l\times p}$ with orthonormal columns such that the system $(E, A, B, UC, UD)$ is outer and further for all $(x, u, y) \in \mathfrak{B}_{(E,A,B,C,D)}$, there exists some $(x, u, \widetilde{y}) \in \mathfrak{B}_{(E,A,B,UC,UD)}$ such that*

$$\|y(t)\| = \|\widetilde{y}(t)\| \text{ for almost all } t \geq 0.$$

# 3 Linear-Quadratic Control Theory for Differential-Algebraic Equations

## 3.1 Introduction

In this chapter we will study the infinite time horizon linear-quadratic optimal control problem

Minimize

$$\mathcal{J}(x,u) := \int_0^\infty \begin{pmatrix} x(\tau) \\ u(\tau) \end{pmatrix}^\mathsf{T} \begin{bmatrix} Q & S \\ S^\mathsf{T} & R \end{bmatrix} \begin{pmatrix} x(\tau) \\ u(\tau) \end{pmatrix} \mathrm{d}\tau \tag{3.1}$$

subject to $(x,u) \in \mathfrak{B}_{(E,A,B)}(x_0)$ with $\lim_{t\to\infty} Ex(t) = 0$.

Here, $Q = Q^\mathsf{T} \in \mathbb{R}^{n\times n}$, $S \in \mathbb{R}^{n\times m}$ and $R = R^\mathsf{T} \in \mathbb{R}^{m\times m}$ are given matrices. The main questions we want to answer are the following:

- Is this optimal control problem *feasible*, i. e., is *optimal value function*

  $$V^+(Ex_0) := \inf\left\{\mathcal{J}(x,u) : (x,u) \in \mathfrak{B}_{(E,A,B)}(x_0) \text{ and } \lim_{t\to\infty} Ex(t) = 0\right\}$$

  bounded from below for all $x_0 \in \mathbb{R}^n$?

- If the optimal control problem is feasible, does there exist an *optimal control* $(x_*, u_*) \in \mathfrak{B}_{(E,A,B)}(x_0)$ with $V^+(Ex_0) = \mathcal{J}(x_*, u_*)$?

- If an optimal control exists, is it *unique*?

Linear-quadratic optimal control problems (3.1) can be analyzed by means of various algebraic structures. The first concept we study to answer the above questions is the *Popov function* given by

$$\Phi(s) = \begin{bmatrix} (sE-A)^{-1}B \\ I_m \end{bmatrix}^\sim \begin{bmatrix} Q & S \\ S^\mathsf{T} & R \end{bmatrix} \begin{bmatrix} (sE-A)^{-1}B \\ I_m \end{bmatrix} \in \mathbb{R}(s)^{m\times m}. \tag{3.2}$$

Note that $\Phi(\mathrm{i}\omega)$ is Hermitian for all $\omega \in \mathbb{R}$ with $\mathrm{i}\omega \notin \Lambda(E,A)$. The feasibility of the optimal control problem is related to the pointwise positive semidefiniteness of $\Phi(\mathrm{i}\cdot) : \{\omega \in \mathbb{R} : \mathrm{i}\omega \notin \Lambda(E,A)\} \to \mathbb{C}^{m\times m}$.

In the case of ordinary differential equations (that is, $E = I_n$), this property can be assessed by the famous Kalman-Yakubovich-Popov lemma, see, e. g., [Kal63, Yak62, Pop62, And67] and [Ran96] and references therein. More precisely, under certain assumptions related to controllability, the pointwise positive semidefiniteness of $\Phi(\mathrm{i}\cdot)$ is equivalent to the solvability of the Kalman-Yakubovich-Popov (KYP) inequality, namely there exists a $P \in \mathbb{R}^{n\times n}$ such that

$$\begin{bmatrix} A^\top P + PA + Q & PB + S \\ B^\top P + S^\top & R \end{bmatrix} \geq 0, \quad P = P^\top. \tag{3.3}$$

There are several attempts to generalize this lemma to differential-algebraic equations, e. g., in [Mas06]. However, this paper only treats the case where $sE - A$ is regular and of index at most one. Recently, in [Brü11b, Brü11a], the KYP lemma has been generalized to behavioral systems. This theory is then also applied to descriptor systems under the assumption of complete controllability. There the pointwise positive semidefiniteness of $\Phi(\mathrm{i}\cdot)$ is related to the solvability of the linear matrix inequality

$$\begin{bmatrix} A^\top P_1 + P_1^\top A + Q & A^\top P_2 + P_1^\top B + S \\ B^\top P_1 + P_2^\top A + S^\top & B^\top P_2 + P_2^\top B + R \end{bmatrix} \geq 0, \quad E^\top P_1 = P_1^\top E, \quad E^\top P_2 = 0,$$

for a pair $(P_1, P_2) \in \mathbb{R}^{n\times n} \times \mathbb{R}^{n\times n}$. Other authors generalize a certain modification of this lemma, namely the *positive real lemma* with particular choices of $Q$, $S$, and $R$. This is done for instance in [FJ04]. However, strong artificial assumptions on the Markov parameters are made in order to prove the main result. These assumptions are dropped in [CF07] by considering a linear matrix inequality related to (3.3) on a certain subspace. We employ a similar idea to present a new, more general version of the KYP lemma for differential-algebraic systems, namely we relate pointwise positive semidefiniteness of $\Phi(\mathrm{i}\cdot)$ to the solvability of the *descriptor KYP inequality*. That is, there exists some $P \in \mathbb{R}^{n\times n}$, such that

$$\begin{bmatrix} A^\top P + P^\top A + Q & P^\top B + S \\ B^\top P + S^\top & R \end{bmatrix} \geq_{\mathcal{V}_{\mathrm{sys}}} 0, \quad E^\top P = P^\top E, \tag{3.4}$$

where

$$\mathcal{V}_{\mathrm{sys}} := \left\{ \begin{pmatrix} x \\ u \end{pmatrix} \in \mathbb{R}^{n+m} : Ax + Bu \in \operatorname{im} E \right\} \tag{3.5}$$

is the *system space* of $(E, A, B)$. Note that for impulse controllable systems, $\mathcal{V}_{\mathrm{sys}}$ is the smallest subspace of $\mathbb{R}^{n+m}$ in which the solution trajectories $(x, u) \in \mathfrak{B}_{(E,A,B)}$ pointwisely evolve, i. e., it is the smallest subspace with

$$\begin{pmatrix} x(t) \\ u(t) \end{pmatrix} \in \mathcal{V}_{\mathrm{sys}} \quad \text{for all } (x, u) \in \mathfrak{B}_{(E,A,B)} \text{ and almost all } t \in \mathbb{R}.$$

Note further that for non-impulse controllable systems, the solution trajectories evolve in an even smaller subspace. However, for ease of terminology and since our focus is on impulse controllable systems, we also use the definition of the systems space as in (3.5) for non-impulse controllable systems. However, this slightly differs from the notion used in [RRV14].

In order to solve the optimal control problem one employs algebraic matrix equations that can be derived from the KYP inequality. If one considers ODEs with $R > 0$, then one directly obtains the *algebraic Riccati equation* [Wil71, LR95]

$$A^\mathsf{T} X + XA - (XB + S)R^{-1}(B^\mathsf{T} X + S^\mathsf{T}) + Q = 0, \quad X = X^\mathsf{T}. \tag{3.6}$$

Solvability criteria and solutions of (3.6) can be given in terms of spectral information of the Hamiltonian matrix

$$\mathcal{A}_H = \begin{bmatrix} A - BR^{-1}S & -BR^{-1}B^\mathsf{T} \\ S^\mathsf{T} R^{-1} S - Q & -(A - BR^{-1}S)^\mathsf{T} \end{bmatrix} \in \mathbb{R}^{2n \times 2n}. \tag{3.7}$$

A well-known sufficient solvability criterion for (3.6) is the absence of purely imaginary eigenvalues of $\mathcal{A}_H$, see [LR95]. Then the solutions of (3.6) can be constructed via the invariant subspaces of (3.7). Difficulties in the characterization of solvability of (3.6) arise when purely imaginary eigenvalues are present. Then the spectral structure of $\mathcal{A}_H$ has to be studied in more detail, in particular the sign-characteristics of the purely imaginary eigenvalues.

Another fundamental difficulty arises when the matrix $R$ is not invertible. Then neither the ARE (3.6) nor the associated Hamiltonian matrix can be formulated. Under the assumption that

$$\begin{bmatrix} Q & S \\ S^\mathsf{T} & R \end{bmatrix} = \begin{bmatrix} C^\mathsf{T} \\ D^\mathsf{T} \end{bmatrix} \begin{bmatrix} C & D \end{bmatrix} \geq 0, \tag{3.8}$$

this problem was first rigorously studied in [HS83] with a special focus on the system structure. In particular, it was pointed out, that in contrast to an optimal control problem with $R > 0$, optimal controls might not exist for piecewise continuous inputs $u(\cdot)$ and thus the use of impulsive inputs was suggested. Mathematically, this is reflected by distributions. Due to these observations optimal control problems with singular $R$ are also called *singular*, otherwise they are called *regular*. Further results were later obtained by [Gee89]. There, it is for instance shown that for the case that $E = I_n$, the optimal value function is quadratic and can be characterized by the maximal and rank-minimizing solution of the KYP inequality (3.3). In this context, "rank-minimizing" refers to the minimization of the rank of the left-hand side of (3.3). Moreover, in [Gee89], the optimal control problem is relaxed in the sense that also nonzero terminal conditions are admitted. If one assumes the condition

$\lim_{t\to\infty} \inf_{x_\mathcal{T}\in\mathcal{T}} \|x(t) - x_\mathcal{T}\|_2 = 0$ with an arbitrary subspace $\mathcal{T} \subseteq \mathbb{R}^n$, then the corresponding optimal costs are still quadratic and can be expressed by a rank-minimizing solution of the KYP inequality (3.3).

Further contributions are given in [Rei11], where also the quite restrictive assumption (3.8) is dropped. Instead of the ARE one considers the so-called *Lur'e equation*

$$\begin{bmatrix} A^\mathsf{T}X + XA + Q & XB + S \\ B^\mathsf{T}X + S^\mathsf{T} & R \end{bmatrix} = \begin{bmatrix} K^\mathsf{T} \\ L^\mathsf{T} \end{bmatrix} \begin{bmatrix} K & L \end{bmatrix}, \quad X = X^\mathsf{T}, \tag{3.9}$$

which has to be solved for $(X, K, L) \in \mathbb{R}^{n\times n} \times \mathbb{R}^{q\times n} \times \mathbb{R}^{q\times m}$ with $q$ as small as possible. To obtain solvabilility criteria and construct solutions we replace the Hamiltonian matrix by the even matrix pencil

$$s\mathcal{E} - \mathcal{A} = \begin{bmatrix} 0 & -sI_n + A & B \\ sI_n + A^\mathsf{T} & Q & S \\ B^\mathsf{T} & S^\mathsf{T} & R \end{bmatrix} \in \mathbb{R}[s]^{2n+m\times 2n+m}. \tag{3.10}$$

Similarly as for the generalized Riccati equation, a sufficient condition for the solvability of the Lur'e equation is the absence of purely imaginary and higher-order infinite eigenvalues of the even matrix pencil $s\mathcal{E} - \mathcal{A}$. However, if there are such eigenvalues, the solvability criteria become much more involved. Then, we have to consider the eigenstructure of $s\mathcal{E} - \mathcal{A}$ in more detail. In [Rei11] this is done by evaluating its even Kronecker canonical form. It can also be shown, that the solution of (3.9) can be constructed via the deflating subspaces of (3.10). A pencil structure similar to (3.10) has also already been considered in [IOW99]. However, some of the special structural features of even matrix pencils, in particular the sign-characteristics of purely imaginary and infinite eigenvalues have not been exploited. So in contrast to [Rei11], where equivalent conditions for the solvability of the Lur'e equation in terms of the spectral structure of (3.10) are derived, [IOW99] only gives necessary conditions.

On the other hand, there are also generalizations of the ARE to descriptor systems. In [Meh91], a *generalized ARE* of the form

$$A^\mathsf{T}XE + E^\mathsf{T}XA - \big(E^\mathsf{T}XB + S\big)R^{-1}\big(B^\mathsf{T}XE + S^\mathsf{T}\big) + Q = 0, \quad X = X^\mathsf{T}, \tag{3.11}$$

is considered. In [Meh91] it has been shown how to construct stabilizing solutions of this equation for systems that are strongly stabilizable. Then the main idea consists of a feedback regularization of the system $(E, A, B)$, an approach we will also make use of in this thesis. However, in general, the relationship between the solutions of the generalized ARE (3.11) and the optimal control problem (3.1) is hidden or even lost. For instance, in [BL87] it is pointed out, that (3.11) might not have a solution, even if (3.1) is feasible.

Moreover, a strong focus of [Meh91] is the treatment of two-point boundary value problems resulting from an application of Pontryagin's maximum principle. These attain the form

$$\mathcal{E}\dot{z}(t) = \mathcal{A}z(t), \quad \mathcal{P}_1 z(0) = 0, \quad \lim_{t\to\infty} \mathcal{P}_2 z(t) = 0, \tag{3.12}$$

where

$$s\mathcal{E} - \mathcal{A} = \begin{bmatrix} 0 & -sE + A & B \\ sE^\top + A^\top & Q & S \\ B^\top & S^\top & R \end{bmatrix} \in \mathbb{R}[s]^{2n+m\times 2n+m}, \tag{3.13}$$

and $\mathcal{P}_1$, $\mathcal{P}_2 \in \mathbb{R}^{2n+m\times 2n+m}$. For many of these considerations essential assumptions are (3.8) and the regularity of $s\mathcal{E} - \mathcal{A}$, in particular for the numerical solution of the boundary value problem (3.12) and the associated generalized ARE (3.11) via deflating subspaces of $s\mathcal{E} - \mathcal{A}$.

An alternative for generalizing the algebraic Riccati equation to DAEs is the approach of [KTK99, KK02] which studies a generalized ARE of the form

$$A^\top X + X^\top A - (X^\top B + S)R^{-1}(B^\top X + S^\top) + Q = 0, \quad E^\top X = X^\top E, \tag{3.14}$$

however, an unnatural side condition, namely the solvability of an *algebraic quadratic equation* is necessary to guarantee the existence of a stabilizing solution.

Even more interesting is the case where both $E$ and $R$ are singular. Note that for DAEs it is much more involved to decide whether the optimal control problem under consideration is regular or singular. In particular, the condition $R > 0$ is neither sufficient nor necessary for regularity of the optimal control problem. This also suggests that the invertibility of $R$ is a rather artificial condition. We refer to [Gee93] for a detailed discussion. This problem was already previously considered in [Gee94], even for nonsquare and non-regular descriptor systems, however still an assumption analogous to (3.8) and impulse controllability is assumed. In [Gee94] existence and uniqueness results for the optimal control are derived which are related to zero dynamics of a particular linear system. Moreover, the optimal controls derived there are related to the solutions $P \in \mathbb{R}^{n\times n}$ of the linear matrix inequality

$$\begin{bmatrix} A^\top PE + E^\top PA + Q & E^\top PB + S \\ B^\top PE + S^\top & R \end{bmatrix} \geq 0, \quad P = P^\top,$$

however the rank-minimization property for the optimal $P$ is lost.

In this thesis we choose a different approach, namely by generalizing the results in [Rei11]. We introduce a new type of algebraic matrix equation, namely, the *descriptor Lur'e equation*

$$\begin{bmatrix} A^\top X + X^\top A + Q & X^\top B + S \\ B^\top X + S^\top & R \end{bmatrix} =_{\mathcal{V}_{\text{sys}}} \begin{bmatrix} K^\top \\ L^\top \end{bmatrix} \begin{bmatrix} K & L \end{bmatrix}, \quad E^\top X = X^\top E, \tag{3.15}$$

that has be solved for $(X, K, L) \in \mathbb{R}^{n\times n} \times \mathbb{R}^{q\times n} \times \mathbb{R}^{q\times m}$ where $\mathcal{V}_{\text{sys}}$ is as in (3.5) and $q$ is as small as possible.

The structure of the solution set of this matrix equation is furthermore investigated. In contrast to [Gee94], with $P = X$, solutions of (3.15) are *rank-minimizing solutions* of (3.4), i. e., with a basis matrix $M_{\mathcal{V}_{\text{sys}}}$ of $\mathcal{V}_{\text{sys}}$ it holds that

$$M_{\mathcal{V}_{\text{sys}}}^{\mathsf{T}} \begin{bmatrix} A^{\mathsf{T}}P + P^{\mathsf{T}}A + Q & P^{\mathsf{T}}B + S \\ B^{\mathsf{T}}P + S^{\mathsf{T}} & R \end{bmatrix} M_{\mathcal{V}_{\text{sys}}}$$

has minimal rank among all solutions of the descriptor KYP inequality (3.4). We show that, under some conditions related to controllability of (2.6), there exist *stabilizing* and *anti-stabilizing* solutions that are simultaneously *extremal solutions*, where "extremal" has to be understood in terms of definiteness of $E^{\mathsf{T}}X$.

The foundation of all these considerations will be the eigenstructure of the *associated even matrix pencil* (3.13). We generalize the results from [Rei11] and show that certain solutions of the descriptor Lur'e equation can be constructed from deflating subspaces of this pencil.

In this work, we will drop most of the restrictions of the previously mentioned approaches. For instance, we neither assume a property related to (3.8) nor the regularity of $s\mathcal{E} - \mathcal{A}$. We will mainly assume that $sE - A$ is regular and that the system (2.6) is impulse controllable. No assumptions on the index of the differential-algebraic equation are made. Note, that when $(E, A, B) \in \Sigma_{n,m}$ is not impulse controllable, we can recover the impulse controllable subsystem, for instance by a staircase reduction of the triple $(E, A, B)$, see [BGMN94]. Moreover, in Section 3.12 we briefly describe a way to approach non-impulse controllable systems directly using the so-called *feedback equivalence form* of a system [RRV14].

The solutions of the descriptor Lur'e equations will further be shown to define a *spectral factorization* of the Popov function. That is, we can construct some $W(s) \in \mathbb{R}(s)^{q \times m}$ such that

$$\Phi(s) = W^{\sim}(s)W(s). \tag{3.16}$$

In particular, the matrix function $W(s)$ will be outer, if it is constructed from the stabilizing solution of the descriptor Lur'e equation (3.15).

A special emphasis will be placed on the characterization whether the descriptor KYP inequality (3.4) has a *nonpositive solution*. That is, (3.4) holds with $E^{\mathsf{T}}P = P^{\mathsf{T}}E \leq 0$. We show that, in this case, an optimal control problem is feasible which is stronger than (3.1). We will present consequences for the analysis of dissipativity of differential-algebraic systems. In particular, important special cases such as *contractivity* and *passivity* will be treated. The descriptor KYP inequality (3.4) and the characterization of nonpositivity of solutions will result in differential-algebraic versions of the *bounded real lemma* and *positive real lemma*, which are both well-known for ODEs, see, e. g., [AV73].

As a further application of our presented theory, we show that *normalized coprime factorizations* [MG89, Var98] and *inner-outer factorizations* [Gre88] of general transfer functions $G(s) \in \mathbb{R}(s)^{p\times m}$ can be constructed with the theory that is developed in this chapter.

To complete the discussion we point towards some generalizations into the direction of time-varying and nonlinear differential-algebraic equations. In [KM04], linear-quadratic optimal control problems for time varying DAEs and time-varying weights $Q$, $S$, and $R$ in the cost functional are treated. Then, a time-varying boundary value problem similar to (3.12) is constructed. Necessary and sufficient conditions for feasibility of the optimal control problems are derived via an inherent Hamiltonian ODE system which is obtained by applying appropriate projection operators to the boundary value problem. We also refer to [Bac03, Bac06] where the special case of a DAE of index two is considered.

A different approach for time-varying optimal control problems is chosen in [KM11]. There, one considers two optimality boundary value problems similar to (3.12), one constructed from the original system and another one based on a so-called *strangeness-free* formulation of the DAE system. Then the solvability conditions and solutions of both boundary value problems are studied and they are related to each other. Moreover, in the recent works [KMS14, MS14], structured global condensed forms for the optimality system (analogous to the even staircase form (2.1)) are derived which allow to analyze its properties.

Finally, optimal control problems subject to nonlinear differential-algebraic equations are discussed, e. g., in [KM01, KMR01, KM08] by analyzing local linearizations of the nonlinear DAE which usually result in a time-varying linear DAE and allow the application of the previously mentioned techniques.

This chapter is structured as follows. In Section 3.2 we introduce some basic state and feedback transformations of the system (2.6) and their influence on the above introduced objects. In Section 3.3 we formulate the Kalman-Yakubovich-Popov lemma for differential-algebraic equations in which we relate the nonnegativity of the Popov function on $\mathrm{i}\mathbb{R}$ to the existence of a solution of the descriptor KYP inequality. In Section 3.4 we consider associated even matrix pencils and descriptor Lur'e equations. In particular, we show that there is a one-to-one correspondence between the solutions of the descriptor Lur'e equation and certain deflating subspaces of the even matrix pencil. In Section 3.5 we consider stabilizing, anti-stabilizing, and extremal solutions. In particular, we show that the stabilizing solution corresponds to a semistable deflating subspace of the associated even matrix pencil. The spectral factorization problem is treated thereafter in Section 3.6. We prove that the stabilizing solution defines a spectral factorization in which the spectral factor is an outer transfer function. The problem of existence of nonpositive solutions of the descriptor KYP inequality is treated in Section 3.7. We will introduce the so-called *modified Popov function.* Some criteria for the existence on nonpositive solutions will be formulated by means of this function.

Whereas Sections 3.3–3.7 are of purely linear algebraic nature, we turn to the linear-quadratic optimal control problem in Section 3.8. We prove that a stabilizing solution of the descriptor KYP inequality exists if and only if the optimal control problem (3.1) is feasible for all $x_0 \in \mathbb{R}^n$. Furthermore, if a solution of the descriptor KYP inequality exists, then the system further satisfies a certain dissipation inequality. The latter results will be used in Section 3.9 to formulate equivalent criteria for dissipativity and cyclo-dissipativity of systems. In particular, this gives rise to differential-algebraic versions of the bounded real lemma and the positive real lemma. In Sections 3.10 and 3.11 we further apply our results to construct realizations of inner-outer factorizations and normalized coprime factorizations by means of the solutions of the descriptor Lur'e equations. Finally, in Section 3.12 we summarize the results and discuss open problems as well as possible extensions of our theory.

## 3.2 State and Feedback Transformations

Here we develop several auxiliary results which are based on transformations of the form

$$\boxed{\begin{aligned} E &\in \mathbb{R}^{n\times n}, \\ A &\in \mathbb{R}^{n\times n}, \\ B &\in \mathbb{R}^{n\times m}, \\ Q = Q^\top &\in \mathbb{R}^{n\times n}, \\ S &\in \mathbb{R}^{m\times n}, \\ R = R^\top &\in \mathbb{R}^{m\times m} \end{aligned}} \quad \underset{\substack{W,T \in \mathrm{Gl}_n(\mathbb{R}), \\ V \in \mathrm{Gl}_m(\mathbb{R}), \\ F \in \mathbb{R}^{m\times n}}}{\rightsquigarrow} \quad \boxed{\begin{aligned} E_F &= WET, \\ A_F &= W(A+BF)T, \\ B_F &= WBV, \\ Q_F &= T^\top(Q+SF+F^\top S^\top + F^\top RF)T, \\ S_F &= T^\top(S+F^\top R)V, \\ R_F &= V^\top RV \end{aligned}} \tag{3.17}$$

For a differential-algebraic system (2.6), $W$ describes a transformation of the equations, $T$ a transformation of the state $x(\cdot)$, $V$ a transformation of the input $u(\cdot)$, and $F$ describes a feedback action. It is not difficult to see that a transformation of the form (3.17) is a group action and therefore defines an equivalence relation. In particular, we have reversibility of any transformation.

**Lemma 3.2.1** (Feedback regularization). *Let $(E, A, B) \in \Sigma_{n,m}$ be given. Then with $r = \operatorname{rank} E$, there exist $W$, $T \in \mathrm{Gl}_n(\mathbb{R})$, $F \in \mathbb{R}^{m\times n}$ such that*

$$W(sE - (A+BF))T = \begin{bmatrix} sI_r - A_{11} & 0 \\ 0 & -I_{n-r} \end{bmatrix},$$

*if and only if $(E, A, B)$ is impulse controllable.*

*Proof.* According to [BGMN92, Cor. 7 and p. 59], there exists some $F \in \mathbb{R}^{m\times n}$ such that $sE - (A + BF)$ is regular and of index at most one, if and only if $(E, A, B)$ is impulse controllable. A transformation $W(sE - (A + BF))T$ to QWF then leads to the desired result. □

Lemma 3.2.1 allows, in case of impulse controllability, a transformation of type (3.17) to matrices of the form

$$\begin{aligned} sE_F - A_F &= \begin{bmatrix} sI_r - A_{11} & 0 \\ 0 & -I_{n-r} \end{bmatrix}, & B_F &= \begin{bmatrix} B_1 \\ B_2 \end{bmatrix}, & \\ Q_F &= \begin{bmatrix} Q_{11} & Q_{12} \\ Q_{12}^\mathsf{T} & Q_{22} \end{bmatrix}, & S_F &= \begin{bmatrix} S_1 \\ S_2 \end{bmatrix}, & R_F = R, \end{aligned} \tag{3.18}$$

where $A_{11}, Q_{11} \in \mathbb{R}^{r\times r}$, $B_1, S_1 \in \mathbb{R}^{r\times m}$, $B_2, S_2 \in \mathbb{R}^{n-r\times m}$, $Q_{12} \in \mathbb{R}^{r\times n-r}$, and $Q_{22} \in \mathbb{R}^{n-r\times n-r}$. We now show how the associated matrix pencil, the KYP matrix, and the optimal control problem behave under feedback, state, and input transformations.

**Proposition 3.2.2.** *Let $(E, A, B) \in \Sigma_{n,m}$ with the system space $\mathcal{V}_{\mathrm{sys}}$ be given and let $Q = Q^\mathsf{T} \in \mathbb{R}^{n\times n}$, $S \in \mathbb{R}^{n\times m}$, and $R = R^\mathsf{T} \in \mathbb{R}^{m\times m}$. Further, assume that $F \in \mathbb{R}^{m\times n}$, $W, T \in \mathrm{Gl}_n(\mathbb{R})$ and $V \in \mathrm{Gl}_m(\mathbb{R})$ are given and let $E_F$, $A_F$, $B_F$, $Q_F$, $S_F$, $R_F$ be defined as in (3.17) such that $(E_F, A_F, B_F) \in \Sigma_{n,m}$ with the system space $\mathcal{V}_{\mathrm{sys},F}$. Then the following statements hold:*

*a) The two Popov functions $\Phi(s), \Phi_F(s) \in \mathbb{R}(s)^{m\times m}$ with*

$$\Phi(s) = \begin{bmatrix} (sE-A)^{-1}B \\ I_m \end{bmatrix}^\sim \begin{bmatrix} Q & S \\ S^\mathsf{T} & R \end{bmatrix} \begin{bmatrix} (sE-A)^{-1}B \\ I_m \end{bmatrix},$$

$$\Phi_F(s) = \begin{bmatrix} (sE_F-A_F)^{-1}B_F \\ I_m \end{bmatrix}^\sim \begin{bmatrix} Q_F & S_F \\ S_F^\mathsf{T} & R_F \end{bmatrix} \begin{bmatrix} (sE_F-A_F)^{-1}B_F \\ I_m \end{bmatrix}$$

*are related via*

$$\Phi_F(s) = \Theta_F^\sim(s)\Phi(s)\Theta_F(s) \text{ with } \Theta_F(s) := V + FT(sE_F - A_F)^{-1}B_F.$$

*In particular, $\Phi(\mathrm{i}\omega) \geq 0$ holds for all $\mathrm{i}\omega \notin \Lambda(E, A)$ if and only if $\Phi_F(\mathrm{i}\omega) \geq 0$ holds for all $\mathrm{i}\omega \notin \Lambda(E_F, A_F)$.*

*b) For $P \in \mathbb{R}^{n\times n}$ and $P_F = W^{-\mathsf{T}}PT$ it holds that $E^\mathsf{T}P = P^\mathsf{T}E$ if and only if $E_F^\mathsf{T}P_F = P_F^\mathsf{T}E_F$. Furthermore, it holds that*

$$\begin{aligned} &\begin{bmatrix} A_F^\mathsf{T}P_F + P_F^\mathsf{T}A_F + Q_F & P_F^\mathsf{T}B_F + S_F \\ B_F^\mathsf{T}P_F + S_F^\mathsf{T} & R_F \end{bmatrix} \\ &\qquad = \begin{bmatrix} T^\mathsf{T} & T^\mathsf{T}F^\mathsf{T} \\ 0 & V^\mathsf{T} \end{bmatrix} \cdot \begin{bmatrix} A^\mathsf{T}P + P^\mathsf{T}A + Q & P^\mathsf{T}B + S \\ B^\mathsf{T}P + S^\mathsf{T} & R \end{bmatrix} \cdot \begin{bmatrix} T & 0 \\ FT & V \end{bmatrix}, \end{aligned}$$

*and the system spaces*

$$\mathcal{V}_{\mathrm{sys}} = \left\{ \begin{pmatrix} x \\ u \end{pmatrix} \in \mathbb{R}^{n+m} : Ax + Bu \in \operatorname{im} E \right\},$$

$$\mathcal{V}_{\mathrm{sys},F} = \left\{ \begin{pmatrix} x \\ u \end{pmatrix} \in \mathbb{R}^{n+m} : A_Fx + B_Fu \in \operatorname{im} E_F \right\}$$

*are related by*

$$\mathcal{V}_{\text{sys}} = \begin{bmatrix} T & 0 \\ FT & V \end{bmatrix} \cdot \mathcal{V}_{\text{sys},F}.$$

*In particular, it holds that* $E_F^\mathsf{T} P_F > (\geq, <, \leq)\, 0$ *if and only if* $E^\mathsf{T} P > (\geq, <, \leq)\, 0$ *and, moreover,*

$$\begin{aligned} &\begin{bmatrix} A^\mathsf{T} P + P^\mathsf{T} A + Q & P^\mathsf{T} B + S \\ B^\mathsf{T} P + S^\mathsf{T} & R \end{bmatrix} \geq_{\mathcal{V}_{\text{sys}}} 0, \\ \Leftrightarrow &\begin{bmatrix} A_F^\mathsf{T} P_F + P_F^\mathsf{T} A_F + Q_F & P_F^\mathsf{T} B_F + S_F \\ B_F^\mathsf{T} P_F + S_F^\mathsf{T} & R_F \end{bmatrix} \geq_{\mathcal{V}_{\text{sys},F}} 0. \end{aligned}$$

c) *The triple* $(X, K, L)$ *solves the descriptor Lur'e equation*

$$\begin{bmatrix} A^\mathsf{T} X + X^\mathsf{T} A + Q & X^\mathsf{T} B + S \\ B^\mathsf{T} X + S^\mathsf{T} & R \end{bmatrix} =_{\mathcal{V}_{\text{sys}}} \begin{bmatrix} K^\mathsf{T} \\ L^\mathsf{T} \end{bmatrix} \begin{bmatrix} K & L \end{bmatrix}, \quad E^\mathsf{T} X = X^\mathsf{T} E, \quad (3.19)$$

*if and only if*

$$(X_F, K_F, L_F) = (W^{-\mathsf{T}} X T, KT + LFT, LV)$$

*solves*

$$\begin{bmatrix} A_F^\mathsf{T} X_F + X_F^\mathsf{T} A_F + Q_F & X_F^\mathsf{T} B_F + S_F \\ B_F^\mathsf{T} X_F + S_F^\mathsf{T} & R_F \end{bmatrix} =_{\mathcal{V}_{\text{sys},F}} \begin{bmatrix} K_F^\mathsf{T} \\ L_F^\mathsf{T} \end{bmatrix} \begin{bmatrix} K_F & L_F \end{bmatrix},$$
$$E_F^\mathsf{T} X_F = X_F^\mathsf{T} E_F. \quad (3.20)$$

*Moreover,* $(X, K, L)$ *is a stabilizing (anti-stabilizing) solution of* (3.19) *if and only if the triple* $(X_F, K_F, L_F)$ *is a stabilizing (anti-stabilizing) solution of* (3.20), *see Definition 3.5.1.*

d) *The even matrix pencils* $s\mathcal{E} - \mathcal{A}$, $s\mathcal{E}_F - \mathcal{A}_F \in \mathbb{R}[s]^{2n+m \times 2n+m}$ *with*

$$\begin{aligned} s\mathcal{E} - \mathcal{A} &= \begin{bmatrix} 0 & -sE + A & B \\ sE^\mathsf{T} + A^\mathsf{T} & Q & S \\ B^\mathsf{T} & S^\mathsf{T} & R \end{bmatrix}, \\ s\mathcal{E}_F - \mathcal{A}_F &= \begin{bmatrix} 0 & -sE_F + A_F & B_F \\ sE_F^\mathsf{T} + A_F^\mathsf{T} & Q_F & S_F \\ B_F^\mathsf{T} & S_F^\mathsf{T} & R_F \end{bmatrix} \end{aligned}$$

*are related via*

$$s\mathcal{E}_F - \mathcal{A}_F = U^\mathsf{T} (s\mathcal{E} - \mathcal{A}) U,$$

*where*

$$U = \begin{bmatrix} W^\mathsf{T} & 0 & 0 \\ 0 & T & 0 \\ 0 & FT & V \end{bmatrix}.$$

*e) It holds that* $(x,u) \in \mathfrak{B}_{(E,A,B)}$ *if and only if* $(x_F, u_F) := (T^{-1}x, V^{-1}(u - Fx)) \in \mathfrak{B}_{(E_F,A_F,B_F)}$. *Furthermore, for all* $t_0, t_1 \in \mathbb{R} \cup \{-\infty, \infty\}$ *it holds that*

$$\int_{t_0}^{t_1} \begin{pmatrix} x(\tau) \\ u(\tau) \end{pmatrix}^\top \begin{bmatrix} Q & S \\ S^\top & R \end{bmatrix} \begin{pmatrix} x(\tau) \\ u(\tau) \end{pmatrix} \mathrm{d}\tau = \int_{t_0}^{t_1} \begin{pmatrix} x_F(\tau) \\ u_F(\tau) \end{pmatrix}^\top \begin{bmatrix} Q_F & S_F \\ S_F^\top & R_F \end{bmatrix} \begin{pmatrix} x_F(\tau) \\ u_F(\tau) \end{pmatrix} \mathrm{d}\tau.$$

*Proof.* We only proof statement a), b)–e) follow from simple algebraic manipulations. Define $G := \lambda E - A$ and $G_F := \lambda E_F - A_F$ for a fixed $\lambda \notin \Lambda(E,A) \cup \Lambda(E_F, A_F) \cup \Lambda(-E,A) \cup \Lambda(-E_F, A_F)$. By applying the Sherman-Morrison-Woodbury identity [GL96, p. 50] we obtain

$$\begin{aligned} TG_F^{-1}B_F &= (G - BF)^{-1}\, BV \\ &= \left(G^{-1} + G^{-1}BF(G - BF)^{-1}\right) BV \\ &= G^{-1}B(V + F(G - BF)^{-1}BV) \\ &= G^{-1}B\Theta_F(\lambda). \end{aligned} \tag{3.21}$$

Then we have

$$\begin{aligned} \Phi_F(s) &= \begin{bmatrix} (sE_F - A_F)^{-1}B_F \\ I_m \end{bmatrix}^\sim \begin{bmatrix} Q_F & S_F \\ S_F^\top & R_F \end{bmatrix} \begin{bmatrix} (sE_F - A_F)^{-1}B_F \\ I_m \end{bmatrix} \\ &= \begin{bmatrix} (sE_F - A_F)^{-1}B_F \\ I_m \end{bmatrix}^\sim \begin{bmatrix} T^\top & T^\top F^\top \\ 0 & V^\top \end{bmatrix} \begin{bmatrix} Q & S \\ S^\top & R \end{bmatrix} \begin{bmatrix} T & 0 \\ FT & V \end{bmatrix} \\ &\qquad \cdot \begin{bmatrix} (sE_F - A_F)^{-1}B_F \\ I_m \end{bmatrix} \\ &= \Theta_F^\sim(s)\Phi(s)\Theta_F(s), \end{aligned}$$

where the latter relation follows from (3.21). In particular, we have shown that $\Phi(\mathrm{i}\omega) \geq 0$ for all $\mathrm{i}\omega \notin \Lambda(E,A) \cup \Lambda(E_F, A_F)$ is equivalent to $\Phi_F(\mathrm{i}\omega) \geq 0$ for all $\mathrm{i}\omega \notin \Lambda(E,A) \cup \Lambda(E_F, A_F)$. The result now follows from the continuity of $\Phi(s)$ and $\Phi_F(s)$ in a neighborhood of points $\mathrm{i}\omega \notin \Lambda(E,A)$ and $\mathrm{i}\omega \notin \Lambda(E_F, A_F)$, respectively. □

## 3.3 Kalman-Yakubovich-Popov Lemma

In this section we present a differential-algebraic version of the Kalman-Yakubovich-Popov (KYP) lemma. Thereby we equivalently characterize the positive semidefiniteness of the Popov function on the imaginary axis by the solvability of a linear matrix inequality.

The main result of this section is presented below, see also [RRV14, Thm 4.1]. We will later present some facts on the structure of the solution set of the occurring linear matrix inequality.

**Theorem 3.3.1** (KYP lemma for differential-algebraic systems). *Let $(E, A, B) \in \Sigma_{n,m}$ with the system space $\mathcal{V}_{\text{sys}}$ be given, and let $Q = Q^\top \in \mathbb{R}^{n\times n}$, $S \in \mathbb{R}^{n\times m}$, and $R = R^\top \in \mathbb{R}^{m\times m}$. Then the following statements hold true:*

*a) Assume that there exists some $P \in \mathbb{R}^{n\times n}$ such that*

$$\begin{bmatrix} A^\top P + P^\top A + Q & P^\top B + S \\ B^\top P + S^\top & R \end{bmatrix} \geq_{\mathcal{V}_{\text{sys}}} 0, \quad E^\top P = P^\top E. \tag{3.22}$$

*Then it holds that*

$$\Phi(\mathrm{i}\omega) \geq 0 \quad \forall \omega \in \mathbb{R} \text{ with } \mathrm{i}\omega \notin \Lambda(E, A). \tag{3.23}$$

*b) Assume that at least one of the following two assumptions holds true:*

*i) $(E, A, B)$ is strongly sign-controllable and the Popov function (3.2) satisfies*

$$\operatorname{rank}_{\mathbb{R}(s)} \Phi(s) = m;$$

*ii) $(E, A, B)$ is strongly controllable.*

*Further, assume that the Popov function fulfills (3.23). Then there exists some $P \in \mathbb{R}^{n\times n}$ that solves the linear matrix inequality (3.22).*

*Proof.* First, we prove statement a). For all $\lambda \in \mathbb{C} \setminus \Lambda(E, A)$ it holds that

$$\begin{aligned} \begin{bmatrix} A & B \end{bmatrix} \begin{bmatrix} (\lambda E - A)^{-1} B \\ I_m \end{bmatrix} &= A(\lambda E - A)^{-1} B + B \\ &= (A + \lambda E - A)(\lambda E - A)^{-1} B \\ &= \lambda E (\lambda E - A)^{-1} B \in \operatorname{im} E. \end{aligned}$$

Therefore, we have

$$\operatorname{im} \begin{bmatrix} (\lambda E - A)^{-1} B \\ I_m \end{bmatrix} \subseteq \mathcal{V}_{\text{sys}} \quad \forall \lambda \in \mathbb{C} \setminus \Lambda(E, A). \tag{3.24}$$

Let $P \in \mathbb{R}^{n\times n}$ such that (3.22) is fulfilled. Then we obtain that

$$\begin{aligned}
&\begin{bmatrix}(sE-A)^{-1}B\\ I_m\end{bmatrix}^{\sim}\begin{bmatrix}A^{\mathsf T}P+P^{\mathsf T}A & P^{\mathsf T}B\\ B^{\mathsf T}P & 0\end{bmatrix}\begin{bmatrix}(sE-A)^{-1}B\\ I_m\end{bmatrix}\\
&= B^{\mathsf T}\big(-sE^{\mathsf T}-A^{\mathsf T}\big)^{-1}\big(A^{\mathsf T}P+P^{\mathsf T}A\big)\big(sE-A\big)^{-1}B\\
&\qquad + B^{\mathsf T}\big(-sE^{\mathsf T}-A^{\mathsf T}\big)^{-1}P^{\mathsf T}B + B^{\mathsf T}P\big(sE-A\big)^{-1}B\\
&= B^{\mathsf T}\big(-sE^{\mathsf T}-A^{\mathsf T}\big)^{-1}\Big(\big(sE^{\mathsf T}+A^{\mathsf T}\big)P+P^{\mathsf T}\big(-sE+A\big)\Big)\big(sE-A\big)^{-1}B\\
&\qquad + B^{\mathsf T}\big(-sE^{\mathsf T}-A^{\mathsf T}\big)^{-1}P^{\mathsf T}B + B^{\mathsf T}P\big(sE-A\big)^{-1}B\\
&= -B^{\mathsf T}P\big(sE-A\big)^{-1}B - B^{\mathsf T}\big(-sE^{\mathsf T}-A^{\mathsf T}\big)^{-1}P^{\mathsf T}B\\
&\qquad + B^{\mathsf T}\big(-sE^{\mathsf T}-A^{\mathsf T}\big)^{-1}P^{\mathsf T}B + B^{\mathsf T}P\big(sE-A\big)^{-1}B\\
&= 0.
\end{aligned} \tag{3.25}$$

This gives rise to the fact that for all $\omega \in \mathbb{R}$ with $\mathrm{i}\omega \notin \Lambda(E,A)$ it holds that

$$\begin{aligned}
\Phi(\mathrm{i}\omega) &= \begin{bmatrix}(\mathrm{i}\omega E-A)^{-1}B\\ I_m\end{bmatrix}^{\mathsf H}\begin{bmatrix}Q & S\\ S^{\mathsf T} & R\end{bmatrix}\begin{bmatrix}(\mathrm{i}\omega E-A)^{-1}B\\ I_m\end{bmatrix}\\
&\overset{(3.22)\&(3.24)}{\geq} -\begin{bmatrix}(\mathrm{i}\omega E-A)^{-1}B\\ I_m\end{bmatrix}^{\mathsf H}\begin{bmatrix}A^{\mathsf T}P+P^{\mathsf T}A & P^{\mathsf T}B\\ B^{\mathsf T}P & 0\end{bmatrix}\begin{bmatrix}(\mathrm{i}\omega E-A)^{-1}B\\ I_m\end{bmatrix}\overset{(3.25)}{=}0.
\end{aligned} \tag{3.26}$$

Now we prove statement b). This part is based on feedback and the KYP lemma for ODE systems, i. e., Theorem 3.3.1 holds for $E = I_n$, see, e. g., [CALM97]. Note that for ODE systems, impulse controllability is trivially fulfilled, the space $\mathcal{V}_{\text{sys}}$ is not anymore a proper subspace of $\mathbb{R}^{n+m}$ and, moreover, the equation $E^{\mathsf T}P = P^{\mathsf T}E$ is equivalent to $P$ being symmetric.

Let $r = \operatorname{rank} E$ and assume that i) or ii) holds true. By Lemma 3.2.1 and Proposition 3.2.2 a) and b), it suffices to prove the statement for the case where

$$sE-A = \begin{bmatrix}sI_r - A_{11} & 0\\ 0 & -I_{n-r}\end{bmatrix},\quad B = \begin{bmatrix}B_1\\ B_2\end{bmatrix},\quad Q = \begin{bmatrix}Q_{11} & Q_{12}\\ Q_{12}^{\mathsf T} & Q_{22}\end{bmatrix},\quad S = \begin{bmatrix}S_1\\ S_2\end{bmatrix},$$

and $A_{11}, Q_{11} \in \mathbb{R}^{r\times r}$, $B_1, S_1 \in \mathbb{R}^{r\times m}$, $B_2, S_2 \in \mathbb{R}^{n-r\times m}$, $Q_{12} \in \mathbb{R}^{r\times n-r}$, and $Q_{22} \in \mathbb{R}^{n-r\times n-r}$. Now we can observe the following facts:

1) The system $(E,A,B)$ is strongly sign-controllable (strongly controllable) if and only if $(I_r, A_{11}, B_1)$ is strongly sign-controllable (strongly controllable), respectively.

2) We may rewrite the Popov function as

$$\begin{aligned}\Phi(s) &= \begin{bmatrix}(sE-A)^{-1}B\\ I_m\end{bmatrix}^{\sim}\begin{bmatrix}Q & S\\ S^\top & R\end{bmatrix}\begin{bmatrix}(sE-A)^{-1}B\\ I_m\end{bmatrix}\\
&= \begin{bmatrix}(sI_r-A_{11})^{-1}B_1\\ -B_2\\ I_m\end{bmatrix}^{\sim}\begin{bmatrix}Q_{11} & Q_{12} & S_1\\ Q_{12}^\top & Q_{22} & S_2\\ S_1^\top & S_2^\top & R\end{bmatrix}\begin{bmatrix}(sI_r-A_{11})^{-1}B_1\\ -B_2\\ I_m\end{bmatrix}\\
&= \begin{bmatrix}(sI_r-A_{11})^{-1}B_1\\ I_m\end{bmatrix}^{\sim}\begin{bmatrix}Q_{11} & S_1-Q_{12}B_2\\ S_1^\top-B_2^\top Q_{12}^\top & B_2^\top Q_{22}B_2-B_2^\top S_2-S_2^\top B_2+R\end{bmatrix}\\
&\qquad\cdot\begin{bmatrix}(sI_r-A_{11})^{-1}B_1\\ I_m\end{bmatrix}.\end{aligned}$$

3) A basis matrix for $\mathcal{V}_{\text{sys}}$ is given by

$$M_{\mathcal{V}_{\text{sys}}} := \begin{bmatrix}I_r & 0\\ 0 & -B_2\\ 0 & I_{n-r}\end{bmatrix}. \tag{3.27}$$

In particular, we obtain from 1) and 2) that assumptions i) and ii) are fulfilled by the sextuple $(E,A,B,Q,S,R)$ if and only if

$$\left(I_r, A_{11}, B_1, Q_{11}, S_1-Q_{12}B_2, B_2^\top Q_{22}B_2-B_2^\top S_2-S_2^\top B_2+R\right)$$

has the respective property. Assuming that (3.23) holds true, the KYP lemma for ODE systems implies that there exists a symmetric $P_{11}\in\mathbb{K}^{r\times r}$, such that

$$\begin{bmatrix}A_{11}^\top P_{11}+P_{11}A_{11}+Q_{11} & P_{11}B_1+S_1-Q_{12}B_2\\ B_1^\top P_{11}+S_1^\top-B_2^\top Q_{12}^\top & B_2^\top Q_{22}B_2-B_2^\top S_2-S_2^\top B_2+R\end{bmatrix}\geq 0. \tag{3.28}$$

Defining

$$P=\begin{bmatrix}P_{11} & 0\\ 0 & 0\end{bmatrix}\in\mathbb{R}^{n\times n},$$

we obtain $E^\top P = P^\top E$. Assume that $x\in\mathbb{R}^n$, $u\in\mathbb{R}^m$ such that

$$\begin{pmatrix}x\\ u\end{pmatrix}\in\mathcal{V}_{\text{sys}}.$$

and partition

$$x=\begin{pmatrix}x_1\\ x_2\end{pmatrix}$$

with $x_1 \in \mathbb{R}^r$ and $x_2 \in \mathbb{R}^{n-r}$. Now the structure of $\mathcal{V}_{\text{sys}}$ given by (3.27) implies $x_2 = -B_2 u$. Therefore, it holds that

$$\begin{aligned}
&\begin{pmatrix} x \\ u \end{pmatrix}^\top \begin{bmatrix} A^\top P + P^\top A + Q & P^\top B + S \\ B^\top P + S^\top & R \end{bmatrix} \begin{pmatrix} x \\ u \end{pmatrix} \\
&= \begin{pmatrix} x_1 \\ -B_2 u \\ u \end{pmatrix}^\top \begin{bmatrix} A_{11}^\top P_{11} + P_{11} A_{11} + Q_{11} & Q_{12} & P_{11} B_1 + S_1 \\ Q_{12}^\top & Q_{22} & S_2 \\ B_1^\top P_{11} + S_1^\top & S_2^\top & R \end{bmatrix} \begin{pmatrix} x_1 \\ -B_2 u \\ u \end{pmatrix} \\
&= \begin{pmatrix} x_1 \\ u \end{pmatrix}^\top \begin{bmatrix} A_{11}^\top P_{11} + P_{11} A_{11} + Q_{11} & P_{11} B_1 + S_1 - Q_{12} B_2 \\ B_1^\top P_{11} + S_1^\top - B_2^\top Q_{12}^\top & B_2^\top Q_{22} B_2 - B_2^\top S_2 - S_2^\top B_2 + R \end{bmatrix} \begin{pmatrix} x_1 \\ u \end{pmatrix} \geq 0.
\end{aligned}$$

□

*Remark* 3.3.2.

a) We obtain from Theorem 3.3.1 that, in the case where at least one of the assumptions i) or ii) is fulfilled, the feasibility of the linear matrix inequality (3.22) is equivalent to the nonnegativity property (3.23) of the Popov function.

b) If the Popov function fulfills (3.23), then the linear matrix inequality (3.22) might have an empty solution set. Counter-examples exist already in the ODE case, see [Sch91b, p. 88]. Further, note that impulse controllability (which is included in i) and ii)) has to be assumed. As a counter-example, consider the matrices

$$E = \begin{bmatrix} 1 & 0 \\ 0 & 0 \end{bmatrix}, \quad A = \begin{bmatrix} 0 & 1 \\ 1 & 0 \end{bmatrix}, \quad B = S = \begin{bmatrix} 0 \\ 0 \end{bmatrix}, \quad Q = \begin{bmatrix} 0 & 0 \\ 0 & -1 \end{bmatrix}, \quad R = 1.$$

The system $(E, A, B)$ is not impulse controllable. The Popov function is constant one, i. e., $\Phi(s) = 1$. In particular, (3.23) is fulfilled. The space $\mathcal{V}_{\text{sys}}$ reads

$$\mathcal{V}_{\text{sys}} = \left\{ \begin{pmatrix} 0 \\ x_2 \\ u \end{pmatrix} : x_2, u \in \mathbb{R} \right\}.$$

Now assume that $P \in \mathbb{R}^{2\times 2}$ solves (3.22). By $E^\top P = P^\top E$, we can infer that there exist $p_{11}, p_{21}, p_{22} \in \mathbb{R}$ with

$$P = \begin{bmatrix} p_{11} & 0 \\ p_{21} & p_{22} \end{bmatrix}.$$

Then we have

$$\begin{bmatrix} A^\top P + P^\top A + Q & P^\top B + S \\ B^\top P + S^\top & R \end{bmatrix} = \begin{bmatrix} 2p_{21} & p_{11} + p_{22} & 0 \\ p_{11} + p_{22} & -1 & 0 \\ 0 & 0 & 1 \end{bmatrix}.$$

However, it holds that

$$v^{\mathsf{T}} \begin{bmatrix} A^{\mathsf{T}}P + P^{\mathsf{T}}A + Q & P^{\mathsf{T}}B + S \\ B^{\mathsf{T}}P + S^{\mathsf{T}} & R \end{bmatrix} v = -1 \not\geq 0 \text{ for } v = \begin{pmatrix} 0 \\ 1 \\ 0 \end{pmatrix} \in \mathcal{V}_{\text{sys}}.$$

Consequently, the linear matrix inequality (3.22) does not have any solutions.

## 3.4 Even Matrix Pencils and Descriptor Lur'e Equations

With the matrices as in the previous section, we now consider the even matrix pencil

$$s\mathcal{E} - \mathcal{A} = \begin{bmatrix} 0 & -sE + A & B \\ sE^{\mathsf{T}} + A^{\mathsf{T}} & Q & S \\ B^{\mathsf{T}} & S^{\mathsf{T}} & R \end{bmatrix} \in \mathbb{R}[s]^{2n+m \times 2n+m}. \tag{3.29}$$

This section is devoted to the analysis of relations between $s\mathcal{E}-\mathcal{A}$, the semidefiniteness of $\Phi(s)$ on $\mathrm{i}\mathbb{R}\setminus\Lambda(E,A)$, the solution, and the solution structure of the descriptor Lur'e equation (3.15). In particular, we show that their solutions can be constructed by using deflating subspaces of $s\mathcal{E} - \mathcal{A}$.

The following theorem relates the eigenstructure of $s\mathcal{E} - \mathcal{A}$ to the positive semidefiniteness of the Popov function $\Phi(s)$ on $\mathrm{i}\mathbb{R} \setminus \Lambda(E, A)$. Note that this result generalizes [Rei11, Thm. 3.1] in two ways: First, the differential-algebraic case is considered instead of only the standard case $E = I_n$. Second, neither condition i) nor ii) from Theorem 3.3.1 has to be presumed. Instead we employ the weaker condition of absence of uncontrollable modes on the imaginary axis. First, we present an auxiliary lemma.

**Lemma 3.4.1.** *For all $\omega \in \mathbb{R}$ with $\det(\mathrm{i}\omega\mathcal{E} - \mathcal{A}) \neq 0$ it holds that*

$$U(\mathrm{i}\omega)^{\mathsf{H}}(\mathrm{i}\omega\mathcal{E} - \mathcal{A})U(\mathrm{i}\omega) = \begin{bmatrix} 0 & -\mathrm{i}\omega E + A & 0 \\ \mathrm{i}\omega E^{\mathsf{T}} + A^{\mathsf{T}} & Q & 0 \\ 0 & 0 & \Phi(\mathrm{i}\omega) \end{bmatrix}$$

*with* $U(\mathrm{i}\omega) = \begin{bmatrix} I_n & 0 & \left(\mathrm{i}\omega E^{\mathsf{T}} + A^{\mathsf{T}}\right)^{-1}\left(Q(-\mathrm{i}\omega E + A)^{-1}B - S\right) \\ 0 & I_n & -(-\mathrm{i}\omega E + A)^{-1}B \\ 0 & 0 & I_m \end{bmatrix}.$

*Proof.* The fact can be verified by simple matrix calculations. □

**Theorem 3.4.2.** *Let the control system $(E, A, B) \in \Sigma_{n,m}$ be given. Further, let $Q = Q^{\mathsf{T}} \in \mathbb{R}^{n\times n}$, $S \in \mathbb{R}^{n\times m}$, and $R = R^{\mathsf{T}} \in \mathbb{R}^{m\times m}$. Assume that $(E, A, B)$ has no uncontrollable modes on the imaginary axis. Further, let $s\mathcal{E} - \mathcal{A}$ as in (3.29) be the associated even matrix pencil, and $\Phi(s)$ be the Popov function as in (3.2). Define $r = \operatorname{rank} E$ and $q = \operatorname{rank}_{\mathbb{R}(s)} \Phi(s)$. Then the following statements are equivalent:*

*a) For all* $\omega \in \mathbb{R}$ *with* $\mathrm{i}\omega \notin \Lambda(E, A)$ *it holds that* $\Phi(\mathrm{i}\omega) \geq 0$.

*b) The EKCF of* $s\mathcal{E} - \mathcal{A}$ *has the following structure:*

*i) All blocks of type E2 have even size and negative sign-characteristic.*

*ii) There exist exactly* $2(n - r) + q$ *blocks of type E3.*

*iii) There exist* $q$ *blocks of type E3 with odd size and positive sign-characteristic.*

*iv) The remaining* $2(n - r)$ *blocks of type E3 are either of even size; or of odd size and the number of odd-sized blocks with positive and negative sign-characteristic is equal.*

*v) There exist exactly* $m - q$ *blocks of type E4.*

*c) The EKCF of* $s\mathcal{E} - \mathcal{A}$ *has the following structure:*

*i') All blocks of type E2 have even size.*

- *The properties b) ii)–v) are valid.*

*Proof.* The absence of uncontrollable modes on the imaginary axis implies that there exists some feedback that removes all imaginary modes. By Proposition 3.2.2 a) and c), it is therefore no loss of generality to assume that $\mathrm{i}\omega \notin \Lambda(E, A)$ for all $\omega \in \mathbb{R}$ and, consequently, the Popov function has no poles on the imaginary axis.

We show that a) implies b): By $\operatorname{rank}_{\mathbb{R}(s)} \Phi(s) = q$ we can conclude that there exists some function $a : \mathbb{R} \to \mathbb{N}_0$ with finite support, such that for all $\omega \in \mathbb{R}$, it holds that

$$\operatorname{In}(\Phi(\mathrm{i}\omega)) = (q - a(\omega), m - q + a(\omega), 0).$$

By applying Lemma 3.4.1, Sylvester's law of inertia [GL96, p. 403] and, by the absence of purely imaginary eigenvalues of $sE - A$, we obtain

$$\operatorname{In}\left(\begin{bmatrix} 0 & -\mathrm{i}\omega E + A \\ \mathrm{i}\omega E^{\mathsf{T}} + A^{\mathsf{T}} & Q \end{bmatrix}\right) = (n, 0, n) \quad \forall \omega \in \mathbb{R}, \tag{3.30}$$

and thus we deduce that

$$\operatorname{In}(\mathrm{i}\omega\mathcal{E} - \mathcal{A}) = (n + q - a(\omega), m - q + a(\omega), n). \tag{3.31}$$

Sylvester's law of inertia further implies that the inertia of $\mathrm{i}\omega\mathcal{E} - \mathcal{A}$ coincides with that of the EKCF of $s\mathcal{E} - \mathcal{A}$ at $\mathrm{i}\omega$. The strategy of the proof consists of analyzing which combination of blocks of an EKCF generates such an inertia pattern. Identified blocks will then be removed to further consider the remaining subpencil.

First, we see from Lemma 3.4.1 that $s\mathcal{E} - \mathcal{A} \in \mathbb{R}[s]^{2n+m \times 2n+m}$, $\operatorname{rank}_{\mathbb{R}(s)}(s\mathcal{E} - \mathcal{A}) = 2n + q$ and $\operatorname{rank} \mathcal{E} = 2r$. From that we conclude that the EKCF has exactly $m - q$ blocks of type E4, i. e., v) holds true. By removing these blocks and using

Lemma 2.1.15 f), we obtain a pencil in EKCF, denoted by $s\mathcal{E}_1 - \mathcal{A}_1 \in \mathbb{C}[s]^{2n_1+q\times 2n_1+q}$, with $\operatorname{rank}\mathcal{E}_1 = 2n_1 - 2(n-r)$ and

$$\operatorname{In}(\mathrm{i}\omega\mathcal{E}_1 - \mathcal{A}_1) = (n_1 + q - a(\omega), a(\omega), n_1).$$

Using again Lemma 2.1.15, we conclude that $s\mathcal{E}_1 - \mathcal{A}_1$ has exactly $q$ blocks of type E3 with positive sign-characteristic and odd size. In other words, iii) holds true.

We remove these blocks and denote the emerging pencil by $s\mathcal{E}_2 - \mathcal{A}_2 \in \mathbb{C}[s]^{2n_2\times 2n_2}$. This pencil fulfills $\operatorname{rank}\mathcal{E}_1 = 2n_2 - 2(n-r)$ and

$$\operatorname{In}(\mathrm{i}\omega\mathcal{E}_2 - \mathcal{A}_2) = (n_2 - a(\omega), a(\omega), n_2).$$

By Lemma 2.1.15, this structure reveals that all blocks of type E2 have even size and negative sign-characteristic. This shows statement i).

Removing these blocks, we obtain a pencil $s\mathcal{E}_3 - \mathcal{A}_3 \in \mathbb{C}[s]^{2n_3\times 2n_3}$ with $\operatorname{rank}\mathcal{E}_1 = 2n_3 - 2(n-r)$ and

$$\operatorname{In}(\mathrm{i}\omega\mathcal{E}_3 - \mathcal{A}_3) = (n_3, 0, n_3).$$

From the rank deficiency of $\mathcal{E}_3$ we can conclude that $s\mathcal{E}_3 - \mathcal{A}_3$ has $2(n-r)$ blocks of type E3, hence the number of blocks of type E3 in the EKCF of $s\mathcal{E} - \mathcal{A}$ is $2(n-r)+q$. Thus, ii) holds true.

The inertial properties of $\mathrm{i}\omega\mathcal{E}_3 - \mathcal{A}_3$ give rise to the fact that the number of odd-sized blocks of type E3 with positive sign-characteristic is equal to the number of odd-sized blocks of type E3 with negative sign-characteristic. This shows iv).

The proof that b) implies c) is trivial.

Now we show that c) implies a): Assume that the EKCF of the associated pencil $s\mathcal{E} - \mathcal{A}$ has the properties i') and ii)–v). Then Lemma 2.1.15 implies that there exist functions $a, b : \mathbb{R} \to \mathbb{N}_0$ with finite support and

$$\operatorname{In}(\mathrm{i}\omega\mathcal{E} - \mathcal{A}) = (n + q - a(\omega), m - q + a(\omega) + b(\omega), n - b(\omega)).$$

In particular, the quantity $\sum_{\omega\in\mathbb{R}} b(\omega)$ is the number of even-sized blocks of type E2 with positive sign-characteristic.

Since we have, without loss of generality, assumed that $\Lambda(E, A) \cap \mathrm{i}\mathbb{R} = \emptyset$, the relation (3.30) holds. Sylvester's law of inertia together with Lemma 3.4.1 then implies

$$\operatorname{In}(\Phi(\mathrm{i}\omega)) = (q - a(\omega), m - q + a(\omega) + b(\omega), -b(\omega)) \quad \forall \omega \in \mathbb{R}. \tag{3.32}$$

Assume that $b$ is not the constant zero function. Then for an $\omega$ with $b(\omega) > 0$, the number of negative eigenvalues of $\Phi(\mathrm{i}\omega)$ would be negative which is a contradiction.

Consequently, $b \equiv 0$, and (3.32) implies that $\Phi(s)$ is pointwise positive semidefinite on the imaginary axis. In other words, a) is satisfied. □

*Remark* 3.4.3 (Spectral structure for impulse controllable systems). If the system $(E, A, B)$ is impulse controllable, we can make a more precise statement about the spectral structure of $s\mathcal{E} - \mathcal{A}$. In this case we can reduce the system to the form (3.18). Then instead of $s\mathcal{E} - \mathcal{A}$ we consider the reduced even matrix pencil

$$s\widehat{\mathcal{E}} - \widehat{\mathcal{A}} := \begin{bmatrix} 0 & -sI_r + A_{11} & B_1 \\ sI_r + A_{11}^{\mathsf{T}} & Q_{11} & -Q_{12}B_2 + S_1 \\ B_1^{\mathsf{T}} & -B_2^{\mathsf{T}}Q_{12}^{\mathsf{T}} + S_1^{\mathsf{T}} & B_2^{\mathsf{T}}Q_{22}B_2 - B_2^{\mathsf{T}}S_2 - S_2^{\mathsf{T}}B_2 + R \end{bmatrix} \in \mathbb{R}[s]^{2r+m \times 2r+m}. \tag{3.33}$$

Then Theorem 3.4.2 yields that $\Phi(\mathrm{i}\omega) \geq 0$ for all $\mathrm{i}\omega \notin \Lambda(E, A)$ is equivalent to the EKCF of $s\widehat{\mathcal{E}} - \widehat{\mathcal{A}}$ having the following structure:

- All blocks of type E2 have even size and negative sign-characteristic.
- There exist exactly $q$ blocks of type E3 with odd size and positive sign-characteristic.
- There exist exactly $m - q$ blocks of type E4.

Since $s\mathcal{E} - \mathcal{A}$ has $2(n - r)$ more rows and columns and also $2(n - r)$ more blocks of type E3, statement iv) of Theorem 3.4.2 can be replaced by

iv') The remaining $2(n - r)$ blocks of type E3 are all of size $1 \times 1$, and $n - r$ of them have positive sign-characteristic and $n - r$ have negative sign-characteristic.

**Example 3.4.4.** To illustrate the results of Theorem 3.4.2 and Remark 3.4.3 consider the matrices

$$E = \begin{bmatrix} 1 & 0 \\ 0 & 0 \end{bmatrix}, \quad A = \begin{bmatrix} 1 & 0 \\ 0 & 1 \end{bmatrix}, \quad B = S = \begin{bmatrix} 0 \\ 0 \end{bmatrix}, \quad Q = \begin{bmatrix} 0 & 0 \\ 0 & 0 \end{bmatrix}, \quad R = 1.$$

The system $(E, A, B) \in \Sigma_{2,1}$ is impulse controllable and the associated Popov function is $\Phi(s) = 1$. Using the notation of Theorem 3.4.2, we have $n = 2$ and $m = r = q = 1$. Let $s\mathcal{E} - \mathcal{A} \in \mathbb{R}[s]^{5\times 5}$ be the associated even matrix pencil as in (3.29). With

$$U = \begin{bmatrix} 1 & 0 & 0 & 0 & 0 \\ 0 & 0 & \frac{1}{\sqrt{2}} & -\frac{1}{\sqrt{2}} & 0 \\ 0 & 1 & 0 & 0 & 0 \\ 0 & 0 & \frac{1}{\sqrt{2}} & \frac{1}{\sqrt{2}} & 0 \\ 0 & 0 & 0 & 0 & 1 \end{bmatrix},$$

the pencil $s\mathcal{E} - \mathcal{A}$ is transformed to EKCF, i. e.,

$$U^{\mathsf{H}}(s\mathcal{E} - \mathcal{A})U = \left[\begin{array}{cc|c|c|c} 0 & -s+1 & 0 & 0 & 0 \\ s+1 & 0 & 0 & 0 & 0 \\ \hline 0 & 0 & -1 & 0 & 0 \\ \hline 0 & 0 & 0 & 1 & 0 \\ \hline 0 & 0 & 0 & 0 & 1 \end{array}\right]. \tag{3.34}$$

Indeed, the right-hand side of (3.34) contains one block of type E1, no blocks of type E2, but $2(n-r)+q = 3$ blocks of type E3 of size $1 \times 1$ (two with positive sign-characteristic and one with negative sign-characteristic) and $m - q = 0$ blocks of type E4. All this is in agreement with Theorem 3.4.2 and Remark 3.4.3.

In the following we use the so far developed results on even matrix pencils to study solutions of the descriptor Lur'e equation

$$\begin{bmatrix} A^{\mathsf{T}}X + X^{\mathsf{T}}A + Q & X^{\mathsf{T}}B + S \\ B^{\mathsf{T}}X + S^{\mathsf{T}} & R \end{bmatrix} =_{\mathcal{V}_{\mathrm{sys}}} \begin{bmatrix} K^{\mathsf{T}} \\ L^{\mathsf{T}} \end{bmatrix} \begin{bmatrix} K & L \end{bmatrix}, \quad E^{\mathsf{T}}X = X^{\mathsf{T}}E. \tag{3.35}$$

Obviously, $P = X$ is a solution of the descriptor KYP inequality (3.22). However, solutions of the descriptor Lur'e equation are special in the following sense.

**Definition 3.4.5** (Solution of a descriptor Lur'e equation)**.** [RRV14, Sect. 5] A triple $(X, K, L) \in \mathbb{R}^{n\times n} \times \mathbb{R}^{q\times n} \times \mathbb{R}^{q\times n}$ is called *solution of the descriptor Lur'e equation* (3.35), if it fulfills (3.35) and

$$\operatorname{rank}_{\mathbb{R}(s)} \begin{bmatrix} -sE + A & B \\ K & L \end{bmatrix} = n + q. \tag{3.36}$$

*Remark* 3.4.6. If $E = I_n$, then $\mathcal{V}_{\mathrm{sys}} = \mathbb{R}^{n+m}$ and the descriptor Lur'e equation (3.35) reduces to a standard Lur'e equation of the form (3.9) [Rei11] with

$$\operatorname{rank}_{\mathbb{R}(s)} \begin{bmatrix} -sI_n + A & B \\ K & L \end{bmatrix} = n + q.$$

If, further, $R > 0$, then $L \in \mathrm{Gl}_m(\mathbb{R})$. In this case, the unknowns $K$ and $L$ in the descriptor Lur'e equation can be eliminated, such that an algebraic Riccati equation

$$A^{\mathsf{T}}X + XA + Q - (XB + S)R^{-1}(XB + S)^{\mathsf{T}} = 0$$

is obtained.

Now we highlight a correspondence between solutions of the descriptor Lur'e equations and certain deflating subspaces of the associated even matrix pencil. For the corresponding theorem and its proof, we need the following two lemmas. The first one states relations between solutions of the descriptor Lur'e equation to solutions of a certain standard Lur'e equation, see also [RRV14, Lem. 5.4].

**Lemma 3.4.7.** *Let $(E, A, B) \in \Sigma_{n,m}$ with the system space $\mathcal{V}_{\text{sys}}$ be impulse controllable and assume that $Q = Q^{\mathsf{T}} \in \mathbb{R}^{n \times n}$, $S \in \mathbb{R}^{n \times m}$, $R = R^{\mathsf{T}} \in \mathbb{R}^{m \times m}$ are given. Let $W, T \in \mathrm{Gl}_n(\mathbb{R})$ and $F \in \mathbb{R}^{m \times n}$ be given such that*

$$sE_F - A_F = W(sE - (A + BF))T = \begin{bmatrix} sI_r - A_{11} & 0 \\ 0 & -I_{n-r} \end{bmatrix}, \quad B_F = WB = \begin{bmatrix} B_1 \\ B_2 \end{bmatrix}, \tag{3.37}$$

*and let*

$$\begin{aligned} Q_F &= T^{\mathsf{T}}(Q + SF + F^{\mathsf{T}}S^{\mathsf{T}} + F^{\mathsf{T}}RF)T = \begin{bmatrix} Q_{11} & Q_{12} \\ Q_{12}^{\mathsf{T}} & Q_{22} \end{bmatrix}, \\ S_F &= T^{\mathsf{T}}(S + F^{\mathsf{T}}R) = \begin{bmatrix} S_1 \\ S_2 \end{bmatrix}, \quad R_F = R \end{aligned} \tag{3.38}$$

*be accordingly partitioned. Then the following statements hold true:*

*a) If $(X, K, L) \in \mathbb{R}^{n \times n} \times \mathbb{R}^{q \times n} \times \mathbb{R}^{q \times m}$ is a solution of the descriptor Lur'e equation (3.35), then for*

$$X_F = W^{-\mathsf{T}}XT = \begin{bmatrix} X_{11} & 0 \\ X_{21} & X_{22} \end{bmatrix}, \quad K_F = (K + LF)T = \begin{bmatrix} K_1 & K_2 \end{bmatrix}, \tag{3.39}$$

*partitioned according to the block structure of (3.37), it holds that*

$$\begin{aligned} &\begin{bmatrix} A_{11}^{\mathsf{T}}X_{11} + X_{11}A_{11} + Q_{11} & X_{11}B_1 + S_1 - Q_{12}B_2 \\ B_1^{\mathsf{T}}X_{11} + S_1^{\mathsf{T}} - B_2^{\mathsf{T}}Q_{12}^{\mathsf{T}} & B_2^{\mathsf{T}}Q_{22}B_2 - B_2^{\mathsf{T}}S_2 - S_2^{\mathsf{T}}B_2 + R \end{bmatrix} \\ &\qquad = \begin{bmatrix} K_1^{\mathsf{T}} \\ (L - K_2B_2)^{\mathsf{T}} \end{bmatrix} \begin{bmatrix} K_1 & L - K_2B_2 \end{bmatrix}, \quad X_{11} = X_{11}^{\mathsf{T}} \end{aligned} \tag{3.40}$$

*with*

$$\operatorname{rank}_{\mathbb{R}(s)} \begin{bmatrix} -sI_r + A_{11} & B_1 \\ K_1 & L - K_2B_2 \end{bmatrix} = r + q. \tag{3.41}$$

*b) If $(X_{11}, K_1, L_1) \in \mathbb{R}^{r \times r} \times \mathbb{R}^{q \times r} \times \mathbb{R}^{q \times m}$ solves the Lur'e equation*

$$\begin{aligned} &\begin{bmatrix} A_{11}^{\mathsf{T}}X_{11} + X_{11}A_{11} + Q_{11} & X_{11}B_1 + S_1 - Q_{12}B_2 \\ B_1^{\mathsf{T}}X_{11} + S_1^{\mathsf{T}} - B_2^{\mathsf{T}}Q_{12}^{\mathsf{T}} & B_2^{\mathsf{T}}Q_{22}B_2 - B_2^{\mathsf{T}}S_2 - S_2^{\mathsf{T}}B_2 + R \end{bmatrix} \\ &\qquad = \begin{bmatrix} K_1^{\mathsf{T}} \\ L_1^{\mathsf{T}} \end{bmatrix} \begin{bmatrix} K_1 & L_1 \end{bmatrix}, \quad X_{11} = X_{11}^{\mathsf{T}}, \end{aligned} \tag{3.42}$$

*then for*

$$X = W^{\mathsf{T}} \begin{bmatrix} X_{11} & 0 \\ 0 & 0 \end{bmatrix} T^{-1}, \quad K = \begin{bmatrix} K_1 & 0 \end{bmatrix} T^{-1} - L_1F, \quad L = L_1,$$

*the triple $(X, K, L) \in \mathbb{R}^{n \times n} \times \mathbb{R}^{q \times n} \times \mathbb{R}^{q \times m}$ solves the descriptor Lur'e equation (3.35).*

*Proof.* First we show a): With the basis matrix $M_{\mathcal{V}_{\text{sys}}}$ of the space $\mathcal{V}_{\text{sys}} \subseteq \mathbb{R}^{n+m}$ as in (3.27), the relation

$$M_{\mathcal{V}_{\text{sys}}}^{\mathsf{T}} \begin{bmatrix} A^{\mathsf{T}}X + X^{\mathsf{T}}A + Q & X^{\mathsf{T}}B + S \\ B^{\mathsf{T}}X + S^{\mathsf{T}} & R \end{bmatrix} M_{\mathcal{V}_{\text{sys}}} = M_{\mathcal{V}_{\text{sys}}}^{\mathsf{T}} \begin{bmatrix} K^{\mathsf{T}} \\ L^{\mathsf{T}} \end{bmatrix} \begin{bmatrix} K & L \end{bmatrix} M_{\mathcal{V}_{\text{sys}}} \tag{3.43}$$

directly yields (3.40). Furthermore, from

$$\begin{aligned} n + q &= \operatorname{rank}_{\mathbb{R}(s)} \begin{bmatrix} -sE + A & B \\ K & L \end{bmatrix} \\ &= \operatorname{rank}_{\mathbb{R}(s)} \begin{bmatrix} W & 0 \\ 0 & I_q \end{bmatrix} \begin{bmatrix} -sE + A & B \\ K & L \end{bmatrix} \begin{bmatrix} T & 0 \\ FT & I_m \end{bmatrix} \\ &= \operatorname{rank}_{\mathbb{R}(s)} \begin{bmatrix} -sI_r + A_{11} & 0 & B_1 \\ 0 & I_{n-r} & B_2 \\ K_1 & K_2 & L \end{bmatrix} \begin{bmatrix} I_r & 0 & 0 \\ 0 & -B_2 & I_{n-r} \\ 0 & I_m & 0 \end{bmatrix} \\ &= \operatorname{rank}_{\mathbb{R}(s)} \begin{bmatrix} -sI_r + A_{11} & B_1 & 0 \\ 0 & 0 & I_{n-r} \\ K_1 & L - K_2B_2 & K_2 \end{bmatrix} \\ &= n - r + \operatorname{rank}_{\mathbb{R}(s)} \begin{bmatrix} -sI_r + A_{11} & B_1 \\ K_1 & L - K_2B_2 \end{bmatrix}, \end{aligned} \tag{3.44}$$

we directly obtain (3.41).

Next, we show b): Assume that $(X_{11}, K_1, L_1)$ is a solution of the Lur'e equation (3.42). Assume further that $(X, K, L)$ is not a solution of the descriptor Lur'e equation (3.35). If we have

$$\begin{bmatrix} A^{\mathsf{T}}X + X^{\mathsf{T}}A + Q & X^{\mathsf{T}}B + S \\ B^{\mathsf{T}}X + S^{\mathsf{T}} & R \end{bmatrix} \neq_{\mathcal{V}_{\text{sys}}} \begin{bmatrix} K^{\mathsf{T}} \\ L^{\mathsf{T}} \end{bmatrix} \begin{bmatrix} K & L \end{bmatrix}, \tag{3.45}$$

then (3.43) is not satisfied which, by a simple calculation, is a contradiction to (3.42). On the other hand, if $(X, K, L)$ does not satisfy (3.36), we obtain

$$\begin{aligned} n + q &= \operatorname{rank}_{\mathbb{R}(s)} \begin{bmatrix} -sI_r + A_{11} & B_1 \\ K_1 & L_1 \end{bmatrix} + n - r \\ &= \operatorname{rank}_{\mathbb{R}(s)} \begin{bmatrix} W^{-1} & 0 \\ 0 & I_q \end{bmatrix} \begin{bmatrix} -sI_r + A_{11} & 0 & B_1 \\ 0 & I_{n-r} & B_2 \\ K_1 & 0 & L_1 \end{bmatrix} \begin{bmatrix} T^{-1} & 0 \\ -F & I_m \end{bmatrix} \\ &= \operatorname{rank}_{\mathbb{R}(s)} \begin{bmatrix} -sE + A & B \\ K & L \end{bmatrix} \\ &\neq n + q, \end{aligned}$$

which is a contradiction as well. Thus $(X, K, L)$ is indeed a solution of the descriptor Lur'e equation (3.35). □

The next technical lemma is the basis for an alternative formulation of the descriptor Lur'e equation.

**Lemma 3.4.8.** *Let $\mathcal{A} \in \mathbb{R}^{n \times n}$ be symmetric and assume that an $m$-dimensional subspace $\mathcal{V} \subseteq \mathbb{R}^n$ is given. Then the following statements hold true:*

*a) If there exist matrices $\mathcal{R}_1 \in \mathbb{R}^{k \times n}$, $\mathcal{R}_2 \in \mathbb{R}^{n-m \times n}$ with $\ker \mathcal{R}_2 = \mathcal{V}$, and a signature matrix $\Sigma_{22} \in \mathbb{R}^{n-m \times n-m}$ (that is a diagonal matrix with only $-1$, $0$, and $+1$ on the diagonal) such that*

$$\mathcal{A} = \mathcal{R}_1^\mathsf{T} \mathcal{R}_1 + \mathcal{R}_2^\mathsf{T} \Sigma_{22} \mathcal{R}_2,$$

*then it holds that $\mathcal{A} =_\mathcal{V} \mathcal{R}_1^\mathsf{T} \mathcal{R}_1$.*

*b) If $\mathcal{A} =_\mathcal{V} \mathcal{R}_1^\mathsf{T} \mathcal{R}_1$ for $\mathcal{R}_1 \in \mathbb{R}^{k \times n}$, then there exist matrices $\widetilde{\mathcal{R}}_1 \in \mathbb{R}^{k \times n}$, $\widetilde{\mathcal{R}}_2 \in \mathbb{R}^{n-m \times n}$ with $\ker \widetilde{\mathcal{R}}_2 = \mathcal{V}$, and a signature matrix $\widetilde{\Sigma}_{22} \in \mathbb{R}^{n-m \times n-m}$ such that*

$$\mathcal{A} = \widetilde{\mathcal{R}}_1^\mathsf{T} \widetilde{\mathcal{R}}_1 + \widetilde{\mathcal{R}}_2^\mathsf{T} \widetilde{\Sigma}_{22} \widetilde{\mathcal{R}}_2.$$

*Proof.* Statement a) is trivial, since for all $v \in \mathcal{V}$ it holds that $v^\mathsf{T} \mathcal{A} v = v^\mathsf{T} \mathcal{R}_1^\mathsf{T} \mathcal{R}_1 v$.

Now we show b): By the assumptions there exists a matrix $\mathcal{R}_2 \in \mathbb{R}^{n-m \times n}$ with $\ker \mathcal{R}_2 = \mathcal{V}$ such that

$$\mathcal{A} = \begin{bmatrix} \mathcal{R}_1^\mathsf{T} & \mathcal{R}_2^\mathsf{T} \end{bmatrix} \begin{bmatrix} I_k & \Sigma_{12} \\ \Sigma_{12}^\mathsf{T} & \Sigma_{22} \end{bmatrix} \begin{bmatrix} \mathcal{R}_1 \\ \mathcal{R}_2 \end{bmatrix}$$

with $\Sigma_{12} \in \mathbb{R}^{k \times n-m}$ and $\Sigma_{22} \in \mathbb{R}^{n-m \times n-m}$. Moreover, it holds that

$$\begin{bmatrix} I_k & \Sigma_{12} \\ \Sigma_{12}^\mathsf{T} & \Sigma_{22} \end{bmatrix} = \begin{bmatrix} I_k & 0 \\ \Sigma_{12}^\mathsf{T} & I_{n-m} \end{bmatrix} \begin{bmatrix} I_k & 0 \\ 0 & \Sigma_{22} - \Sigma_{12}^\mathsf{T} \Sigma_{12} \end{bmatrix} \begin{bmatrix} I_k & \Sigma_{12} \\ 0 & I_{n-m} \end{bmatrix}.$$

By Sylvester's law of inertia there exists a matrix $\mathcal{H} \in \mathrm{Gl}_{n-m}(\mathbb{R})$ such that $\Sigma_{22} - \Sigma_{12}^\mathsf{T} \Sigma_{12} = \mathcal{H}^\mathsf{T} \widetilde{\Sigma}_{22} \mathcal{H}$ with a signature matrix $\widetilde{\Sigma}_{22}$. Thus the claim holds true for

$$\widetilde{\mathcal{R}}_1 = \mathcal{R}_1 + \Sigma_{12} \mathcal{R}_2, \quad \widetilde{\mathcal{R}}_2 = \mathcal{H} \mathcal{R}_2. \tag{3.46}$$

□

Lemma 3.4.8 suggests that the descriptor Lur'e equation (3.35) has a solution $(X, K, L)$ if and only if with $r = \operatorname{rank} E$, the *alternative descriptor Lur'e equation*

$$\begin{bmatrix} A^\mathsf{T} X + X^\mathsf{T} A + Q & X^\mathsf{T} B + S \\ B^\mathsf{T} X + S^\mathsf{T} & R \end{bmatrix} = \begin{bmatrix} \widetilde{K}^\mathsf{T} \\ \widetilde{L}^\mathsf{T} \end{bmatrix} \begin{bmatrix} \widetilde{K} & \widetilde{L} \end{bmatrix} + \begin{bmatrix} H^\mathsf{T} \\ J^\mathsf{T} \end{bmatrix} \Sigma \begin{bmatrix} H & J \end{bmatrix},$$

$$E^\mathsf{T} X = X^\mathsf{T} E \tag{3.47}$$

is solved by $\left(X, \widetilde{K}, \widetilde{L}, H, J, \Sigma\right) \in \mathbb{R}^{n\times n} \times \mathbb{R}^{q\times n} \times \mathbb{R}^{q\times m} \times \mathbb{R}^{n-r\times n} \times \mathbb{R}^{n-r\times m} \times \mathbb{R}^{n-r\times n-r}$, where $\ker \begin{bmatrix} H & J \end{bmatrix} = \mathcal{V}_{\mathrm{sys}}$, $\Sigma$ is a signature matrix, and

$$\operatorname{rank}_{\mathbb{R}(s)} \begin{bmatrix} -sE+A & B \\ \widetilde{K} & \widetilde{L} \end{bmatrix} = n+q.$$

Now we state the main theorem of this section, see also [RRV14, Thm. 6.2] for a similar result.

**Theorem 3.4.9.** *Assume that $(E, A, B) \in \Sigma_{n,m}$ with the system space $\mathcal{V}_{\mathrm{sys}}$ and $r = \operatorname{rank} E$ are given and let the even matrix pencil $s\mathcal{E} - \mathcal{A}$ be as in (3.29) and the Popov function $\Phi(s) \in \mathbb{R}(s)^{m\times m}$ be as in (3.2) with $q = \operatorname{rank}_{\mathbb{R}(s)} \Phi(s)$. Consider the following two statements:*

*1) The descriptor Lur'e equation (3.35) is solvable.*

*2) It holds that $\Phi(\mathrm{i}\omega) \geq 0$ for all $\mathrm{i}\omega \in \mathrm{i}\mathbb{R} \setminus \Lambda(E, A)$ and there exist $Y_\mu$, $Y_x \in \mathbb{R}^{n\times n+m}$, $Y_u \in \mathbb{R}^{m\times n+m}$, $Z_\mu$, $Z_x \in \mathbb{R}^{n\times n+q}$, $Z_u \in \mathbb{R}^{m\times n+q}$ such that for*

$$Y = \begin{bmatrix} Y_\mu \\ Y_x \\ Y_u \end{bmatrix}, \quad Z = \begin{bmatrix} Z_\mu \\ Z_x \\ Z_u \end{bmatrix}, \tag{3.48}$$

*the following holds true:*

*i) the space $\operatorname{im} Y$ is $n+m$-dimensional and $\mathcal{E}$-neutral;*

*ii) $\mathcal{V}_{\mathrm{sys}} \subseteq \operatorname{im} \begin{bmatrix} Y_x \\ Y_u \end{bmatrix}$;*

*iii) $\operatorname{rank} EY_x = r$;*

*iv) there exist $\widetilde{\mathcal{E}}, \widetilde{\mathcal{A}} \in \mathbb{R}^{n+q\times n+m}$ with $\operatorname{rank}_{\mathbb{R}(s)} \left(s\widetilde{\mathcal{E}} - \widetilde{\mathcal{A}}\right) = n+q$, such that*

$$(s\mathcal{E} - \mathcal{A})Y = Z\left(s\widetilde{\mathcal{E}} - \widetilde{\mathcal{A}}\right). \tag{3.49}$$

*Then the following implications hold:*

*a) Statement 1) implies 2).*

*b) If $(E, A, B)$ is impulse controllable, then 2) implies 1).*

*In the case where 2) holds true, then there exists a matrix $Y$ that fulfills 2) with $\operatorname{rank} Y_x = n$. Moreover, there exists a matrix $Y_x^- \in \mathbb{R}^{n+m\times n}$ with $Y_x Y_x^- = I_n$ such that*

$$X = Y_\mu Y_x^-. \tag{3.50}$$

*Proof.* First we prove a): Since the descriptor Lur'e equation (3.35) is solvable, also the alternative descriptor Lur'e equation (3.47) has a solution $(X, \widetilde{K}, \widetilde{L}, H, J, \Sigma)$. As a preliminary observation, we have

$$\ker \begin{bmatrix} H & J \end{bmatrix} = \mathcal{V}_{\mathrm{sys}} = \ker T_\infty^\mathsf{T} \begin{bmatrix} A & B \end{bmatrix},$$

where $T_\infty \in \mathbb{R}^{n \times n-r}$ with $\operatorname{im} T_\infty = \ker E^\mathsf{T}$. Therefore, there exists some $M \in \mathrm{Gl}_{n-r}(\mathbb{R})$ with

$$\begin{bmatrix} H & J \end{bmatrix} = M T_\infty^\mathsf{T} \begin{bmatrix} A & B \end{bmatrix}. \tag{3.51}$$

Therefore, a solution of the alternative descriptor Lur'e equation fulfills

$$\begin{aligned} &\begin{bmatrix} 0 & -sE + A & B \\ sE^\mathsf{T} + A^\mathsf{T} & Q & S \\ B^\mathsf{T} & S^\mathsf{T} & R \end{bmatrix} \begin{bmatrix} X & 0 \\ I_n & 0 \\ 0 & I_m \end{bmatrix} \\ &\qquad = \begin{bmatrix} I_n & 0 \\ -X^\mathsf{T} + H^\mathsf{T} \Sigma M T_\infty^\mathsf{T} & \widetilde{K}^\mathsf{T} \\ J^\mathsf{T} \Sigma M T_\infty^\mathsf{T} & \widetilde{L}^\mathsf{T} \end{bmatrix} \begin{bmatrix} -sE + A & B \\ \widetilde{K} & \widetilde{L} \end{bmatrix}. \end{aligned} \tag{3.52}$$

Using (3.36), we see from (3.52) that solutions of the alternative descriptor Lur'e equation define deflating subspaces of the associated even matrix pencil. The equation $E^\mathsf{T} X = X^\mathsf{T} E$ further implies that this deflating subspace is $\mathcal{E}$-neutral.

Now we prove b) which is more difficult. Since $(E, A, B)$ is impulse controllable, there exist $W, T \in \mathrm{Gl}_n(\mathbb{R})$ and $F \in \mathbb{R}^{m \times n}$ such that $E_F$, $A_F$, $B_F$, as well as $Q_F$, $S_F$, $R_F$ are in the form (3.37) and (3.38), respectively. Moreover, let $X_F$ and $K_F$ be as in (3.39).

Now we define the matrices

$$U := \begin{bmatrix} W^\mathsf{T} & 0 & 0 \\ 0 & T & 0 \\ 0 & FT & I_m \end{bmatrix}, \quad \widehat{U} := \begin{bmatrix} I_r & 0 & 0 & 0 & 0 \\ 0 & -Q_{12}^\mathsf{T} & Q_{22}B_2 - S_2 & -\frac{1}{2}Q_{22} & I_{n-r} \\ 0 & I_r & 0 & 0 & 0 \\ 0 & 0 & -B_2 & I_{n-r} & 0 \\ 0 & 0 & I_m & 0 & 0 \end{bmatrix}$$

and obtain

$$\begin{aligned} &s\widehat{\mathcal{E}}_F - \widehat{\mathcal{A}}_F := \widehat{U}^\mathsf{T} U^\mathsf{T} (s\mathcal{E} - \mathcal{A}) U \widehat{U} \\ &= \begin{bmatrix} 0 & -sI_r + A_{11} & B_1 & 0 & 0 \\ sI_r + A_{11}^\mathsf{T} & Q_{11} & -Q_{12}B_2 + S_1 & 0 & 0 \\ B_1^\mathsf{T} & -B_2^\mathsf{T} Q_{12}^\mathsf{T} + S_1^\mathsf{T} & B_2^\mathsf{T} Q_{22} B_2 - B_2^\mathsf{T} S_2 - S_2^\mathsf{T} B_2 + R & 0 & 0 \\ 0 & 0 & 0 & 0 & I_{n-r} \\ 0 & 0 & 0 & I_{n-r} & 0 \end{bmatrix}. \end{aligned}$$

Furthermore, by setting

$$U^{-1}Y_\mu =: \begin{bmatrix} Y_{\mu,1} \\ Y_{\mu,2} \end{bmatrix}, \quad U^{-1}Y_x =: \begin{bmatrix} Y_{x,1} \\ Y_{x,2} \end{bmatrix}, \quad U^{-1}Y_u =: Y_{u,1}$$

partitioned according to the block structure in (3.37) and (3.38), we obtain

$$\widehat{Y}_F := \widehat{U}^{-1}U^{-1}Y = \begin{bmatrix} I_r & 0 & 0 & 0 & 0 \\ 0 & 0 & I_{n-r} & 0 & 0 \\ 0 & 0 & 0 & 0 & I_m \\ 0 & 0 & 0 & I_{n-r} & B_2 \\ 0 & I_r & Q_{12}^\mathsf{T} & \frac{1}{2}Q_{22} & -\frac{1}{2}Q_{22}B_2 + S_2 \end{bmatrix} \begin{bmatrix} Y_{\mu,1} \\ Y_{\mu,2} \\ Y_{x,1} \\ Y_{x,2} \\ Y_{u,1} \end{bmatrix} = \begin{bmatrix} Y_{\mu,1} \\ Y_{x,1} \\ Y_{u,1} \\ \widehat{Y}_{\mu,2} \\ \widehat{Y}_{x,2} \end{bmatrix}. \tag{3.53}$$

Finally, by setting $\widehat{Z}_F := \widehat{U}^\mathsf{T}U^\mathsf{T}Z$ and using Proposition 3.2.2 d) we obtain $(s\widehat{\mathcal{E}}_F - \widehat{\mathcal{A}}_F)\widehat{Y}_F = \widehat{Z}_F(s\widetilde{\mathcal{E}} - \widetilde{\mathcal{A}})$, where we assume w. l. o. g. that $s\widetilde{\mathcal{E}} - \widetilde{\mathcal{A}}$ is in KCF.

Since $\operatorname{im}\widehat{Y}_F$ is $\widehat{\mathcal{E}}_F$-neutral, the space $\operatorname{im}\begin{bmatrix} Y_{\mu,1} \\ Y_{x,1} \end{bmatrix}$ is $\begin{bmatrix} 0 & -I_r \\ I_r & 0 \end{bmatrix}$-neutral and thus its dimension is at most $r$. On the other hand, by Proposition 3.2.2 b) we have

$$\mathcal{V}_{\mathrm{sys},F} \subseteq \operatorname{im}\begin{bmatrix} Y_{x,1} \\ Y_{x,2} \\ Y_{u,1} \end{bmatrix} \quad \text{with} \quad \mathcal{V}_{\mathrm{sys},F} = \begin{bmatrix} T^{-1} & 0 \\ -F & I_m \end{bmatrix}\mathcal{V}_{\mathrm{sys}}.$$

Since $\dim \mathcal{V}_{\mathrm{sys},F} = r + m$, it follows that $\operatorname{rank}\begin{bmatrix} Y_{x,1} \\ Y_{u,1} \end{bmatrix} = r + m$. By property iii) in 2) we have $\operatorname{rank} Y_{x,1} = r$. This yields $\operatorname{rank}\begin{bmatrix} Y_{\mu,1}^\mathsf{T} & Y_{x,1}^\mathsf{T} & Y_{u,1}^\mathsf{T} \end{bmatrix}^\mathsf{T} = r + m$. Moreover, due to the block-diagonal structure of $s\widehat{\mathcal{E}}_F - \widehat{\mathcal{A}}_F$, it holds that $\operatorname{rank}\begin{bmatrix} \widehat{Y}_{\mu,2} \\ \widehat{Y}_{x,2} \end{bmatrix} = n - r$.

From these facts it follows that there exists a matrix $V \in \mathrm{Gl}_{n+m}(\mathbb{R})$ such that

$$\begin{bmatrix} Y_{\mu,1} \\ Y_{x,1} \\ Y_{u,1} \\ \widehat{Y}_{\mu,2} \\ \widehat{Y}_{x,2} \end{bmatrix} V = \begin{bmatrix} X_{11} & 0 & 0 \\ I_{n_1} & 0 & 0 \\ 0 & I_m & 0 \\ 0 & 0 & \widetilde{Y}_{\mu,2} \\ 0 & 0 & \widetilde{Y}_{x,2} \end{bmatrix}, \tag{3.54}$$

where $\operatorname{rank}\begin{bmatrix} \widetilde{Y}_{\mu,2} \\ \widetilde{Y}_{x,2} \end{bmatrix} = n - r$.

Now we can make use of this theorem for ODE systems [Rei11, Thm. 11], i. e., there exist $K_1 \in \mathbb{K}^{q\times r}$ and $L_1 \in \mathbb{K}^{q\times m}$ such that

$$\begin{bmatrix} 0 & -sI_r + A_{11} & B_1 \\ sI_r + A_{11}^\mathsf{T} & Q_{11} & -Q_{12}B_2 + S_1 \\ B_1^\mathsf{T} & -B_2^\mathsf{T} Q_{12}^\mathsf{T} + S_1^\mathsf{T} & B_2^\mathsf{T} Q_{22} B_2 - B_2^\mathsf{T} S_2 - S_2^\mathsf{T} B_2 + R \end{bmatrix} \begin{bmatrix} X_{11} & 0 \\ I_r & 0 \\ 0 & I_m \end{bmatrix} = \begin{bmatrix} I_r & 0 \\ -X_{11} & K_1^\mathsf{T} \\ 0 & L_1^\mathsf{T} \end{bmatrix} \begin{bmatrix} -sI_r + A_{11} & B_1 \\ K_1 & L_1 \end{bmatrix}, \quad (3.55)$$

where

$$\operatorname{rank}_{\mathbb{R}(s)} \begin{bmatrix} -sI_r + A_{11} & B_1 \\ K_1 & L_1 \end{bmatrix} = r + q.$$

Thus $(X_{11}, K_1, L_1)$ solves the Lur'e equation (3.42). Then, by Lemma 3.4.7 b), we obtain a solution of the descriptor Lur'e equation (3.35).

If Statement 2) holds true, we obtain from (3.52) that we can choose $Y_x$ and $Y_u$ such that $\begin{bmatrix} Y_x \\ Y_u \end{bmatrix}$ is invertible. Hence, with $\begin{bmatrix} Y_x^- & Y_u^- \end{bmatrix} := \begin{bmatrix} Y_x \\ Y_u \end{bmatrix}^{-1}$, we obtain a representation for $X$ as in (3.50). □

*Remark* 3.4.10. If $(X, K, L)$ solves the descriptor Lur'e equation (3.35), then we can choose $Y$ and $Z$ in (3.49) such that

$$s\widetilde{\mathcal{E}} - \widetilde{\mathcal{A}} = \begin{bmatrix} -sE + A & B \\ K & L \end{bmatrix}.$$

This can be seen from (3.52). Moreover, by (3.46), for a tuple $\left(X, \widetilde{K}, \widetilde{L}, H, J, \Sigma\right)$ solving the alternative descriptor Lur'e equation, there exists a $\Sigma_{12} \in \mathbb{R}^{n\times m}$ such that

$$\begin{bmatrix} \widetilde{K} & \widetilde{L} \end{bmatrix} = \begin{bmatrix} K & L \end{bmatrix} + \Sigma_{12} \begin{bmatrix} H & J \end{bmatrix}.$$

Together with (3.51) this yields

$$\begin{aligned} \begin{bmatrix} -sE + A & B \\ \widetilde{K} & \widetilde{L} \end{bmatrix} &= \begin{bmatrix} -sE + A & B \\ K + \Sigma_{12} H & L + \Sigma_{12} J \end{bmatrix} \\ &= \begin{bmatrix} -sE + A & B \\ K + \Sigma_{12} M T_\infty^\mathsf{T} (-sE + A) & L + \Sigma_{12} M T_\infty^\mathsf{T} B \end{bmatrix} \\ &= \begin{bmatrix} I_n & 0 \\ \Sigma_{12} M T_\infty^\mathsf{T} & I_m \end{bmatrix} \begin{bmatrix} -sE + A & B \\ K & L \end{bmatrix}. \end{aligned}$$

**Theorem 3.4.11.** *Assume that $(E, A, B) \in \Sigma_{n,m}$ with the system space $\mathcal{V}_{\mathrm{sys}}$ is impulse controllable and let the descriptor Lur'e equation (3.35) with associated even* *matrix pencil* $s\mathcal{E} - \mathcal{A}$ *as in* (3.29) *be given. Furthermore, assume that the descriptor KYP inequality* (3.22) *is feasible. Let an* $n+m$*-dimensional* $\mathcal{E}$*-neutral space* $\operatorname{im} Y$ *with* $Y$ *as in* (3.48) *be given such that* $\mathcal{V}_{\mathrm{sys}} \subseteq \operatorname{im} \begin{bmatrix} Y_x \\ Y_u \end{bmatrix}$ *and* (3.49) *holds for some* $Z \in \mathbb{R}^{2n+m \times n+q}$, $\widetilde{\mathcal{E}}, \widetilde{\mathcal{A}} \in \mathbb{R}^{n+q \times n+m}$. *If for all finite eigenvalues* $\lambda$ *of the pencil* $s\widetilde{\mathcal{E}} - \widetilde{\mathcal{A}}$, *the number* $-\overline{\lambda}$ *is not an uncontrollable mode of* $(E, A, B)$, *then* $\operatorname{rank} EY_x = r$.

*Proof.* We consider again the subspace relation (3.55). Since all assumptions of [Rei11, Thm. 14] are fulfilled, the result immediately follows. □

Note that Theorem 3.4.2 requires that $\Phi(\mathrm{i}\omega) \geq 0$ for all $\omega \in \mathbb{R}$ with $\mathrm{i}\omega \notin \Lambda(E, A)$, whereas in Theorem 3.4.11 we assume that that the descriptor KYP inequality (3.22) is feasible to ensure $\operatorname{rank} EY_x = r$. However, this condition is, by Theorem 3.3.1, slightly stronger than the nonnegativity of $\Phi(s)$ on $\mathrm{i}\mathbb{R} \setminus \Lambda(E, A)$.

In the sequel of this section we discuss how to construct $\mathcal{E}$-neutral deflating subspaces with the properties of Theorem 3.4.9. First, we consider the case of a single block $\mathcal{D}(s)$ as in Theorem 2.1.13.

**Lemma 3.4.12.** *Consider a block* $\mathcal{D}(s) = s\mathcal{E} - \mathcal{A}$ *from Theorem 2.1.13. Then* $\operatorname{im} \mathbf{Y}$ *is an* $\mathcal{E}$*-neutral deflating subspace of* $\mathcal{D}(s)$ *with*

$$\mathcal{D}(s)\mathbf{Y} = \mathbf{Z}\widetilde{\mathcal{D}}(s)$$

*and an associated block* $\widetilde{\mathcal{D}}(s)$ equivalent *to a block from Theorem 2.1.2 as summarized in Table 3.1.*

*Proof.* The proof can be carried out by direct calculation. □

Finally, the next lemma shows that the deflating subspaces might not be unique if $s\widetilde{\mathcal{E}} - \widetilde{\mathcal{A}}$ can be constructed such that its KCF contains multiple blocks $\widetilde{\mathcal{D}}(s)$ from Theorem 2.1.2 of the same kind an size.

**Lemma 3.4.13.** *Let the even matrix pencil* $s\mathcal{E} - \mathcal{A}$ *as in* (3.29) *be given. Assume that for all* $j \in \{1, \ldots, \widetilde{\ell}\}$ *it holds that*

$$(s\mathcal{E} - \mathcal{A})Y_j = Z_j\widetilde{\mathcal{D}}(s),$$

*where* $\operatorname{im} Y_j$ *are* $\mathcal{E}$*-neutral deflating subspaces with* $Y_j \in \mathbb{C}^{2n+m \times k_1}$, $Z_j \in \mathbb{C}^{2n+m \times k_2}$, *and* $\widetilde{\mathcal{D}}(s) \in \mathbb{C}[s]^{k_2 \times k_1}$ *as in Theorem 2.1.2. Moreover, assume that* $\operatorname{im} Y_i \cap \operatorname{im} Y_j = \{0\}$ *for all* $1 \leq i, j \leq \widetilde{\ell}$ *with* $i \neq j$. *Then also*

$$\operatorname{im} \left( \sum_{j=1}^{\widetilde{\ell}} \alpha_j Y_j \right)$$

*is an* $\mathcal{E}$*-neutral deflating subspace for all* $0 \neq (\alpha_1, \ldots, \alpha_{\widetilde{\ell}}) \in \mathbb{R}^{\widetilde{\ell}}$.

Table 3.1: Classification of $\mathcal{E}$-neutral deflating subspaces of a single block $\mathcal{D}(s)$

| | $\mathcal{D}(s)$ | | | | | $\widetilde{\mathcal{D}}(s)$ | | |
|---|---|---|---|---|---|---|---|---|
| # | type | size | parameters | **Y** | **Z** | type | size | parameters |
| 1 | E1 | $2k \times 2k$ | $\lambda = \lambda_0 \in \mathbb{C}^+$ | $\begin{bmatrix} 0_{2k-l \times l} \\ I_l \end{bmatrix}$ | $\begin{bmatrix} I_l \\ 0_{2k-l \times l} \end{bmatrix}$ | K1 | $l \times l$ | $\lambda = \lambda_0$, $l \leq k$ |
| 2 | E1 | $2k \times 2k$ | $\lambda = \lambda_0 \in \mathbb{C}^+$ | $\begin{bmatrix} I_l \\ 0_{2k-l \times l} \end{bmatrix}$ | $\begin{bmatrix} 0_{2k-l \times l} \\ I_l \end{bmatrix}$ | K1 | $l \times l$ | $\lambda = -\overline{\lambda}_0$, $l \leq k$ |
| 3 | E2 | $k \times k$ | $\mu = \mu_0 \in \mathbb{R}$, $\varepsilon \in \{-1, 1\}$ | $\begin{bmatrix} 0_{k-l \times l} \\ I_l \end{bmatrix}$ | $\begin{bmatrix} I_l \\ 0_{k-l \times l} \end{bmatrix}$ | K1 | $l \times l$ | $\lambda = \mathrm{i}\mu_0$, $l \leq \frac{k}{2}$ ($k$ even), $l \leq \frac{k-1}{2}$ ($k$ odd) |
| 4 | E3 | $k \times k$ | $\varepsilon \in \{-1, 1\}$ | $\begin{bmatrix} 0_{k-l \times l} \\ I_l \end{bmatrix}$ | $\begin{bmatrix} I_l \\ 0_{k-l \times l} \end{bmatrix}$ | K2 | $l \times l$ | $l \leq \frac{k}{2}$ ($k$ even), $l \leq \frac{k+1}{2}$ ($k$ odd) |
| 5 | E4 | $(2k-1)$ $\times (2k-1)$ | – | $\begin{bmatrix} 0_{2k-l-1 \times l} \\ I_l \end{bmatrix}$ | $\begin{bmatrix} I_{l-1} \\ 0_{2k-l \times l-1} \end{bmatrix}$ | K3 | $(l-1) \times l$ | $l \leq k$ |

*Proof.* The statement follows from

$$(s\mathcal{E} - \mathcal{A}) \sum_{j=1}^{\widetilde{\ell}} \alpha_j Y_j = \sum_{j=1}^{\widetilde{\ell}} \alpha_j Z_j \widetilde{\mathcal{D}}(s),$$

and the fact that $\operatorname{im} Y_i \cap \operatorname{im} Y_j = \{0\}$ for all $1 \leq i, j \leq \widetilde{\ell}$ with $i \neq j$. □

In the next remark we will make some comments on uniqueness of solutions and the solution structure of the descriptor Lur'e equation.

*Remark* 3.4.14. Assume that $(E, A, B) \in \Sigma_{n,m}$ is impulse controllable.

a) To fulfill the inclusion $\mathcal{V}_{\text{sys}} \subseteq \operatorname{im} \begin{bmatrix} Y_x \\ Y_u \end{bmatrix}$, it is necessary to construct a *maximally* $\widehat{\mathcal{E}}$-neutral deflating subspace $\operatorname{im} \begin{bmatrix} Y_{\mu,1}^{\mathsf{T}} & Y_{x,1}^{\mathsf{T}} & Y_{u,1}^{\mathsf{T}} \end{bmatrix}^{\mathsf{T}}$ of the reduced even pencil $s\widehat{\mathcal{E}} - \widehat{\mathcal{A}}$ in (3.33) (see also (3.55)), for instance as in [Rei11, Thm. 13].

b) There is a certain freedom for the choice of $\widehat{Y}_{\mu,2}$ and $\widehat{Y}_{x,2}$ in (3.53). This can be most easily seen in (3.54), since any $(n-r)$-dimensional subspace $\operatorname{im} \begin{bmatrix} \widetilde{Y}_{\mu,2} \\ \widetilde{Y}_{x,2} \end{bmatrix} \subseteq \mathbb{R}^{2(n-r)}$ can be generated by appropriately choosing $\widetilde{Y}_{\mu,2}$ and $\widetilde{Y}_{x,2}$. This freedom also directly follows from Lemma 3.4.13. From Table 3.1, we can choose $2(n-r)$ linearly independent vectors $v_1, \ldots, v_{2(n-r)}$ that span a deflating subspace corresponding to a single block $\widetilde{\mathcal{D}}(s) = 1$. However, for the construction of $\widetilde{Y}_{\mu,2}$ and $\widetilde{Y}_{x,2}$, we only need a set of $n-r$ linearly independent vectors of the form

$$y_i = \sum_{j=1}^{2(n-r)} \alpha_j^{(i)} v_j.$$

c) There is further freedom in choosing the deflating subspaces for the blocks of type E1, following from Table 3.1. In order to guarantee $\operatorname{rank} EY_x = r$, we have to construct the subspace according to type #2 from Table 3.1 if $\lambda$ is an uncontrollable mode and according to type #1 from Table 3.1 if $-\overline{\lambda}$ is an uncontrollable mode of $(E, A, B)$. This criterion implicitly contains strong sign-controllability of $(E, A, B)$ [Rei11]. In particular, in case of a solvable descriptor Lur'e equation, an $\mathcal{E}$-neutral subspace $\mathcal{Y}$ of dimension $n+m$ as in Theorem 3.4.9 can be constructed such that $\operatorname{rank} EY_x = r$, if the triple $(E, A, B)$ is strongly controllable.

d) Let the assumptions of Theorem 3.4.11 be satisfied. Then the following two statements are equivalent:

i) The matrix $E^\mathsf{T}X$ can only attain a finite number of distinct values, where $(X, K, L)$ is a solution of the descriptor Lur'e equation.

ii) The EKCF of the pencil $s\mathcal{E} - \mathcal{A}$ has no multiple blocks of type E1 corresponding to the same parameter $\mu \in \mathbb{C}^+$.

If the EKCF of the pencil $s\mathcal{E} - \mathcal{A}$ has multiple blocks of type E1 corresponding to the same parameter $\mu \in \mathbb{C}^+$, we can construct an infinite amount of other $\mathcal{E}$-neutral deflating subspaces corresponding to the eigenvalues $\mu$ and $-\overline{\mu}$ by appropriately mixing subspaces of types #1 and #2 in Table 3.1 corresponding to different blocks. Otherwise, the number of possible values of $E^\mathsf{T}X$ is at most $2^r$, since for every block of the KCF of $s\widetilde{\mathcal{E}} - \widetilde{\mathcal{A}}$ of type K1 a maximal deflating subspace can be either chosen according to type #1 or #2 in Table 3.1. See also [Wil71, Rem. 18] for a related observation in the ODE case. Note that for real problems, the eigenvalues of $s\widetilde{\mathcal{E}} - \widetilde{\mathcal{A}}$ appear in complex conjugate pairs. So in fact, there might be even less *real* solutions, since the deflating subspaces must then also be picked in such pairs.

Finally, in the next remark, we present some examples illustrating the result of Theorem 3.4.9.

*Remark* 3.4.15.

(a) Even if all properties of statement 2) of Theorem 3.4.9 are fulfilled, the invertibility of $\begin{bmatrix} Y_x \\ Y_u \end{bmatrix}$ *cannot* be guaranteed for an arbitrary choice of $Y$. To see this consider the following simple example. Choose

$$E = \begin{bmatrix} 1 & 0 \\ 0 & 0 \end{bmatrix}, \quad A = \begin{bmatrix} 1 & 0 \\ 0 & 1 \end{bmatrix}, \quad B = S = \begin{bmatrix} 0 \\ 0 \end{bmatrix}, \quad Q = \begin{bmatrix} 0 & 0 \\ 0 & 0 \end{bmatrix}, \quad R = 1.$$

The system $(E, A, B) \in \Sigma_{2,1}$ is impulse controllable. With $X = \begin{bmatrix} x_{11} & 0 \\ x_{21} & x_{22} \end{bmatrix}$ we have

$$\begin{bmatrix} A^\mathsf{T}X + X^\mathsf{T}A + Q & X^\mathsf{T}B + S \\ B^\mathsf{T}X + S^\mathsf{T} & R \end{bmatrix} = \begin{bmatrix} 2x_{11} & x_{21} & 0 \\ x_{21} & 2x_{22} & 0 \\ 0 & 0 & 1 \end{bmatrix}$$

which is positive definite, e. g., for $x_{11} = x_{22} = 1$ and $x_{21} = 0$. Furthermore,

$$\begin{bmatrix} 0 & 0 & -s+1 & 0 & 0 \\ 0 & 0 & 0 & 1 & 0 \\ s+1 & 0 & 0 & 0 & 0 \\ 0 & 1 & 0 & 0 & 0 \\ 0 & 0 & 0 & 0 & 1 \end{bmatrix} \begin{bmatrix} 0 & 0 & 0 \\ 0 & 0 & 0 \\ 1 & 0 & 0 \\ 0 & 1 & 0 \\ 0 & 0 & 1 \end{bmatrix} = \begin{bmatrix} 1 & 0 & 0 \\ 0 & 1 & 0 \\ 0 & 0 & 0 \\ 0 & 0 & 0 \\ 0 & 0 & 1 \end{bmatrix} \begin{bmatrix} -s+1 & 0 & 0 \\ 0 & 1 & 0 \\ 0 & 0 & 1 \end{bmatrix},$$

i. e., $\operatorname{rank}\begin{bmatrix} Y_x \\ Y_u \end{bmatrix} = 3 = n + m$. On the other hand, we also have

$$\begin{bmatrix} 0 & 0 & -s+1 & 0 & 0 \\ 0 & 0 & 0 & 1 & 0 \\ s+1 & 0 & 0 & 0 & 0 \\ 0 & 1 & 0 & 0 & 0 \\ 0 & 0 & 0 & 0 & 1 \end{bmatrix} \begin{bmatrix} 0 & 0 & 0 \\ 0 & 1 & 0 \\ 1 & 0 & 0 \\ 0 & 0 & 0 \\ 0 & 0 & 1 \end{bmatrix} = \begin{bmatrix} 1 & 0 & 0 \\ 0 & 0 & 0 \\ 0 & 0 & 0 \\ 0 & 1 & 0 \\ 0 & 0 & 1 \end{bmatrix} \begin{bmatrix} -s+1 & 0 & 0 \\ 0 & 1 & 0 \\ 0 & 0 & 1 \end{bmatrix},$$

i. e., $\operatorname{rank}\begin{bmatrix} Y_x \\ Y_u \end{bmatrix} = 2 < n + m$, and therefore, no solution can be constructed from this deflating subspace.

(b) It is also important to note that the requirement of impulse controllability is essential for the existence of a $Y$ with $\begin{bmatrix} Y_x \\ Y_u \end{bmatrix} \in \mathrm{Gl}_{n+m}(\mathbb{R})$. As a counter-example consider the matrices

$$E = \begin{bmatrix} 1 & 0 \\ 0 & 0 \end{bmatrix}, \quad A = \begin{bmatrix} 0 & 1 \\ 1 & 0 \end{bmatrix}, \quad B = S = \begin{bmatrix} 0 \\ 0 \end{bmatrix}, \quad Q = \begin{bmatrix} 0 & 0 \\ 0 & 0 \end{bmatrix}, \quad R = 1.$$

Again we have $X = \begin{bmatrix} x_{11} & 0 \\ x_{21} & x_{22} \end{bmatrix}$ and

$$\begin{bmatrix} A^\mathsf{T} X + X^\mathsf{T} A + Q & X^\mathsf{T} B + S \\ B^\mathsf{T} X + S^\mathsf{T} & R \end{bmatrix} = \begin{bmatrix} 2x_{21} & x_{11} + x_{22} & 0 \\ x_{11} + x_{22} & 0 & 0 \\ 0 & 0 & 1 \end{bmatrix},$$

which is positive semidefinite, e. g., for $x_{11} = x_{22} = 0$ and $x_{21} = 1$. However, we have

$$\begin{bmatrix} 0 & 0 & -s & 1 & 0 \\ 0 & 0 & 1 & 0 & 0 \\ s & 1 & 0 & 0 & 0 \\ 1 & 0 & 0 & 0 & 0 \\ 0 & 0 & 0 & 0 & 1 \end{bmatrix} \begin{bmatrix} 0 & 0 & 0 \\ 1 & 0 & 0 \\ 0 & 0 & 0 \\ 0 & 1 & 0 \\ 0 & 0 & 1 \end{bmatrix} = \begin{bmatrix} 0 & 1 & 0 \\ 0 & 0 & 0 \\ 1 & 0 & 0 \\ 0 & 0 & 0 \\ 0 & 0 & 1 \end{bmatrix} \begin{bmatrix} 1 & 0 & 0 \\ 0 & 1 & 0 \\ 0 & 0 & 1 \end{bmatrix},$$

i. e., $\operatorname{rank}\begin{bmatrix} Y_x \\ Y_u \end{bmatrix} = 2 < n+m$. Since in our case, $s\mathcal{E} - \mathcal{A}$ has only infinite eigenvalues, there is also no way to make $\begin{bmatrix} Y_x \\ Y_u \end{bmatrix}$ invertible since otherwise the right-hand side would not be independent of $s$.

## 3.5 Stabilizing, Anti-Stabilizing, and Extremal Solutions

Here we focus on particular solutions of the descriptor Lur'e equation and KYP inequality. We define stabilizing and anti-stabilizing solutions. We prove that these are extremal in the sense of definiteness.

**Definition 3.5.1** (Stabilizing and anti-stabilizing solutions)**.** [RRV14, Def. 5.1] Let $(E, A, B) \in \Sigma_{n,m}$ with the system space $\mathcal{V}_{\mathrm{sys}}$, and $Q = Q^{\mathsf{T}} \in \mathbb{R}^{n\times n}$, $S \in \mathbb{R}^{n\times m}$, and $R = R^{\mathsf{T}} \in \mathbb{R}^{m\times m}$ be given. A solution $(X, K, L)$ of the *descriptor Lur'e equation* (3.35) is called

a) *stabilizing*, if

$$\operatorname{rank}\begin{bmatrix} -\lambda E + A & B \\ K & L \end{bmatrix} = n + q \quad \forall \lambda \in \mathbb{C}^{+}; \tag{3.56}$$

b) *anti-stabilizing*, if

$$\operatorname{rank}\begin{bmatrix} -\lambda E + A & B \\ K & L \end{bmatrix} = n + q \quad \forall \lambda \in \mathbb{C}^{-}. \tag{3.57}$$

*Remark* 3.5.2 (Stabilizing and anti-stabilizing solutions). A solution $(X, K, L)$ is stabilizing if and only if the system $(E, A, B, K, L) \in \Sigma_{n,m,q}$ is outer. Analogously, $(X, K, L)$ is anti-stabilizing if and only if the system $(-E, A, B, K, L) \in \Sigma_{n,m,q}$ is outer.

Now we state a result on the existence of stabilizing and anti-stabilizing solutions of a descriptor Lur'e equation, see also [RRV14, Thm. 5.3].

**Theorem 3.5.3** (Existence of stabilizing and anti-stabilizing solutions)**.** *Let a system $(E, A, B) \in \Sigma_{n,m}$ with the system space $\mathcal{V}_{\mathrm{sys}}$, and $Q = Q^{\mathsf{T}} \in \mathbb{R}^{n\times n}$, $S \in \mathbb{R}^{n\times m}$, and $R = R^{\mathsf{T}} \in \mathbb{R}^{m\times m}$ be given. Assume that the descriptor KYP inequality* (3.22) *is solvable.*

*a) If $(E, A, B)$ is strongly stabilizable, then the descriptor Lur'e equation* (3.35) *has a stabilizing solution.*

*b) If $(E, A, B)$ is strongly anti-stabilizable, then the descriptor Lur'e equation* (3.35) *has an anti-stabilizing solution.*

*Proof.* Since statement b) follows from a) by turning to the backward system, we only prove the first statement. Since the descriptor KYP inequality (3.22) is solvable, by the KYP lemma (Theorem 3.3.1), the associated Popov function $\Phi(s)$ fulfills $\Phi(\mathrm{i}\omega) \geq 0$ for all $\omega \in \mathbb{R}$ with $\mathrm{i}\omega \notin \Lambda(E, A)$. Furthermore, since $(E, A, B)$ is strongly stabilizable, $(E, A, B)$ has no uncontrollable modes on the imaginary axis. Thus by

Theorem 3.4.2 and Remark 3.4.3, the EKCF of the corresponding even matrix pencil $s\mathcal{E} - \mathcal{A}$ has the structure as in Remark 3.4.3. By picking a deflating subspace of $s\mathcal{E} - \mathcal{A}$ according to type #2 in Table 3.1 for *all* blocks of type E1 and applying Theorem 3.4.11 and Theorem 3.4.9, we finally obtain a stabilizing solution of the descriptor Lur'e equation (3.35). □

In the following theorem we will show that the stabilizing and anti-stabilizing solution of the descriptor Lur'e equation are extremal solutions in terms of definiteness of $E^\top X$, see also [RRV14, Thm. 5.5].

**Theorem 3.5.4.** *Let $(E, A, B) \in \Sigma_{n,m}$ with the system space $\mathcal{V}_{\text{sys}}$ be impulse controllable, and let $Q = Q^\top \in \mathbb{R}^{n\times n}$, $S \in \mathbb{R}^{n\times m}$, and $R = R^\top \in \mathbb{R}^{m\times m}$ be given. Assume that $(X, K, L)$ solves the descriptor Lur'e equation* (3.35) *and $P \in \mathbb{R}^{n\times n}$ solves the descriptor KYP inequality* (3.22).

*a) If $(X, K, L)$ is a stabilizing solution, then*

$$E^\top X \geq E^\top P.$$

*b) If $(X, K, L)$ is an anti-stabilizing solution, then*

$$E^\top X \leq E^\top P.$$

*Furthermore, if $(X, K, L)$ and $\big(\widetilde{X}, \widetilde{K}, \widetilde{L}\big)$ are stabilizing (anti-stabilizing) solutions, then*

$$E^\top X = E^\top \widetilde{X}. \tag{3.58}$$

*Proof.* Statement b) follows from a) by turning to the backward system. Therefore, we only prove a): Since $(E, A, B)$ is impulse controllable, there exist $W$, $T \in \mathrm{Gl}_n(\mathbb{R})$ and $F \in \mathbb{R}^{m\times n}$ such that $E_F$, $A_F$, $B_F$ as well as $Q_F$, $S_F$, $R_F$ are in the form (3.37) and (3.38), respectively. Moreover, assume that $X_F$ and $K_F$ are given as in (3.39) and let

$$P_F = W^{-\top} P T = \begin{bmatrix} P_{11} & 0 \\ P_{21} & P_{22} \end{bmatrix}$$

be accordingly partitioned. From Lemma 3.4.7 a) it follows that (3.40) is satisfied. By an argumentation analogous to (3.44), we obtain that

$$n + q = \operatorname{rank} \begin{bmatrix} -\lambda E + A & B \\ K & L \end{bmatrix} = \operatorname{rank} \begin{bmatrix} -\lambda I_r + A_{11} & B_1 \\ K_1 & L - K_2 B_2 \end{bmatrix} + n - r \quad \forall \lambda \in \mathbb{C}^+,$$

i. e., $(X_{11}, K_1, L - K_2 B_2)$ is a stabilizing solution of the Lur'e equation (3.40). From the proof of Theorem 3.3.1 we further see, that $P_{11}$ solves the KYP inequality (3.28).

Now the corresponding maximality result for ODE systems in [Rei11, Thm. 15] yields $X_{11} \geq P_{11}$. Finally, this results in

$$\begin{aligned} E^{\mathsf{T}} X = T^{-\mathsf{T}} E_F^{\mathsf{T}} X_F T^{-1} &= T^{-\mathsf{T}} \begin{bmatrix} X_{11} & 0 \\ 0 & 0 \end{bmatrix} T^{-1} \\ &\geq T^{-\mathsf{T}} \begin{bmatrix} P_{11} & 0 \\ 0 & 0 \end{bmatrix} T^{-1} = T^{-\mathsf{T}} E_F^{\mathsf{T}} P_F T^{-1} = E^{\mathsf{T}} P. \end{aligned}$$

It remains to prove the uniqueness statement in (3.58). Therefore, assume that $(X, K, L)$ and $\left(\widetilde{X}, \widetilde{K}, \widetilde{L}\right)$ are stabilizing solutions of the descriptor Lur'e equation. Then, since $\widetilde{X}$ also solves the descriptor KYP inequality, we obtain from the previously shown result that $E^{\mathsf{T}} X \geq E^{\mathsf{T}} \widetilde{X}$. Further, by reversing the roles of $X$ and $\widetilde{X}$, we can conclude $E^{\mathsf{T}} \widetilde{X} \geq E^{\mathsf{T}} X$. Altogether, this yields $E^{\mathsf{T}} X = E^{\mathsf{T}} \widetilde{X}$. □

## 3.6 Spectral Factorization

In this section the connection of the so far presented theory to spectral factorization of the Popov function will be presented. This factorization appears to play an important role in filtering theory and linear-quadratic optimal control as pointed out in Section 3.8 of this chapter. Early results on this factorization are given, e. g., in [Wie49, You61, And69, Wil71]. The most general approach known up to now is presented in [OV00]. There, a computational method for the spectral factorization of arbitrary Popov functions with (3.8) is discussed. This procedure relies on a sequence of reductions of the Popov function which are based on transformations of a certain Rosenbrock pencil to filter out a weakly minimal subsystem for which an algebraic Riccati equation is solved. The solution of this Riccati equation is then used to construct realizations of the spectral factors. In this work, however, we use a direct approach which instantly relates a spectral factorization to the solution of a descriptor KYP inequality without a restriction like (3.8).

**Definition 3.6.1** (Spectral factorization)**.** A *spectral factorization* of a Popov function $\Phi(s) \in \mathbb{R}(s)^{m \times m}$ is a representation

$$\Phi(s) = W^{\sim}(s) W(s) \tag{3.59}$$

for some $W(s) \in \mathbb{R}(s)^{l \times m}$. The rational function $W(s) \in \mathbb{R}(s)^{l \times m}$ is then called a *spectral factor* of $\Phi(s)$. A realization of a spectral factor is called a *spectral factor system*.

*Remark* 3.6.2. In the literature such as [Wil71, OV00], a spectral factorization is often referred to as a decomposition of the form (3.59), where the spectral factor $W(s)$ is outer. In this work we call any factorization of the form (3.59) a spectral factorization.

Obviously, a necessary condition for the existence of a spectral factorization is that $\Phi(s)$ is Hermitian and attains positive semidefinite values on the imaginary axis. We will now show that each solution of the descriptor KYP inequality (3.22) induces a spectral factorization, see also [RRV14].

**Theorem 3.6.3.** *Let $(E, A, B) \in \Sigma_{n,m}$ with the system space $\mathcal{V}_{\text{sys}}$ be given and assume that $Q = Q^\top \in \mathbb{R}^{n\times n}$, $S \in \mathbb{R}^{n\times m}$, and $R = R^\top \in \mathbb{R}^{m\times m}$. Assume that $P \in \mathbb{R}^{n\times n}$ solves the descriptor KYP inequality (3.22). Let $K \in \mathbb{R}^{l\times n}$ and $L \in \mathbb{R}^{l\times m}$ with*

$$\begin{bmatrix} A^\top P + P^\top A + Q & P^\top B + S \\ B^\top P + S^\top & R \end{bmatrix} =_{\mathcal{V}_{\text{sys}}} \begin{bmatrix} K^\top \\ L^\top \end{bmatrix} \begin{bmatrix} K & L \end{bmatrix} \tag{3.60}$$

*be given. Then $(E, A, B, K, L) \in \Sigma_{n,m,l}$ is a spectral factor system for the Popov function $\Phi(s) \in \mathbb{R}(s)^{m\times m}$ as in (3.2) and the associated spectral factor is given by*

$$W(s) = K(sE - A)^{-1}B + L. \tag{3.61}$$

*In particular, the following statements hold true:*

*a) If $P = X$, where $(X, K, L)$ is a stabilizing solution of the descriptor Lur'e equation (3.35), then $W(s)$ is outer.*

*b) If $P = X$, where $(X, K, L)$ is an anti-stabilizing solution of the descriptor Lur'e equation (3.35), then $W(-s)$ is outer.*

*Proof.* Using (3.24), the result follows from

$$\begin{aligned}
\Phi(s) &= \begin{bmatrix} (sE - A)^{-1}B \\ I_m \end{bmatrix}^\sim \begin{bmatrix} Q & S \\ S^\top & R \end{bmatrix} \begin{bmatrix} (sE - A)^{-1}B \\ I_m \end{bmatrix} \\
&\overset{(3.60)}{=} \begin{bmatrix} (sE - A)^{-1}B \\ I_m \end{bmatrix}^\sim \begin{bmatrix} K^\top \\ L^\top \end{bmatrix} \begin{bmatrix} K & L \end{bmatrix} \begin{bmatrix} (sE - A)^{-1}B \\ I_m \end{bmatrix} \\
&\quad - \begin{bmatrix} (sE - A)^{-1}B \\ I_m \end{bmatrix}^\sim \begin{bmatrix} A^\top P + P^\top A & P^\top B \\ B^\top P & 0 \end{bmatrix} \begin{bmatrix} (sE - A)^{-1}B \\ I_m \end{bmatrix} \\
&\overset{(3.25)}{=} \left( \begin{bmatrix} K & L \end{bmatrix} \begin{bmatrix} (sE - A)^{-1}B \\ I_m \end{bmatrix} \right)^\sim \left( \begin{bmatrix} K & L \end{bmatrix} \begin{bmatrix} (sE - A)^{-1}B \\ I_m \end{bmatrix} \right) \\
&= W^\sim(s)W(s).
\end{aligned}$$

The statements on the spectral factors corresponding to the stabilizing and anti-stabilizing solutions follow from Proposition 2.2.25. □

It follows from the spectral factorization that for $K \in \mathbb{R}^{l\times n}$, $L \in \mathbb{R}^{l\times m}$ with (3.60) it holds that $l \geq q = \operatorname{rank}_{\mathbb{R}(s)} \Phi(s)$. In the following we show that we even have $l = \operatorname{rank}_{\mathbb{R}(s)} \Phi(s)$ for $P = X$, where $(X, K, L)$ is a solution of the descriptor Lur'e equation (3.35). As a consequence, the descriptor Lur'e equation can be seen as a descriptor KYP inequality in which the rank of the left hand side is minimized. In particular, we have $q \leq m$.

**Theorem 3.6.4.** *Let $(E, A, B) \in \Sigma_{n,m}$ with the system space $\mathcal{V}_{\text{sys}}$ be given, and assume that $Q = Q^{\mathsf{T}} \in \mathbb{R}^{n \times n}$, $S \in \mathbb{R}^{n \times m}$, and $R = R^{\mathsf{T}} \in \mathbb{R}^{m \times m}$. Let $\Phi(s) \in \mathbb{R}(s)^{m \times m}$ be the Popov function (3.2), and let $q = \operatorname{rank}_{\mathbb{R}(s)} \Phi(s)$. If $(X, K, L)$ is a solution of the descriptor Lur'e equation (3.35), then $K \in \mathbb{R}^{q \times n}$ and $L \in \mathbb{K}^{q \times m}$.*

*Proof.* Let $W(s)$ be defined as in (3.61). It follows from Theorem 3.6.3 that for all $\omega \in \mathbb{R}$ with $\mathrm{i}\omega \notin \Lambda(E, A)$ it holds that

$$\Phi(\mathrm{i}\omega) = W^{\mathsf{H}}(\mathrm{i}\omega)W(\mathrm{i}\omega),$$

hence $\operatorname{rank}_{\mathbb{R}(s)} W(s) = q$.

On the other hand, Lemma 2.2.26 gives

$$\operatorname{rank}_{\mathbb{R}(s)} \begin{bmatrix} -sE + A & B \\ K & L \end{bmatrix} = n + q.$$

From the definition of the solution of a descriptor Lur'e equation, see Definition 3.4.6, we now directly get $K \in \mathbb{R}^{q \times n}$ and $L \in \mathbb{R}^{q \times m}$. □

## 3.7 Nonpositive Solutions

In this section we investigate the existence of solutions of the descriptor KYP inequality with $E^{\mathsf{T}}P \leq 0$. Such solutions are referred to as *nonpositive solutions.* The existence of nonpositive solutions will be of great importance for linear-quadratic optimal control (see Section 3.8) and the analysis of dissipative systems (see Section 3.9). Criteria will be given in terms of the *modified Popov function*

$$\Psi : \mathbb{C} \setminus \Lambda(E, A) \to \mathbb{C}^{m \times m} \tag{3.62a}$$

with

$$\Psi(\lambda) = \begin{bmatrix} (\lambda E - A)^{-1}B \\ I_m \end{bmatrix}^{\mathsf{H}} \begin{bmatrix} Q & S \\ S^{\mathsf{T}} & R \end{bmatrix} \begin{bmatrix} (\lambda E - A)^{-1}B \\ I_m \end{bmatrix}. \tag{3.62b}$$

Note that the modified Popov function coincides with the Popov function on the imaginary axis. Moreover, note that in general, the modified Popov function is in general neither rational nor meromorphic. The latter fact can be seen by a simple example. Let $\lambda = \gamma + \mathrm{i}\delta$ and consider

$$\begin{aligned}\Psi(\lambda) &= (\lambda - 1)^{-\mathsf{H}} \cdot (\lambda - 1)^{-1} \\ &= \left(|\lambda|^2 - \lambda - \overline{\lambda} + 1\right)^{-1} \\ &= \left(\gamma^2 + \delta^2 - 2\gamma + 1\right)^{-1}.\end{aligned}$$

Obviously, there exist infinitely many non-isolated pairs $(\gamma, \delta) \in \mathbb{R}^2$ for which $\gamma^2 + \delta^2 - 2\gamma + 1 = 0$, so $\Psi(\cdot)$ cannot be meromorphic.

First we show that the nonnegativity of $\Psi(\cdot)$ in the right complex half-plane is necessary for the existence of a nonpositive solution.

**Theorem 3.7.1.** *Let $(E, A, B) \in \Sigma_{n,m}$ with the system space $\mathcal{V}_{\text{sys}}$ be given and assume that $Q = Q^\mathsf{T} \in \mathbb{R}^{n\times n}$, $S \in \mathbb{R}^{n\times m}$, and $R = R^\mathsf{T} \in \mathbb{R}^{m\times m}$. Assume that $P \in \mathbb{R}^{n\times n}$ is a nonpositive solution of the descriptor KYP inequality* (3.22). *Then the modified Popov function* (3.62) *fulfills*

$$\Psi(\lambda) \geq 0 \quad \forall \lambda \in \overline{\mathbb{C}^+} \setminus \Lambda(E, A). \tag{3.63}$$

*Proof.* The statement holds true since, analogous to (3.25), for all $\lambda \in \mathbb{C} \setminus \Lambda(E, A)$ we have

$$\begin{aligned}\begin{bmatrix}(\lambda E - A)^{-1}B \\ I_m\end{bmatrix}^\mathsf{H} \begin{bmatrix}A^\mathsf{T}P + P^\mathsf{T}A & P^\mathsf{T}B \\ B^\mathsf{T}P & 0\end{bmatrix} \begin{bmatrix}(\lambda E - A)^{-1}B \\ I_m\end{bmatrix}& \\ = 2\operatorname{Re}(\lambda) \cdot B^\mathsf{T}\big(\overline{\lambda}E^\mathsf{T} - A^\mathsf{T}\big)^{-1}E^\mathsf{T}P\big(\lambda E - A\big)^{-1}B&.\end{aligned}$$

The descriptor KYP inequality and (3.24) now yield the inequality

$$\Psi(\lambda) \geq -2\operatorname{Re}(\lambda) \cdot B^\mathsf{T}\big(\overline{\lambda}E^\mathsf{T} - A^\mathsf{T}\big)^{-1}E^\mathsf{T}P\big(\lambda E - A\big)^{-1}B \quad \forall \lambda \in \mathbb{C} \setminus \Lambda(E, A).$$

Now plugging in some $\lambda \in \overline{\mathbb{C}^+} \setminus \Lambda(E, A)$ and using that $E^\mathsf{T}P \leq 0$, we obtain the desired result. □

We now show that, under an additional assumption, the nonnegativity of the modified Popov function also implies the existence of a nonpositive solution. For this we need the following two lemmas.

**Lemma 3.7.2.** *Let $(E, A, B, C, D) \in \Sigma_{n,m,p}$ be strongly stabilizable and strongly detectable. Furthermore, assume that the transfer function $G(s) = C(sE - A)^{-1}B + D \in \mathcal{RH}_\infty^{p\times m}$. Then the index of $sE - A$ is at most one. Furthermore, all finite eigenvalues of $sE - A$ are in $\mathbb{C}^-$.*

*Proof.* Since generalized state-space transformations preserve strong stabilizability and strong detectability, we can w. l. o. g. assume that $sE - A$ is given in QWF, i. e.,

$$sE - A = \begin{bmatrix}sI_r - A_{11} & 0 \\ 0 & sE_{22} - I_{n-r}\end{bmatrix}, \quad B = \begin{bmatrix}B_1 \\ B_2\end{bmatrix}, \quad C = \begin{bmatrix}C_1 & C_2\end{bmatrix},$$

where $E_{22} \in \mathbb{R}^{n-r\times n-r}$ is nilpotent, $B_1 \in \mathbb{R}^{r\times m}$, $B_2 \in \mathbb{R}^{n-r\times m}$, $C_1 \in \mathbb{R}^{p\times r}$, and $C_2 \in \mathbb{R}^{p\times n-r}$. As in (2.13) we have a decomposition

$$G(s) = G_{\text{sp}}(s) + G_{\text{poly}}(s)$$

with the strictly proper part $G_{\rm sp}(s)$ and the polynomial part $G_{\rm poly}(s)$.

The assumption that $G \in \mathcal{RH}_\infty^{p\times m}$ then implies $G_{\rm sp}(s) = C_1(sI_r - A_{11})^{-1}B_1 \in \mathcal{RH}_\infty^{p\times m}$. Strong stabilizability and strong detectability of $(E, A, B, C, D)$ imply stabilizability and detectability of $(I_r, A_{11}, B_1, C_1, D)$. Then [BIR12, Lem. 8.3] implies that $\Lambda(A_{11}) \subset \mathbb{C}^-$, hence $\Lambda(E, A) \subset \mathbb{C}^-$.

It remains to show that the index $\nu \in \mathbb{N}_0$ of $sE - A$ is at most one: Assume that $\nu > 1$. Then $E_{22}^\nu = 0$ and $E_{22}^{\nu-1} \neq 0$. The properness of $G(s)$ implies

$$C_2E_{22}B_2 = \ldots = C_2E_{22}^{\nu-1}B_2 = 0.$$

By using Proposition 2.2.6, impulse controllability and impulse observability lead to

$$\ker \begin{bmatrix} E_{22}^\top \\ B_2^\top \end{bmatrix} \cap \operatorname{im} E_{22}^\top = \{0\}, \text{ and} \tag{3.64a}$$

$$\ker \begin{bmatrix} E_{22} \\ C_2 \end{bmatrix} \cap \operatorname{im} E_{22} = \{0\}. \tag{3.64b}$$

By $C_2E_{22}^{\nu-1}B_2 = 0$ and $E_{22}(E_{22}^{\nu-1}B_2) = 0$, (3.64b) gives $E_{22}^{\nu-1}B_2 = 0$. Hence, we have $B_2^\top \left(E_{22}^\top\right)^{\nu-1} = 0$ and $E_{22}\left(E_{22}^\top\right)^{\nu-1} = 0$. Therefore, (3.64a) gives $E_{22}^{\nu-1} = 0$ which is a contradiction. □

**Lemma 3.7.3.**

*a) Let two systems* $(E, A, B, C_1, D_1) \in \Sigma_{n,m,m}$, $(E, A, B, C_2, D_2) \in \Sigma_{n,m,p}$ *with transfer functions* $G_1(s) = C_1(sE - A)^{-1}B + D_1 \in \mathrm{Gl}_m(\mathbb{R}(s))$, $G_2(s) = C_2(sE - A)^{-1}B + D_2 \in \mathbb{R}(s)^{p\times m}$ *be given. Then the transfer function of the system*

$$(E_{\rm e}, A_{\rm e}, B_{\rm e}, C_{\rm e}) := \left( \begin{bmatrix} E & 0 \\ 0 & 0 \end{bmatrix}, \begin{bmatrix} A & B \\ C_1 & D_1 \end{bmatrix}, \begin{bmatrix} 0 \\ -I_m \end{bmatrix}, \begin{bmatrix} C_2 & D_2 \end{bmatrix} \right) \in \Sigma_{n+m,m,p}$$

*is*

$$G_{\rm e}(s) = G_2(s)G_1^{-1}(s).$$

*b) Let two systems* $(E, A, B_1, C, D_1) \in \Sigma_{n,p,m}$, $(E, A, B_2, C, D_2) \in \Sigma_{n,p,p}$ *with transfer functions* $G_1(s) = C(sE - A)^{-1}B_1 + D_1 \in \mathbb{R}(s)^{p\times m}$, $G_2(s) = C(sE - A)^{-1}B_2 + D_2 \in \mathrm{Gl}_p(\mathbb{R}(s))$ *be given. Then the transfer function of the system*

$$(E_{\rm e}, A_{\rm e}, B_{\rm e}, C_{\rm e}) := \left( \begin{bmatrix} E & 0 \\ 0 & 0 \end{bmatrix}, \begin{bmatrix} A & B_2 \\ C & D_2 \end{bmatrix}, \begin{bmatrix} B_1 \\ D_1 \end{bmatrix}, \begin{bmatrix} 0 & -I_p \end{bmatrix} \right) \in \Sigma_{n+p,m,p}$$

*is*

$$G_{\rm e}(s) = G_2^{-1}(s)G_1(s).$$

*Proof.* First we prove a): It can be verified that

$$\begin{aligned}
&\begin{bmatrix} sE-A & -B \\ -C_1 & -D_1 \end{bmatrix}^{-1} \\
&= \begin{bmatrix} (sE-A)^{-1} - (sE-A)^{-1}BG_1^{-1}(s)C_1(sE-A)^{-1} & -(sE-A)^{-1}BG_1^{-1}(s) \\ -G_1^{-1}(s)C_1(sE-A)^{-1} & -G_1^{-1}(s) \end{bmatrix},
\end{aligned}$$

and therefore, it holds that

$$\begin{aligned}
G_{\mathrm{e}}(s) &= C_{\mathrm{e}}(sE_{\mathrm{e}} - A_{\mathrm{e}})^{-1}B_{\mathrm{e}} \\
&= \begin{bmatrix} C_2 & D_2 \end{bmatrix} \begin{bmatrix} (sE-A)^{-1}BG_1^{-1}(s) \\ G_1^{-1}(s) \end{bmatrix} \\
&= C_2(sE-A)^{-1}BG_1^{-1}(s) + D_2G_1^{-1}(s) \\
&= G_2(s)G_1^{-1}(s).
\end{aligned}$$

Statement b) can be proven similarly by checking that

$$\begin{aligned}
&\begin{bmatrix} sE-A & -B_2 \\ -C & -D_2 \end{bmatrix}^{-1} \\
&= \begin{bmatrix} (sE-A)^{-1} - (sE-A)^{-1}B_2G_2^{-1}(s)C(sE-A)^{-1} & -(sE-A)^{-1}B_2G_2^{-1}(s) \\ -G_2^{-1}(s)C(sE-A)^{-1} & -G_2^{-1}(s) \end{bmatrix},
\end{aligned}$$

and thus

$$\begin{aligned}
G_{\mathrm{e}}(s) &= C_{\mathrm{e}}(sE_{\mathrm{e}} - A_{\mathrm{e}})^{-1}B_{\mathrm{e}} \\
&= \begin{bmatrix} G_2^{-1}(s)C(sE-A)^{-1} & G_2^{-1}(s) \end{bmatrix} \begin{bmatrix} B_1 \\ D_1 \end{bmatrix} \\
&= G_2^{-1}(s)C(sE-A)^{-1}B_1 + G_2^{-1}(s)D_1 \\
&= G_2^{-1}(s)G_1(s).
\end{aligned}$$

□

**Theorem 3.7.4.** *Let $(E, A, B) \in \Sigma_{n,m}$ with the system space $\mathcal{V}_{\mathrm{sys}}$ be given and assume that $Q = Q^{\mathsf{T}} \in \mathbb{R}^{n \times n}$, $S \in \mathbb{R}^{n \times m}$, and $R = R^{\mathsf{T}} \in \mathbb{R}^{m \times m}$. Assume that the modified Popov function* (3.62) *is nonnegative, i. e.,* (3.63) *holds true. Further, assume that at least one of the following two assumptions holds true:*

*a) $(E, A, B)$ is strongly stabilizable and the Popov function* (3.2) *satisfies*

$$\operatorname{rank}_{\mathbb{R}(s)} \Phi(s) = m;$$

*b) $(E, A, B)$ is strongly controllable.*

*Let $C_1 \in \mathbb{R}^{m\times n}$, $C_2 \in \mathbb{R}^{p_2\times n}$, $D_1 \in \mathbb{R}^{m\times m}$, and $D_2 \in \mathbb{R}^{p_2\times m}$ be given such that*

$$\begin{bmatrix} Q & S \\ S^\mathsf{T} & R \end{bmatrix} =_{\mathcal{V}_{\mathrm{sys}}} \begin{bmatrix} C_1^\mathsf{T} C_1 & C_1^\mathsf{T} D_1 \\ D_1^\mathsf{T} C_1 & D_1^\mathsf{T} D_1 \end{bmatrix} - \begin{bmatrix} C_2^\mathsf{T} C_2 & C_2^\mathsf{T} D_2 \\ D_2^\mathsf{T} C_2 & D_2^\mathsf{T} D_2 \end{bmatrix}, \tag{3.65}$$

*and*

$$G_1(s) := C_1(sE - A)^{-1}B + D_1 \in \mathrm{Gl}_m(\mathbb{R}(s)). \tag{3.66}$$

*Then the solution set of the descriptor KYP inequality (3.22) is nonempty. Furthermore, the following holds true:*

*a) There exists a nonpositive solution of the descriptor KYP inequality (3.22).*

*b) If, furthermore, the system*

$$\left( \begin{bmatrix} E & 0 \\ 0 & 0 \end{bmatrix}, \begin{bmatrix} A & B \\ C_1 & D_1 \end{bmatrix}, \begin{bmatrix} 0 \\ -I_m \end{bmatrix}, \begin{bmatrix} C_2 & D_2 \end{bmatrix} \right) \in \Sigma_{n+m,m,p_2} \tag{3.67}$$

*is strongly detectable, then all solutions of the descriptor KYP inequality (3.22) are nonpositive.*

*Proof.* The proof is divided into two main steps. In the first step we prove the result for a special case. In the second step we show how one can reduce the general problem to this special case by an appropriate transformation.
*Step 1: We show that Theorem 3.7.4 holds true under the additional assumption that $C_1 = 0$ and $D_1 = I_m$:*
*Step 1.1: We show that, under the additional assumption that the index of $sE - A$ is at most one and $\Lambda(E, A) \subset \mathbb{C}^-$, the solution set of the descriptor KYP inequality (3.22) is nonempty. Moreover, all its solutions are nonpositive:*
Since the modified Popov function and the Popov function coincide on $\mathrm{i}\mathbb{R}$, we obtain

$$\Phi(\mathrm{i}\omega) = \Psi(\mathrm{i}\omega) \geq 0 \quad \forall \omega \in \mathbb{R} \text{ with } \mathrm{i}\omega \notin \Lambda(E, A).$$

Then, by Theorem 3.3.1, the descriptor KYP inequality (3.22) has a solution $P \in \mathbb{R}^{n\times n}$.
By the additional assumption on $sE - A$, there exist $W, T \in \mathrm{Gl}_n(\mathbb{R})$ with

$$W(sE - A)T = \begin{bmatrix} sI_r - A_{11} & 0 \\ 0 & -I_{n-r} \end{bmatrix}, \quad A_{11} \in \mathbb{R}^{r\times r}, \quad \Lambda(A_{11}) \subset \mathbb{C}^-.$$

Since $P$ and $C_2$ can be accordingly transformed, and, in particular, nonpositivity is preserved under this transformation (see Proposition 3.2.2), we see that it is no loss of generality to assume that $W = T = I_n$.

Then $E^\top P = P^\top E$ implies that any solution $P$ of the descriptor KYP inequality (3.22) fulfills

$$P = \begin{bmatrix} P_{11} & 0 \\ P_{21} & P_{22} \end{bmatrix},$$

where $P_{11} \in \mathbb{R}^{r\times r}$ is symmetric. By an accordant partition

$$C_2 = \begin{bmatrix} C_{21} & C_{22} \end{bmatrix},$$

we obtain from the descriptor KYP inequality that

$$A_{11}^\top P_{11} + P_{11}A_{11} - C_{21}^\top C_{21} \geq 0.$$

In particular, there exists some $Q_{11} \geq 0$ with $A_{11}^\top P_{11} + P_{11}A_{11} = Q_{11}$. By $\Lambda(A_{11}) \subset \mathbb{C}^-$, we obtain from [ZDG96, Lem. 3.18] that the matrix $P_{11}$ can be expressed by the integral

$$P_{11} = -\int_0^\infty \mathrm{e}^{A_{11}^\top \tau} Q_{11} \mathrm{e}^{A_{11}\tau} \mathrm{d}\tau \leq 0,$$

and hence we obtain

$$E^\top P = \begin{bmatrix} P_{11} & 0 \\ 0 & 0 \end{bmatrix} \leq 0.$$

*Step 1.2: We prove that statement b) holds true:*
The assumptions $C_1 = 0$ and $D_1 = I_m$ imply that

$$\operatorname{rank}\begin{bmatrix} -\lambda E + A & B \\ C_1 & D_1 \\ C_2 & D_2 \end{bmatrix} = \operatorname{rank}\begin{bmatrix} -\lambda E + A & B \\ 0 & I_m \\ C_2 & D_2 \end{bmatrix} = m + \operatorname{rank}\begin{bmatrix} \lambda E - A \\ C_2 \end{bmatrix} \quad \forall \lambda \in \mathbb{C}.$$

Consequently, the condition that the system (3.67) is strongly detectable is equivalent to strong detectability of $(E, A, B, C_2)$. Furthermore, the assumption $C_1 = 0$ and $D_1 = I_m$ implies $G_1(s) = I_m$. The latter gives that, for $G_2(s) := C_2(sE - A)^{-1}B + D_2 \in \mathbb{R}(s)^{p_2 \times m}$ and all $\lambda \in \overline{\mathbb{C}^+} \setminus \Lambda(E, A)$ it holds that

$$\Psi(\lambda) = I_m - G_2^\mathsf{H}(\lambda)G_2(\lambda) \geq 0.$$

As a consequence, $G_2 \in \mathcal{RH}_\infty^{p\times m}$. Using that $(E, A, B, C_2)$ is strongly detectable, we can use Lemma 3.7.2 to infer that the index of $sE - A$ is at most one, and all finite eigenvalues of $sE - A$ are in $\mathbb{C}^-$. Then assertion b) follows from the findings in Step 1.1.
*Step 1.3: We prove that statement a) holds true:*
By the *Kalman decomposition for observability* [Dai89, Sect. 2.5], we may w.l.o.g. assume that

$$sE - A = \begin{bmatrix} sE_{11} - A_{11} & 0 \\ sE_{21} - A_{21} & sE_{22} - A_{22} \end{bmatrix}, \quad B = \begin{bmatrix} B_1 \\ B_2 \end{bmatrix}, \quad C_2 = \begin{bmatrix} C_{21} & 0 \end{bmatrix}, \tag{3.68}$$

where $(E_{11}, A_{11}, B_1, C_{21}) \in \Sigma_{r,m,p_2}$ is completely observable (and therefore, strongly detectable). Then we have $G_2(s) = C_{21}(sE_{11} - A_{11})^{-1}B_1 + D_2$. By the assumptions and Theorem 3.3.1, there exists some solution $P_{11} \in \mathbb{R}^{r\times r}$ of the descriptor KYP inequality

$$\begin{bmatrix} A_{11}^\top P_{11} + P_{11}^\top A_{11} - C_{21}^\top C_{21} & P_{11}^\top B_1 - C_{21}^\top D_2 \\ B_1^\top P_{11} - D_2^\top C_{21} & I_m - D_2^\top D_2. \end{bmatrix} \geq_{\mathcal{V}_{\text{sys},1}} 0,$$

where

$$\mathcal{V}_{\text{sys},1} = \left\{ \begin{pmatrix} x_1 \\ u \end{pmatrix} \in \mathbb{R}^{r+m} : A_{11}x_1 + B_1 u \in \operatorname{im} E_{11} \right\}. \tag{3.69}$$

Since $(E, A, B, C_2, D_2)$ is strongly stabilizable and strongly detectable, we can make use of the results of Step 1.2 to see that $E_{11}^\top P_{11} \leq 0$.

By the block triangular structure of $E$, $A$, and $B$, we obtain that

$$\mathcal{V}_{\text{sys}} \subseteq \widetilde{\mathcal{V}}_{\text{sys}} := \left\{ \begin{pmatrix} x_1 \\ x_2 \\ u \end{pmatrix} \in \mathbb{R}^{n+m} : \begin{pmatrix} x_1 \\ u \end{pmatrix} \in \mathcal{V}_{\text{sys},1} \right\}.$$

Now define

$$P = \begin{bmatrix} P_{11} & 0 \\ 0 & 0 \end{bmatrix} \in \mathbb{R}^{n\times n}.$$

Then we have $E^\top P = P^\top E \leq 0$, and

$$\begin{bmatrix} A^\top P + P^\top A - C_2^\top C_2 & P^\top B - C_2^\top D_2 \\ B^\top P - D_2^\top C_2 & I_m - D_2^\top D_2. \end{bmatrix}$$
$$= \begin{bmatrix} A_{11}^\top P_{11} + P_{11}^\top A_{11} - C_{21}^\top C_{21} & 0 & P_{11}^\top B_1 - C_{21}^\top D_2 \\ 0 & 0 & 0 \\ B_1^\top P_{11} - D_2^\top C_{21} & 0 & I_m - D_2^\top D_2. \end{bmatrix} \geq_{\widetilde{\mathcal{V}}_{\text{sys}}} 0.$$

Hence, by (3.69), we see that $P$ solves the descriptor KYP inequality (3.22).

*Step 2: We prove the theorem for the general case:*

*Step 2.1: First we show statement a):*

Define the rational function

$$G_2(s) := C_2(sE - A)^{-1}B + D_2 \in \mathbb{R}(s)^{p_2\times m}.$$

Using (3.24), we obtain that the modified Popov function fulfills

$$\begin{aligned}
\Psi(\lambda) &= \begin{bmatrix}(\lambda E - A)^{-1}B\\ I_m\end{bmatrix}^{\mathsf{H}} \begin{bmatrix}Q & S\\ S^{\mathsf{T}} & R\end{bmatrix} \begin{bmatrix}(\lambda E - A)^{-1}B\\ I_m\end{bmatrix}\\
&= \begin{bmatrix}(\lambda E - A)^{-1}B\\ I_m\end{bmatrix}^{\mathsf{H}} \begin{bmatrix}C_1^{\mathsf{T}}C_1 & C_1^{\mathsf{T}}D_1\\ D_1^{\mathsf{T}}C_1 & D_1^{\mathsf{T}}D_1\end{bmatrix} \begin{bmatrix}(\lambda E - A)^{-1}B\\ I_m\end{bmatrix}\\
&\quad - \begin{bmatrix}(\lambda E - A)^{-1}B\\ I_m\end{bmatrix}^{\mathsf{H}} \begin{bmatrix}C_2^{\mathsf{T}}C_2 & C_2^{\mathsf{T}}D_2\\ D_2^{\mathsf{T}}C_2 & D_2^{\mathsf{T}}D_2\end{bmatrix} \begin{bmatrix}(\lambda E - A)^{-1}B\\ I_m\end{bmatrix}\\
&= G_1^{\mathsf{H}}(\lambda)G_1(\lambda) - G_2^{\mathsf{H}}(\lambda)G_2(\lambda).
\end{aligned}$$

In particular, the function $G_{\mathrm{e}}(s) = G_2(s)G_1^{-1}(s) \in \mathbb{R}(s)^{p_2 \times m}$ fulfills

$$I_m - G_{\mathrm{e}}^{\mathsf{H}}(\lambda)G_{\mathrm{e}}(\lambda) \geq 0 \tag{3.70}$$

for all $\lambda \in \overline{\mathbb{C}^+} \setminus \Lambda(E, A)$ which are no transmission zeros of $(E, A, B, C_1, D_1)$. Now define the system

$$(E_{\mathrm{e}}, A_{\mathrm{e}}, B_{\mathrm{e}}, C_{\mathrm{e}}) := \left(\begin{bmatrix}E & 0\\ 0 & 0\end{bmatrix}, \begin{bmatrix}A & B\\ C_1 & D_1\end{bmatrix}, \begin{bmatrix}0\\ -I_m\end{bmatrix}, \begin{bmatrix}C_2 & D_2\end{bmatrix}\right) \in \Sigma_{n+m,m,p_2}. \tag{3.71}$$

Then by Lemma 2.2.26 the pencil $sE_{\mathrm{e}} - A_{\mathrm{e}}$ is regular. Further, by Lemma 3.7.3 we obtain

$$C_{\mathrm{e}}(sE_{\mathrm{e}} - A_{\mathrm{e}})^{-1}B_{\mathrm{e}} = G_2(s)G_1^{-1}(s) = G_{\mathrm{e}}(s).$$

The structure of $E_{\mathrm{e}}$, $A_{\mathrm{e}}$, and $B_{\mathrm{e}}$ yields

$$\operatorname{rank}\begin{bmatrix}\lambda E_{\mathrm{e}} - A_{\mathrm{e}} & B_{\mathrm{e}}\end{bmatrix} = \operatorname{rank}\begin{bmatrix}\lambda E - A & B\end{bmatrix} + m \quad \forall \lambda \in \mathbb{C}.$$

Hence, strong stabilizability of $(E, A, B)$ implies strong stabilizability of $(E_{\mathrm{e}}, A_{\mathrm{e}}, B_{\mathrm{e}})$. Now consider the descriptor KYP inequality

$$\begin{bmatrix}A_{\mathrm{e}}^{\mathsf{T}}P_{\mathrm{e}} + P_{\mathrm{e}}^{\mathsf{T}}A_{\mathrm{e}} + Q_{\mathrm{e}} & P_{\mathrm{e}}^{\mathsf{T}}B_{\mathrm{e}} + S_{\mathrm{e}}\\ B_{\mathrm{e}}^{\mathsf{T}}P_{\mathrm{e}} + S_{\mathrm{e}}^{\mathsf{T}} & R_{\mathrm{e}}\end{bmatrix} \geq_{\mathcal{V}_{\mathrm{sys,e}}} 0, \quad E_{\mathrm{e}}^{\mathsf{T}}P_{\mathrm{e}} = P_{\mathrm{e}}^{\mathsf{T}}E_{\mathrm{e}} \tag{3.72}$$

with

$$Q_{\mathrm{e}} = -\begin{bmatrix}C_2^{\mathsf{T}}C_2 & C_2^{\mathsf{T}}D_2\\ D_2^{\mathsf{T}}C_2 & D_2^{\mathsf{T}}D_2\end{bmatrix}, \quad S_{\mathrm{e}} = 0, \quad R_{\mathrm{e}} = I_m,$$

and

$$\mathcal{V}_{\mathrm{sys,e}} = \left\{\begin{pmatrix}x_{\mathrm{e}}\\ u\end{pmatrix} \in \mathbb{R}^{n+2m} : A_{\mathrm{e}}x_{\mathrm{e}} + B_{\mathrm{e}}u \in \operatorname{im} E_{\mathrm{e}}\right\}.$$

The modified Popov function associated to the descriptor KYP inequality (3.72) then reads $\Psi_{\mathrm{e}}(\lambda) = I_m - G_{\mathrm{e}}^{\mathsf{H}}(\lambda)G_{\mathrm{e}}(\lambda)$. By (3.70) and the results from Step 1, there exists a nonpositive solution $P_{\mathrm{e}} \in \mathbb{R}^{n+m\times n+m}$ of (3.72). Partition

$$P_{\mathrm{e}} = \begin{bmatrix} P & P_{\mathrm{e},12} \\ P_{\mathrm{e},21} & P_{\mathrm{e},22} \end{bmatrix}$$

with $P \in \mathbb{R}^{n\times n}$. The equation $E_{\mathrm{e}}^{\mathsf{T}} P_{\mathrm{e}} = P_{\mathrm{e}}^{\mathsf{T}} E_{\mathrm{e}} \leq 0$ gives rise to $E^{\mathsf{T}} P = P^{\mathsf{T}} E \leq 0$ and $E^{\mathsf{T}} P_{\mathrm{e},12} = 0$. It suffices for the desired statement a) to prove that $P$ solves the original descriptor KYP inequality (3.22).

By definition of $\mathcal{V}_{\mathrm{sys,e}}$, $E_{\mathrm{e}}$, $A_{\mathrm{e}}$ and $B_{\mathrm{e}}$, we have

$$\mathcal{V}_{\mathrm{sys,e}} = \left\{ \begin{pmatrix} x \\ u \\ z \end{pmatrix} \in \mathbb{R}^{n+2m} : \begin{pmatrix} x \\ u \end{pmatrix} \in \mathcal{V}_{\mathrm{sys}} \text{ and } z = C_1 x + D_1 u \right\}.$$

Thus, the matrix

$$T_{\mathrm{e}} := \begin{bmatrix} I_n & 0 \\ 0 & I_m \\ C_1 & D_1 \end{bmatrix}$$

fulfills $T_{\mathrm{e}}\mathcal{V}_{\mathrm{sys}} = \mathcal{V}_{\mathrm{sys,e}}$. Further, the relation $P_{\mathrm{e},12}^{\mathsf{T}} E = 0$ implies

$$\begin{bmatrix} 0 & A^{\mathsf{T}} P_{\mathrm{e},12} \\ P_{\mathrm{e},12}^{\mathsf{T}} A & P_{\mathrm{e},12}^{\mathsf{T}} B + B^{\mathsf{T}} P_{\mathrm{e},12} \end{bmatrix} =_{\mathcal{V}_{\mathrm{sys}}} 0.$$

Thereby, we obtain

$$\begin{aligned}
0 &\leq_{\mathcal{V}_{\mathrm{sys}}} T_{\mathrm{e}}^{\mathsf{T}} \begin{bmatrix} A_{\mathrm{e}}^{\mathsf{T}} P_{\mathrm{e}} + P_{\mathrm{e}}^{\mathsf{T}} A_{\mathrm{e}} + Q_{\mathrm{e}} & P_{\mathrm{e}}^{\mathsf{T}} B_{\mathrm{e}} + S_{\mathrm{e}} \\ B_{\mathrm{e}}^{\mathsf{T}} P_{\mathrm{e}} + S_{\mathrm{e}}^{\mathsf{T}} & R_{\mathrm{e}} \end{bmatrix} T_{\mathrm{e}} \\
&=_{\mathcal{V}_{\mathrm{sys}}} \begin{bmatrix} A^{\mathsf{T}} P + P^{\mathsf{T}} A + C_1^{\mathsf{T}} C_1 - C_2^{\mathsf{T}} C_2 & P^{\mathsf{T}} B + C_1^{\mathsf{T}} D_1 - C_2^{\mathsf{T}} D_2 + A^{\mathsf{T}} P_{\mathrm{e},12} \\ B^{\mathsf{T}} P + D_1^{\mathsf{T}} C_1 - D_2^{\mathsf{T}} C_2 + P_{\mathrm{e},12}^{\mathsf{T}} A & D_1^{\mathsf{T}} D_1 - D_2^{\mathsf{T}} D_2 + P_{\mathrm{e},12}^{\mathsf{T}} B + B^{\mathsf{T}} P_{\mathrm{e},12} \end{bmatrix} \\
&=_{\mathcal{V}_{\mathrm{sys}}} \begin{bmatrix} A^{\mathsf{T}} P + P^{\mathsf{T}} A + Q & P^{\mathsf{T}} B + S \\ B^{\mathsf{T}} P + S^{\mathsf{T}} & R \end{bmatrix} + \begin{bmatrix} 0 & A^{\mathsf{T}} P_{\mathrm{e},12} \\ P_{\mathrm{e},12}^{\mathsf{T}} A & P_{\mathrm{e},12}^{\mathsf{T}} B + B^{\mathsf{T}} P_{\mathrm{e},12} \end{bmatrix} \\
&=_{\mathcal{V}_{\mathrm{sys}}} \begin{bmatrix} A^{\mathsf{T}} P + P^{\mathsf{T}} A + Q & P^{\mathsf{T}} B + S \\ B^{\mathsf{T}} P + S^{\mathsf{T}} & R \end{bmatrix}.
\end{aligned}$$

*Step 2.2: We show statement b):*

It remains to show that all solutions of the descriptor KYP inequality are nonpositive, if the system (3.67) is strongly detectable. This, however, follows directly from

$$\begin{bmatrix} -\lambda E_{\mathrm{e}} + A_{\mathrm{e}} \\ C_{\mathrm{e}} \end{bmatrix} = \begin{bmatrix} -\lambda E + A & B \\ C_1 & D_1 \\ C_2 & D_2 \end{bmatrix} \quad \forall \lambda \in \mathbb{C}.$$

Now let $P \in \mathbb{R}^{n\times n}$ be a solution of the descriptor KYP inequality (3.22). Then

$$P_{\mathrm{e}} = \begin{bmatrix} P & 0 \\ 0 & 0 \end{bmatrix}$$

solves the descriptor KYP inequality (3.72). Using the results of Step 1, we obtain that

$$E_{\mathrm{e}}^{\top} P_{\mathrm{e}} = \begin{bmatrix} E^{\top} P & 0 \\ 0 & 0 \end{bmatrix} \leq 0,$$

and therefore, $E^{\top} P \leq 0$. This completes the proof. □

*Remark* 3.7.5.

a) It follows from Sylvester's law of inertia [GL96, p. 403] that a decomposition (3.65) always exists. The only restriction we face is the invertibility of $G_1(s) \in \mathbb{R}(s)^{m\times m}$.

b) For controllable systems governed by ordinary differential equations, it has been initially claimed by Willems in his seminal article [Wil71] that positivity of the modified Popov function is also sufficient for the existence of a nonpositive solution, see also [Tre89, Tre99]. The same author disproved this claim by giving a counterexample in an erratum [Wil74] shortly after that. This article also contains statement a) of Theorem 3.7.4 for controllable systems governed by ordinary differential equations (without a proof). Further note that a similar problem has been considered for image representations of behaviors in [WT98].

Finally in this section, we investigate a further case where the descriptor KYP inequality a nonpositive solution. Namely, we will show that nonpositive solutions exist if

$$\begin{bmatrix} Q & S \\ S^{\top} & R \end{bmatrix} \geq_{\mathcal{V}_{\mathrm{sys}}} 0. \tag{3.73}$$

This is by far simpler to prove than the case treated in Theorem 3.7.4.

**Proposition 3.7.6.** *Let $(E, A, B) \in \Sigma_{n,m}$ with the system space $\mathcal{V}_{\mathrm{sys}}$ be given and assume that $Q = Q^{\top} \in \mathbb{R}^{n\times n}$, $S \in \mathbb{R}^{n\times m}$, and $R = R^{\top} \in \mathbb{R}^{m\times m}$. Let* (3.73) *be satisfied. Then the descriptor KYP inequality* (3.22) *has a nonpositive solution.*

*Proof.* This is a simple consequence of that fact that $P = 0$ is already a solution of the descriptor KYP inequality. □

## 3.8 Linear-Quadratic Optimal Control

In this section we discuss the main application of the so far developed theory, namely the linear-quadratic optimal control problem. This section is divided into two parts. The first part deals with optimal control problems with *zero terminal conditions*, i. e., we look for controls that stabilize (or anti-stabilize) the system while minimizing a certain cost functional. These problems will be directly related to stabilizing and anti-stabilizing solutions of associated descriptor Lur'e equations. The second part will be devoted to optimal control problems with *free terminal conditions*. These are much more involved and could yet not be solved completely. However, there are some relations to nonpositive solutions of the descriptor KYP inequality that will be pointed out.

### 3.8.1 Optimal Control with Zero Terminal Condition

For an interval $[t_0, t_1] \subseteq \mathbb{R} \cup \{-\infty, \infty\}$ and matrices $Q = Q^{\mathsf{T}} \in \mathbb{R}^{n \times n}$, $S \in \mathbb{R}^{n \times m}$, and $R = R^{\mathsf{T}} \in \mathbb{R}^{m \times m}$, consider the *cost functional*

$$\mathcal{J}(x, u, t_0, t_1) = \int_{t_0}^{t_1} \begin{pmatrix} x(\tau) \\ u(\tau) \end{pmatrix}^{\mathsf{T}} \begin{bmatrix} Q & S \\ S^{\mathsf{T}} & R \end{bmatrix} \begin{pmatrix} x(\tau) \\ u(\tau) \end{pmatrix} \mathrm{d}\tau. \tag{3.74}$$

Let $(E, A, B) \in \Sigma_{n,m}$ and $x_0 \in \mathbb{R}^n$ be given. Recall from (2.7) that $\mathfrak{B}_{(E,A,B)}(x_0)$ is the set of $(x, u) \in \mathfrak{B}_{(E,A,B)}$ with $Ex(0) = Ex_0$. First we consider two optimal control problems, namely the respective minimization of the cost functional on the negative and positive time horizon with constraints $(x, u) \in \mathfrak{B}_{(E,A,B)}(x_0)$ and $\lim_{t \to \pm\infty} Ex(t) = 0$. These are given by

a) $$V^+(Ex_0) = \inf \left\{ \mathcal{J}(x, u, 0, \infty) : (x, u) \in \mathfrak{B}_{(E,A,B)}(x_0) \text{ and } \lim_{t \to \infty} Ex(t) = 0 \right\}, \tag{3.75}$$

b) $$V^-(Ex_0) = -\inf \left\{ \mathcal{J}(x, u, -\infty, 0) : (x, u) \in \mathfrak{B}_{(E,A,B)}(x_0) \text{ and } \lim_{t \to -\infty} Ex(t) = 0 \right\}. \tag{3.76}$$

We aim to characterize finiteness of $V^+(Ex_0)$ and $V^-(Ex_0)$ for all $x_0 \in \mathbb{R}^n$. Note that strong stabilizability and strong anti-stabilizability of $(E, A, B)$ are, respectively, equivalent to the fact that for all $x_0 \in \mathbb{R}^n$, the sets

$$\begin{aligned} &\left\{ (x, u) \in \mathfrak{B}_{(E,A,B)}(x_0) : \lim_{t \to \infty} Ex(t) = 0 \right\}, \text{ and} \\ &\left\{ (x, u) \in \mathfrak{B}_{(E,A,B)}(x_0) : \lim_{t \to -\infty} Ex(t) = 0 \right\} \end{aligned}$$

are nonempty. Consequently, strong stabilizability and strong anti-stabilizability of $(E, A, B)$ are, respectively, equivalent to $V^{+}(Ex_0) < \infty$ and $V^{-}(Ex_0) > -\infty$ for all $x_0 \in \mathbb{R}^n$.

To present equivalent criteria for the cost functional and the system, we consider a class of *dissipation functions* $V : \operatorname{im} E \to \mathbb{R}$. That is, $V$ is continuous, $V(0) = 0$, and it satisfies the *dissipation inequality*

$$\mathcal{J}(x, u, t_0, t_1) + V(Ex(t_1)) \geq V(Ex(t_0)) \quad \forall (x, u) \in \mathfrak{B}_{(E,A,B)},\ t_0,\ t_1 \in \mathbb{R} \text{ with } t_0 \leq t_1. \tag{3.77}$$

Assume moreover that $V$ is differentiable. Then the inequality is equivalent to

$$\begin{pmatrix} x(t) \\ u(t) \end{pmatrix}^{\top} \begin{bmatrix} Q & S \\ S^{\top} & R \end{bmatrix} \begin{pmatrix} x(t) \\ u(t) \end{pmatrix} \geq -\nabla V(Ex(t))E\dot{x}(t) = -\nabla V(Ex(t))(Ax(t) + Bu(t))$$
$$\forall (x, u) \in \mathfrak{B}_{(E,A,B)} \text{ and almost all } t \in \mathbb{R}, \tag{3.78}$$

where $\nabla V(Ex(t)) \in \mathbb{R}^{1 \times n}$ is the gradient of $V$ in $Ex(t)$.

For quadratic dissipation functions we can make the ansatz

$$V(Ex_0) = x_0^{\top} P^{\top} E x_0, \tag{3.79}$$

where $P \in \mathbb{R}^{n \times n}$ is a matrix with $P^{\top}E = E^{\top}P$ (the latter property makes $V(Ex_0)$ well-defined). Using that

$$(\nabla V(Ex_0))z = 2x_0^{\top} P^{\top} z \quad \forall z \in \mathbb{R}^n,$$

the dissipation inequality (3.78) is now equivalent to the property that for all $(x, u) \in \mathfrak{B}_{(E,A,B)}$ it holds that

$$\begin{pmatrix} x(t) \\ u(t) \end{pmatrix}^{\top} \begin{bmatrix} A^{\top}P + P^{\top}A + Q & P^{\top}B + S \\ B^{\top}P + S^{\top} & R \end{bmatrix} \begin{pmatrix} x(t) \\ u(t) \end{pmatrix} \geq 0 \text{ for almost all } t \in \mathbb{R}.$$

The relation to the descriptor KYP inequality (3.22) is then based on the fact that $(x, u) \in \mathfrak{B}_{(E,A,B)}$ pointwisely evolves in $\mathcal{V}_{\text{sys}}$ almost everywhere. In particular, each solution of the descriptor KYP inequality induces a dissipation function via (3.79). This and the relation to linear-quadratic optimal control will be made more precise throughout this section.

In the following we show that, under the assumption of impulse controllability, the existence of stabilizing and anti-stabilizing solutions of the descriptor Lur'e equation is an equivalent criterion for the finiteness of $V^{+}$ and $V^{-}$, respectively. We also refer to [IR14] for a similar consideration in the ODE case. For the proof we make use of the following two auxiliary results.

**Lemma 3.8.1.** *Let $(E, A, B) \in \Sigma_{n,m}$. Assume that the index of $sE - A$ is at most one, and $\Lambda(E, A) \subset \mathbb{C}^-$. Then for all $x_0 \in \mathbb{R}^n$ and $u \in \mathcal{L}^2([0, \infty), \mathbb{R}^m)$, the following holds true:*

*a) there exists some unique $(x, u) \in \mathfrak{B}_{(E,A,B)}(x_0)$;*

*b) this trajectory fulfills $\lim_{t\to\infty} Ex(t) = 0$.*

*Proof.* Since for all $W, T \in \mathrm{Gl}_n(\mathbb{R})$ it holds that

$$(x, u) \in \mathfrak{B}_{(E,A,B)}(x_0) \Leftrightarrow \left(T^{-1}x, u\right) \in \mathfrak{B}_{(WET,WAT,WB)}\left(T^{-1}x_0\right),$$

we can assume that

$$sE - A = \begin{bmatrix} sI_r - A_{11} & 0 \\ 0 & -I_{n-r} \end{bmatrix}, \quad B = \begin{bmatrix} B_1 \\ B_2 \end{bmatrix},$$

where $B_1 \in \mathbb{R}^{r\times m}$, $B_2 \in \mathbb{R}^{n-r\times m}$ and $A_{11} \in \mathbb{R}^{r\times r}$ with $\Lambda(A_{11}) \subset \mathbb{C}^-$. Partition

$$T^{-1}x(t) = \begin{pmatrix} x_1(t) \\ x_2(t) \end{pmatrix}, \quad T^{-1}x_0 = \begin{pmatrix} x_{10} \\ x_{20} \end{pmatrix}$$

according to the block structure of $W(sE - A)T$. Then the result follows from the fact that the solution of the ordinary differential equation $\dot{x}_1(t) = A_{11}x_1(t) + B_1u(t)$, $x_1(0) = x_{10}$ is unique. Thus

$$T^{-1}x(t) = \begin{pmatrix} \mathrm{e}^{A_{11}t}x_{10} + \int_0^t \mathrm{e}^{A_{11}(t-\tau)}B_1u(\tau)\mathrm{d}\tau \\ -B_2u(t) \end{pmatrix}$$

is unique and tends to zero since $\Lambda(A_{11}) \subset \mathbb{C}^-$ and $u \in \mathcal{L}^2([0, \infty), \mathbb{R}^m)$. □

**Lemma 3.8.2.** *Let $(E, A, B) \in \Sigma_{n,m}$ be impulse controllable. Let $(x_0, u_0) \in \mathbb{R}^n \times \mathbb{R}^m$ be such that $Ax_0 + Bu_0 \in \operatorname{im} E$. Then there exists some infinitely often differentiable $(x, u) \in \mathfrak{B}_{(E,A,B)}$ with $x(0) = x_0$ and $u(0) = u_0$.*

*Proof.* By Lemma 3.2.1 and Proposition 3.2.2 a) and c) it suffices to show the statement for

$$sE - A = \begin{bmatrix} sI_r - A_{11} & 0 \\ 0 & -I_{n-r} \end{bmatrix}, \quad B = \begin{bmatrix} B_1 \\ B_2 \end{bmatrix},$$

where $B_1 \in \mathbb{R}^{r\times m}$, $B_2 \in \mathbb{R}^{n-r\times m}$. Then the structure of $E$, $A$, and $B$ gives rise to the existence of some $x_{01} \in \mathbb{R}^r$ with

$$x_0 = \begin{pmatrix} x_{01} \\ -B_2u_0 \end{pmatrix}.$$

Now define

$$u(t) = u_0, \quad x(t) = \begin{pmatrix} e^{A_{11}t}x_{10} + \int_0^t e^{A_{11}(t-\tau)}B_1 u_0 \mathrm{d}\tau \\ -B_2 u_0 \end{pmatrix}.$$

Then the trajectory $(x, u)$ has the desired properties. □

**Theorem 3.8.3.** *Let $(E, A, B) \in \Sigma_{n,m}$ with the system space $\mathcal{V}_{\mathrm{sys}}$ be given and assume that $Q = Q^\top \in \mathbb{R}^{n\times n}$, $S \in \mathbb{R}^{n\times m}$, and $R = R^\top \in \mathbb{R}^{m\times m}$. Let the cost functional $\mathcal{J}$ be defined as in (3.74), and let the function $V^+ : \operatorname{im} E \to \mathbb{R}$ be defined as in (3.75). Then the following two assertions are equivalent:*

*a) $V^+(Ex_0) \in \mathbb{R}$ for all $x_0 \in \mathbb{R}^n$.*

*b) The system $(E, A, B)$ is impulse controllable, has no uncontrollable modes on the imaginary axis, and the descriptor Lur'e equation (3.35) has a stabilizing solution $(X, K, L)$.*

*In the case where the above statements are valid, we have:*

*i) $V^+(Ex_0) = x_0^\top X^\top E x_0$ for all $x_0 \in \mathbb{R}^n$.*

*ii) For all $x_0 \in \mathbb{R}^n$ and $(x, u) \in \mathfrak{B}_{(E,A,B)}(x_0)$ such that $\lim_{t\to\infty} Ex(t) = 0$ and $\mathcal{J}(x, u, 0, \infty) \in \mathbb{R}$ it holds that*

$$\mathcal{J}(x, u, 0, \infty) = V^+(Ex_0) + \int_0^\infty \|Kx(\tau) + Lu(\tau)\|_2^2 \mathrm{d}\tau. \tag{3.80}$$

*Proof.* We first show that b) implies a), i), and ii):

Assume that $(E, A, B)$ is impulse controllable, and $(X, K, L)$ is a stabilizing solution of the descriptor Lur'e equation (3.35). Then, since

$$n + q = \operatorname{rank} \begin{bmatrix} -\lambda E + A & B \\ K & L \end{bmatrix} = \operatorname{rank} \begin{bmatrix} -\lambda E + A & B \end{bmatrix} + q \quad \forall \lambda \in \mathbb{C}^+,$$

and $(E, A, B)$ has no uncontrollable modes on the imaginary axis, we obtain that $(E, A, B)$ is strongly stabilizable. Let $x_0 \in \mathbb{R}^n$ and $(x, u) \in \mathfrak{B}_{(E,A,B)}(x_0)$ be such that $\lim_{t\to\infty} Ex(t) = 0$ and $\mathcal{J}(x,u,0,\infty) < \infty$. Then we obtain

$$\begin{aligned}
x_0^\top X^\top E x_0 &= -\int_0^\infty \frac{\mathrm{d}}{\mathrm{d}\tau} x(\tau)^\top X^\top E x(\tau) \mathrm{d}\tau \\
&= -\int_0^\infty \dot{x}(\tau)^\top E^\top X x(\tau) + x(\tau)^\top X^\top E \dot{x}(\tau) \mathrm{d}\tau \\
&= -\int_0^\infty (Ax(\tau) + Bu(\tau))^\top X x(\tau) + x(\tau)^\top X^\top (Ax(\tau) + Bu(\tau)) \mathrm{d}\tau \\
&= \int_0^\infty \begin{pmatrix} x(\tau) \\ u(\tau) \end{pmatrix}^\top \begin{bmatrix} -A^\top X - X^\top A & -X^\top B \\ -B^\top X & 0 \end{bmatrix} \begin{pmatrix} x(\tau) \\ u(\tau) \end{pmatrix} \mathrm{d}\tau \\
&= \int_0^\infty \begin{pmatrix} x(\tau) \\ u(\tau) \end{pmatrix}^\top \left( \begin{bmatrix} Q & S \\ S^\top & R \end{bmatrix} - \begin{bmatrix} K^\top K & K^\top L \\ L^\top K & L^\top L \end{bmatrix} \right) \begin{pmatrix} x(\tau) \\ u(\tau) \end{pmatrix} \mathrm{d}\tau \\
&= \mathcal{J}(x,u,0,\infty) - \int_0^\infty \|Kx(\tau) + Lu(\tau)\|_2^2 \, \mathrm{d}\tau.
\end{aligned}$$

This yields

$$x_0^\top X^\top E x_0 + \|Kx + Lu\|^2_{\mathcal{L}^2([0,\infty),\mathbb{R}^q)} = \mathcal{J}(x,u,0,\infty). \tag{3.81}$$

This leads to

$$x_0^\top X^\top E x_0 \le \mathcal{J}(x,u,0,\infty) \quad \forall (x,u) \in \mathfrak{B}_{(E,A,B)}(x_0) \text{ with } \lim_{t\to\infty} Ex(t) = 0,$$

and thus

$$x_0^\top X^\top E x_0 \le V^+(Ex_0) \quad \forall x_0 \in \mathbb{R}^n. \tag{3.82}$$

This inequality together with strong stabilizability of $(E,A,B)$ implies that a) holds true. Now we prove that i) and ii) are valid:

In view of (3.81) and (3.82), it suffices to prove that $x_0 X^\top E x_0 \ge V^+(Ex_0)$ for all $x_0 \in \mathbb{R}^n$. Since the triple $(X,K,L)$ is a stabilizing solution of the descriptor Lur'e equation (3.35), the system $(E,A,B,K,L) \in \Sigma_{n,m,q}$ is outer according to Definition 2.2.27. Proposition 2.2.29 then implies that there exists some sequence $(u_k(\cdot))_{k\in\mathbb{N}}$ in $\mathcal{L}^2([0,\infty),\mathbb{R}^m)$ such that

1) for all $k \in \mathbb{N}$, there exists a $(x_k,u_k,y_k) \in \mathfrak{B}_{(E,A,B,K,L)}(x_0)$ with $\lim_{t\to\infty} Ex_k(t) = 0$;

2) the sequence $(Kx_k(\cdot) + Lu_k(\cdot))_{k\in\mathbb{N}}$ tends to zero in $\mathcal{L}^2([0,\infty),\mathbb{R}^q)$.

Therefore, we obtain

$$\begin{aligned}
V^+(Ex_0) &\le \lim_{k\to\infty} \mathcal{J}(x_k,u_k,0,\infty) \\
&= x_0^\top X^\top E x_0 + \lim_{k\to\infty} \|Kx_k + Lu_k\|^2_{\mathcal{L}^2([0,\infty),\mathbb{R}^q)} \\
&= x_0^\top X^\top E x_0.
\end{aligned}$$

This, together with (3.82) yields statement i) and thus with (3.81) we obtain statement ii).

Now we show that a) implies b):

The assumption that $V^+(Ex_0) < \infty$ for all $x_0 \in \mathbb{R}^n$ implies that $(E, A, B)$ is strongly stabilizable. In particular, it is impulse controllable. Using Lemma 3.2.1, we obtain that there exist $W, T \in \mathrm{Gl}_n(\mathbb{R})$, $F \in \mathbb{R}^{m\times n}$ such that the index of $W(sE - (A+BF)T$ is at most one, and further, $\Lambda(WET, W(A+BF)T) \subset \mathbb{C}^-$. Hence, by Proposition 3.2.2 c) and e), we can w. l. o. g. assume that $\Lambda(E, A) \subset \mathbb{C}^-$, and the index of $sE - A$ is bounded from above by one.

For our proof we first show that $V^+$ is a quadratic functional. For this part we face several technical difficulties. First, in Step 1, we restrict our considerations to *square-integrable* solution trajectories and consider an accordingly modified functional $\widetilde{V}^+$. Due to the restriction of the behavior, the cost functional $\mathcal{J}(\cdot,\cdot,0,\infty)$ is homogeneous and fulfills the parallelogram identity. From these properties we can show that $\widetilde{V}^+$ is quadratic. In Steps 2 and 3 we show that the restrictions of the behavior have no influence the costs, in other words, we prove that $V^+ = \widetilde{V}^+$ and hence, $V^+$ is quadratic as well, i. e., $V^+(Ex_0) = x_0^\top E^\top Y E x_0$ with a symmetric $Y \in \mathbb{R}^{n\times n}$ for all $x_0 \in \mathbb{R}^n$. In Step 4 we show how this representation of $V^+$ can be used to construct a solution of the descriptor KYP inequality. Finally, in Step 5, we show that this solution even defines a stabilizing solution of the descriptor Lur'e equation by making use of the previously introduced theory of outer systems.

*Step 1: We consider the functional*

$$\widetilde{V}^+(Ex_0) = \inf\big\{\mathcal{J}(x,u,0,\infty) : \\ (x,u) \in \mathfrak{B}_{(E,A,B)}(x_0) \cap \big(\mathcal{L}^2([0,\infty),\mathbb{R}^n) \times \mathcal{L}^2([0,\infty),\mathbb{R}^m)\big)\big\},$$

*and show that there exists some symmetric $Y \in \mathbb{R}^{n\times n}$ with $\widetilde{V}^+(Ex_0) = x_0^\top E^\top Y E x_0$ for all $x_0 \in \mathbb{R}^n$:*

First note that, by construction of $\widetilde{V}^+$, we have $\widetilde{V}^+(Ex_0) \geq V^+(Ex_0)$ for all $x_0 \in \mathbb{R}^n$. Further, by $\Lambda(E, A) \subset \mathbb{C}^-$ and the index of $sE - A$ being not greater than one, we obtain that $\widetilde{V}^+(Ex_0) < \infty$ for all $x_0 \in \mathbb{R}^n$.

The assumptions on $sE - A$ together with Lemma 3.8.1 imply that for all $u \in \mathcal{L}^2([0,\infty),\mathbb{R}^m)$, there exists a unique $(x,u) \in \mathfrak{B}_{(E,A,B)}(x_0)$. This trajectory moreover fulfills

$$x \in \mathcal{L}^2([0,\infty),\mathbb{R}^n) \text{ and } \lim_{t\to\infty} Ex(t) = 0.$$

The functional $\widetilde{V}^+$ hence reads

$$\widetilde{V}^+(Ex_0) = \inf\big\{\mathcal{J}(x,u,0,\infty) : (x,u) \in \mathfrak{B}_{(E,A,B)}(x_0) \text{ with } u \in \mathcal{L}^2([0,\infty),\mathbb{R}^m)\big\}.$$

By simple calculations, we obtain that for all $\lambda \in \mathbb{R}$ and $(x,u),(x_1,u_1),(x_2,u_2) \in \mathfrak{B}_{(E,A,B)}$ with $u$, $u_1$, $u_2 \in \mathcal{L}^2([0,\infty),\mathbb{R}^m)$ it holds that

$$\begin{aligned} 2\mathcal{J}(x_1,u_1,0,\infty) + 2\mathcal{J}(x_2,u_2,0,\infty) &= \mathcal{J}(x_1+x_2,u_1+u_2,0,\infty) \\ &\quad + \mathcal{J}(x_1-x_2,u_1-u_2,0,\infty), \\ \mathcal{J}(\lambda x,\lambda u,0,\infty) &= \lambda^2 \mathcal{J}(x,u,0,\infty). \end{aligned} \tag{3.83}$$

Now we prove that $\widetilde{V}^+$ is a quadratic functional:

We have $\widetilde{V}^+(0) = 0$, since $\mathcal{J}(0,0,0,\infty) = 0$. Further, the existence of $(x,u) \in \mathfrak{B}_{(E,A,B)}(0)$ with $\mathcal{J}(x,u,0,\infty) < 0$ would imply, by taking scalar multiples of $(x,u)$, that $\widetilde{V}^+(0) = -\infty$. This means that $\widetilde{V}^+(E(0 \cdot x_0)) = 0 = 0 \cdot \widetilde{V}^+(Ex_0)$. On the other hand, for all $\lambda \in \mathbb{R} \setminus \{0\}$, $x_0 \in \mathbb{R}^n$, and $\varepsilon > 0$, the definition of $\widetilde{V}^+$ implies that there exists a trajectory $(x,u) \in \mathfrak{B}_{(E,A,B)}(x_0)$ with $u \in \mathcal{L}^2([0,\infty),\mathbb{R}^m)$ and

$$\mathcal{J}(x,u,0,\infty) \le \widetilde{V}^+(Ex_0) + \frac{\varepsilon}{\lambda^2},$$

and hence it holds that

$$\begin{aligned} \widetilde{V}^+(E(\lambda x_0)) &\le \mathcal{J}(\lambda x,\lambda u,0,\infty) \\ &= \lambda^2 \mathcal{J}(x,u,0,\infty) \\ &\le \lambda^2 \left(\widetilde{V}^+(Ex_0) + \frac{\varepsilon}{\lambda^2}\right) \\ &= \lambda^2 \widetilde{V}^+(Ex_0) + \varepsilon. \end{aligned}$$

Since the above inequality holds for all $\varepsilon > 0$ it follows that

$$\widetilde{V}^+(E(\lambda x_0)) \le \lambda^2 \widetilde{V}^+(Ex_0). \tag{3.84}$$

The reverse inequality follows from (3.84) and

$$\widetilde{V}^+(Ex_0) = \widetilde{V}^+\left(E\left(\frac{1}{\lambda}\cdot \lambda x_0\right)\right) \le \frac{1}{\lambda^2}\widetilde{V}^+(E(\lambda x_0)).$$

Assuming that $x_{01}, x_{02} \in \mathbb{R}^n$ and $\varepsilon > 0$, the definition of $\widetilde{V}^+$ implies that there exist $(x_1,u_1) \in \mathfrak{B}_{(E,A,B)}(x_{01})$, $(x_2,u_2) \in \mathfrak{B}_{(E,A,B)}(x_{02})$ with $u_1$, $u_2 \in \mathcal{L}^2([0,\infty),\mathbb{R}^m)$ and

$$\begin{aligned} \mathcal{J}(x_1,u_1,0,\infty) &\le \widetilde{V}^+(Ex_{01}) + \frac{\varepsilon}{4}, \\ \mathcal{J}(x_2,u_2,0,\infty) &\le \widetilde{V}^+(Ex_{02}) + \frac{\varepsilon}{4}. \end{aligned}$$

Then, by using (3.83), we obtain

$$\begin{aligned}
&\widetilde{V}^+\left(E(x_{01}+x_{02})\right)+\widetilde{V}^+\left(E(x_{01}-x_{02})\right)\\
&\leq \mathcal{J}(x_1+x_2,u_1+u_2,0,\infty)+\mathcal{J}(x_1-x_2,u_1-u_2,0,\infty)\\
&=2\mathcal{J}(x_1,u_1,0,\infty)+2\mathcal{J}(x_2,u_2,0,\infty)\\
&\leq 2\widetilde{V}^+(Ex_{01})+2\widetilde{V}^+(Ex_{02})+\varepsilon.
\end{aligned}$$

Since the above inequality holds for all $\varepsilon > 0$ we get

$$\widetilde{V}^+\left(E(x_{01}+x_{02})\right)+\widetilde{V}^+\left(E(x_{01}-x_{02})\right)\leq 2\widetilde{V}^+(Ex_{01})+2\widetilde{V}^+(Ex_{02}). \tag{3.85}$$

The reverse inequality follows from (3.85) as follows: For

$$\widetilde{x}_{01}=\frac{1}{2}(x_{01}+x_{02}),\quad \widetilde{x}_{02}=\frac{1}{2}(x_{01}-x_{02}) \tag{3.86}$$

it holds that $\widetilde{x}_{01}+\widetilde{x}_{02}=x_{01}$, $\widetilde{x}_{01}-\widetilde{x}_{02}=x_{02}$. Then we obtain

$$\begin{aligned}
2\widetilde{V}^+(Ex_{01})+2\widetilde{V}^+(Ex_{02})&=2\widetilde{V}^+(E(\widetilde{x}_{01}+\widetilde{x}_{02}))+2\widetilde{V}^+(E(\widetilde{x}_{01}-\widetilde{x}_{02}))\\
&\leq 4\widetilde{V}^+(E\widetilde{x}_{01})+4\widetilde{V}^+(E\widetilde{x}_{02})\\
&=4\widetilde{V}^+\left(E\left(\frac{1}{2}(x_{01}+x_{02})\right)\right)\\
&\quad+4\widetilde{V}^+\left(E\left(\frac{1}{2}(x_{01}-x_{02})\right)\right)\\
&=\widetilde{V}^+(E(x_{01}+x_{02}))+\widetilde{V}^+(E(x_{01}-x_{02})).
\end{aligned}$$

Altogether, we can conclude that $\widetilde{V}^+$ is again quadratic, i. e., for all $x_{01}$, $x_{02}$, $x_0 \in \mathbb{R}^n$ and $\lambda \in \mathbb{R}$ it holds that

$$\begin{aligned}
2\widetilde{V}^+(Ex_{01})+2\widetilde{V}^+(Ex_{02})&=\widetilde{V}^+(E(x_{01}+x_{02}))+\widetilde{V}^+(E(x_{01}-x_{02})),\\
\widetilde{V}^+(E(\lambda x_0))&=\lambda^2\widetilde{V}^+(Ex_0).
\end{aligned}$$

This gives rise to the existence of some symmetric $Y \in \mathbb{R}^{n\times n}$ with

$$\widetilde{V}^+(Ex_0)=(Ex_0)^\top Y(Ex_0)\quad \forall x_0\in\mathbb{R}^n.$$

*Step 2: We prove that $\widetilde{V}^+$ is a dissipation function, i. e., it fulfills the dissipation inequality* (3.77)*:*

Assume that $t \geq 0$ and $(x,u) \in \mathfrak{B}_{(E,A,B)}(x_0)$. By definition of $\widetilde{V}^+$, there exists some $(\widetilde{x},\widetilde{u}) \in \mathfrak{B}_{(E,A,B)}(x(t))$ with $(\widetilde{x},\widetilde{u}) \in \mathcal{L}^2([0,\infty),\mathbb{R}^n)\times\mathcal{L}^2([0,\infty),\mathbb{R}^m)$ and

$$\mathcal{J}(\widetilde{x},\widetilde{u},0,\infty)\leq \widetilde{V}^+(Ex(t))+\varepsilon.$$

Consider the concatenation

$$(\overline{x}, \overline{u}) = (x, u) \underset{t}{\diamondsuit} (\widetilde{x}, \widetilde{u}) \in \mathfrak{B}_{(E,A,B)}(x_0),$$

Then, by using time-invariance, we obtain

$$\begin{aligned}
\widetilde{V}^+(Ex_0) \leq \mathcal{J}(\overline{x}, \overline{u}, 0, \infty) &= \mathcal{J}(\overline{x}, \overline{u}, 0, t) + \mathcal{J}(\overline{x}, \overline{u}, t, \infty) \\
&= \mathcal{J}(\overline{x}, \overline{u}, 0, t) + \mathcal{J}(\overline{x}(\cdot + t), \overline{u}(\cdot + t), 0, \infty) \\
&= \mathcal{J}(x, u, 0, t) + \mathcal{J}(\widetilde{x}, \widetilde{u}, 0, \infty) \\
&\leq \mathcal{J}(x, u, 0, t) + \widetilde{V}^+(Ex(t)) + \varepsilon.
\end{aligned}$$

The result follows, since $\varepsilon > 0$ can be made arbitrarily small.
*Step 3: We show that* $\widetilde{V}^+ = V^+$*:*

By construction of $\widetilde{V}^+$, we have $\widetilde{V}^+ \geq V^+$. Assume that $x_0 \in \mathbb{R}^n$ and $(x, u) \in \mathfrak{B}_{(E,A,B)}(x_0)$ is given such that $\lim_{t\to\infty} Ex(t) = 0$ and $\mathcal{J}(x, u, 0, \infty) \in \mathbb{R}$. Then, since $\widetilde{V}^+$ is a dissipation function, we obtain that for all $t \geq 0$ it holds that

$$\widetilde{V}^+(Ex_0) - \widetilde{V}^+(Ex(t)) \leq \mathcal{J}(x, u, 0, t).$$

Taking the limit $t \to \infty$ and using $\lim_{t\to\infty} Ex(t) = 0$, we obtain

$$\widetilde{V}^+(Ex_0) \leq \mathcal{J}(x, u, 0, \infty).$$

This implies $\widetilde{V}^+ \leq V^+$.
*Step 4: We prove that the matrix* $X := YE \in \mathbb{K}^{n\times n}$*, where* $Y \in \mathbb{R}^{n\times n}$ *is obtained in Step 1, solves the descriptor KYP inequality:*

Let $(x_0, u_0) \in \mathbb{R}^n \times \mathbb{R}^m$ be such that $Ax_0 + Bu_0 \in \operatorname{im} E$. Using Lemma 3.8.2, we obtain that there exists some continuous $(x, u) \in \mathfrak{B}_{(E,A,B)}$ with $x(0) = x_0$ and $u(0) = u_0$. The dissipation inequality implies that for all $h > 0$ we have

$$\frac{x(0)^\top X^\top Ex(0) - x(h)^\top X^\top Ex(h)}{h} \leq \frac{1}{h} \cdot \mathcal{J}(x, u, 0, h).$$

For $h \to 0$, the right hand side converges to

$$\begin{pmatrix} x_0 \\ u_0 \end{pmatrix}^\top \begin{bmatrix} Q & S \\ S^\top & R \end{bmatrix} \begin{pmatrix} x_0 \\ u_0 \end{pmatrix}.$$

The product rule and $E\dot{x}(0) = Ax_0 + Bu_0$ imply that the left hand side converges to

$$\begin{pmatrix} x_0 \\ u_0 \end{pmatrix}^\top \begin{bmatrix} -A^\top X - X^\top A & -X^\top B \\ -B^\top X & 0 \end{bmatrix} \begin{pmatrix} x_0 \\ u_0 \end{pmatrix}.$$

Altogether this shows that $X$ solves the descriptor KYP inequality (3.22).
*Step 5: We prove that the matrix $X$ induces a stabilizing solution of the descriptor Lur'e equation:*

Since by the findings of Step 4, $X$ is a solution of the descriptor KYP inequality, there exist $\widetilde{K} \in \mathbb{R}^{l\times n}$, $\widetilde{L} \in \mathbb{R}^{l\times m}$ with

$$\begin{bmatrix} A^\top X + X^\top A + Q & X^\top B + S \\ B^\top X + S^\top & R \end{bmatrix} =_{\mathcal{V}_{\mathrm{sys}}} \begin{bmatrix} \widetilde{K}^\top \\ \widetilde{L}^\top \end{bmatrix} \begin{bmatrix} \widetilde{K} & \widetilde{L} \end{bmatrix}.$$

Assume that $(x,u) \in \mathfrak{B}_{(E,A,B)}(x_0)$. Then we obtain for all $t \geq 0$ that

$$\begin{aligned}
&x(t)^\top X^\top E x(t) - x_0^\top X^\top E x_0 \\
&= \int_0^t \frac{\mathrm{d}}{\mathrm{d}\tau} x(\tau)^\top X^\top E x(\tau) \mathrm{d}\tau \\
&= \int_0^t \dot{x}(\tau)^\top E^\top X x(\tau) + x(\tau)^\top X^\top E \dot{x}(\tau) \mathrm{d}\tau \\
&= \int_0^t (Ax(\tau) + Bu(\tau))^\top X x(\tau) + x(\tau)^\top X^\top (Ax(\tau) + Bu(\tau)) \mathrm{d}\tau \\
&= \int_0^t \begin{pmatrix} x(\tau) \\ u(\tau) \end{pmatrix}^\top \begin{bmatrix} A^\top X + X^\top A & X^\top B \\ B^\top X & 0 \end{bmatrix} \begin{pmatrix} x(\tau) \\ u(\tau) \end{pmatrix} \mathrm{d}\tau \\
&= \int_0^t \begin{pmatrix} x(\tau) \\ u(\tau) \end{pmatrix}^\top \left( \begin{bmatrix} \widetilde{K}^\top \widetilde{K} & \widetilde{K}^\top \widetilde{L} \\ \widetilde{L}^\top \widetilde{K} & \widetilde{L}^\top \widetilde{L} \end{bmatrix} - \begin{bmatrix} Q & S \\ S^\top & R \end{bmatrix} \right) \begin{pmatrix} x(\tau) \\ u(\tau) \end{pmatrix} \mathrm{d}\tau \\
&= \int_0^t \left\| \widetilde{K} x(\tau) + \widetilde{L} u(\tau) \right\|_2^2 \mathrm{d}\tau - \mathcal{J}(x,u,0,t).
\end{aligned}$$

By taking the limit $t \to \infty$, we obtain

$$x_0^\top X^\top E x_0 + \left\| \widetilde{K} x + \widetilde{L} u \right\|^2_{\mathcal{L}^2([0,\infty),\mathbb{R}^l)} = \mathcal{J}(x,u,0,\infty). \tag{3.87}$$

By Step 1 and Step 4, we have that for all $x_0 \in \mathbb{R}^n$, there exists a sequence $(u_k(\cdot))_{k\in\mathbb{N}}$ in $\mathcal{L}^2([0,\infty),\mathbb{R}^m)$ such that for $(u_k, x_k) \in \mathfrak{B}_{(E,A,B)}(x_0)$ we get

$$\lim_{k\to\infty} \mathcal{J}(x_k, u_k, 0, \infty) = x_0^\top X^\top E x_0.$$

Consequently, by (3.87), we have

$$\lim_{k\to\infty} \left\| \widetilde{K} x_k + \widetilde{L} u_k \right\|_{\mathcal{L}^2([0,\infty),\mathbb{R}^l)} = 0.$$

By Proposition 2.2.30 there exists some matrix $U \in \mathbb{K}^{q\times l}$ with orthogonal columns such that, for $K := U\widetilde{K}$ and $L := U\widetilde{L}$, the system $(E,A,B,K,L) \in \Sigma_{n,m,q}$ is outer and, further, for all $(x,u) \in \mathfrak{B}_{(E,A,B)}$ it holds that

$$\left\| Kx(t) + Lu(t) \right\|_2 = \left\| \widetilde{K} x(t) + \widetilde{L} u(t) \right\|_2 \text{ for almost all } t \geq 0.$$

By the same argumentation as in Step 4, we can show that

$$\begin{bmatrix} A^\top X + X^\top A + Q & X^\top B + S \\ B^\top X + S^\top & R \end{bmatrix} =_{\mathcal{V}_{\text{sys}}} \begin{bmatrix} K^\top \\ L^\top \end{bmatrix} \begin{bmatrix} K & L \end{bmatrix}.$$

Since $(E, A, B, K, L)$ is outer, we have

$$\operatorname{rank} \begin{bmatrix} -\lambda E + A & B \\ K & L \end{bmatrix} = n + q \quad \forall \lambda \in \mathbb{C}^+.$$

Hence, $(X, K, L)$ is a stabilizing solution of the descriptor Lur'e equation. □

In view of the concept introduced in Definition 2.2.21, we can now present structural properties of the infimizing controls. In particular, we can characterize whether some $(x, u) \in \mathfrak{B}_{(E,A,B)}(x_0)$ exists with $\mathcal{J}(x, u, 0, \infty) = V^+(Ex_0)$. Such an infimizer is typically called an *optimal control*. We also refer to [IR14] for a similar consideration in the ODE case.

**Proposition 3.8.4.** *Let $(E, A, B) \in \Sigma_{n,m}$ with the system space $\mathcal{V}_{\text{sys}}$ be impulse controllable and $Q = Q^\top \in \mathbb{R}^{n\times n}$, $S \in \mathbb{R}^{n\times m}$, $R = R^\top \in \mathbb{R}^{m\times m}$. Let the cost functional $\mathcal{J}$ be defined as in (3.74) and let the function $V^+ : \operatorname{im} E \to \mathbb{R}$ be defined as in (3.75). Let $(X, K, L)$ be a stabilizing solution of the descriptor Lur'e equation (3.35). Assume that $(x, u) \in \mathcal{ZD}_{(E,A,B,K,L)}(x_0)$ with $\mathcal{J}(x, u, 0, \infty) < \infty$ and $\lim_{t\to\infty} Ex(t) = 0$. Then it holds that*

$$\mathcal{J}(x, u, 0, \infty) = V^+(Ex_0). \tag{3.88}$$

*Furthermore, $(x, u) \in \mathfrak{B}_{(E,A,B)}(x_0)$ is uniquely described by this equation if and only if the zero dynamics of $(E, A, B, K, L)$ are autonomous.*

In view of (3.80), we see that $(x, u) \in \mathfrak{B}_{(E,A,B)}(x_0)$ fulfills $V^+(Ex_0) = \mathcal{J}(x, u, 0, \infty)$ if and only if we have

$$\begin{aligned} E\dot{x}(t) &= Ax(t) + Bu(t), \quad Ex(0) = Ex_0, \quad \lim_{t\to\infty} Ex(t) = 0, \\ 0 &= Kx(t) + Lu(t). \end{aligned} \tag{3.89}$$

Equivalently, we have $(x, u) \in \mathcal{ZD}_{(E,A,B,K,L)}(x_0)$ with $\lim_{t\to\infty} Ex(t) = 0$.

Hence we can use Definitions 2.2.21 and 2.2.22 to characterize existence and structure of optimal controls for all $x_0 \in \mathbb{R}^n$.

**Corollary 3.8.5.** *Let $(E, A, B) \in \Sigma_{n,m}$ with the system space $\mathcal{V}_{\text{sys}}$ be impulse controllable and let $Q = Q^T \in \mathbb{R}^{n\times n}$, $S \in \mathbb{R}^{n\times m}$, and $R = R^T \in \mathbb{R}^{m\times m}$ be given. Let the cost functional $\mathcal{J}$ be defined as in (3.74) and let the function $V^+ : \operatorname{im} E \to \mathbb{R}$ be defined as in (3.75). Let $(X, K, L)$ be a stabilizing solution of the descriptor Lur'e equation (3.35). Then the following statements hold true:*

a) *For all* $x_0 \in \mathbb{R}^n$*, there exists a* $(x,u) \in \mathfrak{B}_{(E,A,B)}(x_0)$ *with* $V^+(Ex_0) = \mathcal{J}(x,u,0,\infty)$ *if and only if the zero dynamics of* $(E,A,B,K,L)$ *are strongly stabilizable.*

b) *For all* $x_0 \in \mathbb{R}^n$*, there exists a unique* $(x,u) \in \mathfrak{B}_{(E,A,B)}(x_0)$ *with* $V^+(Ex_0) = \mathcal{J}(x,u,0,\infty)$ *if and only if the zero dynamics of* $(E,A,B,K,L)$ *are strongly asymptotically stable.*

*Remark* 3.8.6 (Optimal control, Popov function, even matrix pencil).

a) Let $R(s)$ be the Rosenbrock pencil of $(E,A,B,K,L) \in \Sigma_{n,m,q}$. Then we can use the structural characterization of the zero dynamics in Proposition 2.2.24 to see that the following holds true:

   i) The zero dynamics of $(E,A,B,K,L)$ are strongly stabilizable if and only if $\operatorname{rank} R(\lambda) = n+p$ for all $\lambda \in \overline{\mathbb{C}^+}$ and, furthermore, the index of $R(s)$ is at most one.

   ii) The zero dynamics of $(E,A,B,K,L)$ are strongly asymptotically stable if and only if $R(\lambda) \in \mathrm{Gl}_{n+m}(\mathbb{C})$ for all $\lambda \in \overline{\mathbb{C}^+}$ and, furthermore, the index of $R(s)$ is at most one.

b) The two conditions in item ii) are equivalent to the function $\lambda \mapsto R(\lambda)^{-1}$ being bounded in $\overline{\mathbb{C}^+}$. Since, for $W(s) = K(sE-A)^{-1}B + L \in \mathbb{R}(s)^{q\times m}$ it holds that

$$R^{-1}(s) = \begin{bmatrix} -(sE-A)^{-1} + (sE-A)^{-1}BW^{-1}(s)K(sE-A)^{-1} & (sE-A)^{-1}BW^{-1}(s) \\ W^{-1}(s)K(sE-A)^{-1} & W^{-1}(s) \end{bmatrix},$$

we obtain that this condition implies that $W^{-1}(s)$ is bounded in $\overline{\mathbb{C}^+}$. Theorem 3.6.3 implies that this condition is equivalent to the fact that the Popov function $\Phi(s)$ in boundedly invertible on $\mathrm{i}\mathbb{R}$.

c) Glancing at the proofs of Theorem 3.4.9 and Theorem 3.5.3, we obtain that the set of invariant zeros of $(E,A,B,K,L)$ is in the set of the finite eigenvalues of the even matrix pencil $s\mathcal{E}-\mathcal{A}$ as in (3.29). Invertibility of $R(s)$ (and thus invertibility of $W(s)$) corresponds to regularity of $s\mathcal{E}-\mathcal{A}$. Furthermore, by comparing with Table 3.1 the following holds for the Rosenbrock pencil $R(s)$ of an outer system $(E,A,B,K,L)$:

   - each block of type K1 and size $k \times k$ in the KCF of $R(s)$ corresponds to a block of type E1 and size $2k \times 2k$ in the EKCF of $s\mathcal{E}-\mathcal{A}$;
   - each block of type K2 and size $k \times k$ in the KCF of $R(s)$ corresponds to a block of type E3 and size $(2k-1)\times(2k-1)$ in the EKCF of $s\mathcal{E}-\mathcal{A}$.

Moreover, the latter statement yields that $R(s)$ has index at most one if and only if $s\mathcal{E} - \mathcal{A}$ has index at most one.

d) The transmission zeros of $(E, A, B, K, L)$ are, by Proposition 2.2.25, invariant zeros of $(E, A, B, K, L)$. Hence, they are finite eigenvalues of $s\mathcal{E} - \mathcal{A}$.

e) Assume that for all $x_0 \in \mathbb{R}^n$, there exists a unique $(x, u) \in \mathfrak{B}_{(E,A,B)}(x_0)$ with $\lim_{t\to\infty} Ex(t) = 0$ and $V^+(Ex_0) = \mathcal{J}(x, u, 0, \infty)$. The previous statements yield that $W(s)$ is boundedly invertible in $\overline{\mathbb{C}^+}$ and, further, the Popov function $\Phi(s)$ is boundedly invertible on $\mathrm{i}\mathbb{R}$.

f) A Laplace transform of (3.89) with $U(s) := \mathcal{L}\{u\}(s)$ yields

$$0 = W(s)U(s) + K(sE - A)^{-1}Ex_0.$$

A multiplication from the left with $W^{\sim}(s)$ together with $W^{\sim}(s)W(s) = \Phi(s)$ gives

$$0 = \Phi(s)U(s) + W^{\sim}(s)K(sE - A)^{-1}Ex_0.$$

If $\Phi(s) \in \mathrm{Gl}_m(\mathbb{R}(s))$, we obtain

$$U(s) = -\Phi^{-1}(s)W^{\sim}(s)K(sE - A)^{-1}Ex_0.$$

g) Assume that $(x, u) \in \mathfrak{B}_{(E,A,B)}(x_0)$ with $\lim_{t\to\infty} Ex(t) = 0$ and $V^+(Ex_0) = \mathcal{J}(x, u, 0, \infty)$. Then, by using (3.52) and (3.89), we can formally write

$$\begin{bmatrix} 0 & -\frac{\mathrm{d}}{\mathrm{d}t}E + A & B \\ \frac{\mathrm{d}}{\mathrm{d}t}E^\top + A^\top & Q & S \\ B^\top & S^\top & R \end{bmatrix} \begin{bmatrix} X & 0 \\ I_n & 0 \\ 0 & I_m \end{bmatrix} \begin{pmatrix} x(\cdot) \\ u(\cdot) \end{pmatrix}$$
$$= \begin{bmatrix} I_n & 0 \\ -X^\top + H^\top \Sigma M T_\infty^\top & K^\top \\ J^\top \Sigma M T_\infty^\top & L^\top \end{bmatrix} \underbrace{\begin{bmatrix} -\frac{\mathrm{d}}{\mathrm{d}t}E + A & B \\ K & L \end{bmatrix} \begin{pmatrix} x(\cdot) \\ u(\cdot) \end{pmatrix}}_{\equiv 0}.$$

In particular, the function $\mu(\cdot) = Xx(\cdot)$ is part of a solution of the boundary value problem

$$\begin{bmatrix} 0 & E & 0 \\ -E^\top & 0 & 0 \\ 0 & 0 & 0 \end{bmatrix} \begin{pmatrix} \dot{\mu}(t) \\ \dot{x}(t) \\ \dot{u}(t) \end{pmatrix} = \begin{bmatrix} 0 & A & B \\ A^\top & Q & S \\ B^\top & S^\top & R \end{bmatrix} \begin{pmatrix} \mu(t) \\ x(t) \\ u(t) \end{pmatrix}, \tag{3.90}$$

with $Ex(0) = Ex_0$ and $\lim_{t\to\infty} E^\top \mu(t) = 0$.

On the other hand, if the boundary value problem (3.90) is solvable, then the solution $(\mu(\cdot)^\top, x(\cdot)^\top, u(\cdot)^\top)^\top$ pointwisely evolves in a deflating subspace corresponding to the eigenvalues in $\mathbb{C}^-$. This however implies $\mu(\cdot) = Xx(\cdot)$ and hence the solution of (3.90) is an optimal control. Note that similar statements can also be found in [Meh91], however in a less general situation.

h) In the case where the system is governed by an ordinary differential equation, the pencil

$$\begin{bmatrix} -sI_n + A & B \\ K & L \end{bmatrix} \tag{3.91}$$

is invertible and its index is at most one if and only if $L$ is nonsingular. By simple row operations, we see that the finite eigenvalues of the pencil in (3.91) are the eigenvalues of the *closed-loop matrix* $A - BR^{-1}\left(B^\top X + S\right)$. Consequently, our concept of stabilizing solutions reduces to stabilizing solutions of algebraic Riccati equations [ZDG96, Sec. 13.2] in the case where $R > 0$, $E = I_n$ and the even matrix pencil $s\mathcal{E} - \mathcal{A}$ as in (3.29) has no eigenvalues on $\mathrm{i}\mathbb{R}$.

Further, note that in this case, a multiplication of the equation $0 = Kx(t) + Lu(t)$ from the left with $L^\top$ yields the optimal control $u(t) = -R^{-1}\left(B^\top X + S\right)x(t)$. Therefore, the state trajectory of the optimal control fulfills

$$\dot{x}(t) = \left(A - BR^{-1}(B^\top X + S)\right)x(t).$$

Now we formulate a result for the characterization of the functional $V^-$. That is, we formulate a statement for infimization of the cost functional on the negative time horizon analogously to Theorem 3.8.3. We will see that a transformation to the backward system will lead to an equivalent optimization problem for the positive time horizon.

**Theorem 3.8.7.** *Let* $(E, A, B) \in \Sigma_{n,m}$ *with the system space* $\mathcal{V}_{\mathrm{sys}}$ *be given and assume that* $Q = Q^\top \in \mathbb{R}^{n\times n}$, $S \in \mathbb{R}^{n\times m}$, *and* $R = R^\top \in \mathbb{R}^{m\times m}$. *Let the cost functional* $\mathcal{J}$ *be defined as in* (3.74), *and let the function* $V^- : \operatorname{im} E \to \mathbb{R}$ *be defined as in* (3.76). *Then the following two assertions are equivalent:*

*a)* $V^-(Ex_0) \in \mathbb{R}$ *for all* $x_0 \in \mathbb{R}^n$.

*b) The system* $(E, A, B)$ *is impulse controllable, has no uncontrollable modes on the imaginary axis, and the descriptor Lur'e equation* (3.35) *has an anti-stabilizing solution* $(X, K, L)$.

*In the case where the above statements are valid, we have:*

*i)* $V^-(Ex_0) = x_0^\top X^\top E x_0$ *for all* $x_0 \in \mathbb{R}^n$.

*ii) For all* $x_0 \in \mathbb{R}^n$ *and* $(x, u) \in \mathfrak{B}_{(E,A,B)}(x_0)$ *such that* $\lim_{t\to-\infty} Ex(t) = 0$ *and* $\mathcal{J}(x, u, -\infty, 0) \in \mathbb{R}$ *it holds that*

$$-\mathcal{J}(x, u, -\infty, 0) = V^-(Ex_0) - \int_{-\infty}^{0} \|Kx(t) + Lu(t)\|_2^2 \mathrm{d}t. \tag{3.92}$$

*Proof.* This problem can be lead back to Theorem 3.8.3 by considering the following facts:

1) It holds that

$$(x(\cdot), u(\cdot)) \in \mathfrak{B}_{(E,A,B)}(x_0) \Leftrightarrow (x(-\cdot), u(-\cdot)) \in \mathfrak{B}_{(-E,A,B)}(x_0).$$

2) The triple $(X, K, L)$ is an anti-stabilizing solution of (3.35) if and only if it is a stabilizing solution of the descriptor Lur'e equation (3.35) in which $E$ is replaced by $-E$.

3) We have

$$V^-(Ex_0) = -\inf\left\{\mathcal{J}(x,u,0,\infty) : (x,u) \in \mathfrak{B}_{(-E,A,B)}(x_0) \text{ and } \lim_{t\to\infty} Ex(t) = 0\right\}.$$

□

Finally, we present some equivalences for the finiteness of the cost functionals $V^+$ and $V^-$ for strongly controllable systems. In this situation we are able to obtain a stronger result which relates the feasibility of optimal control problems with zero terminal conditions to equivalent conditions expressed by the dissipation inequality, the Popov function, the descriptor KYP inequality, and the descriptor Lur'e equation. The proof requires the following lemma.

**Lemma 3.8.8.** *Let $(E, A, B) \in \Sigma_{n,m}$ be strongly controllable and assume that $Q = Q^\top \in \mathbb{R}^{n\times n}$, $S \in \mathbb{R}^{n\times m}$, and $R = R^\top \in \mathbb{R}^{m\times m}$ are given. Let the cost functional $\mathcal{J}$ be defined as in (3.74). Moreover, let $x_0 \in \mathbb{R}^n$, $(x,u) \in \mathfrak{B}_{(E,A,B)}(x_0)$, and $\varepsilon > 0$ be given. Then the following statements hold true:*

*a) If the integral $\mathcal{J}(x,u,0,\infty)$ converges and $\lim_{t\to\infty} Ex(t) = 0$, then there exists some $(\overline{x},\overline{u}) \in \mathfrak{B}_{(E,A,B)}(x_0)$ with compact support and*

$$|\mathcal{J}(x,u,0,\infty) - \mathcal{J}(\overline{x},\overline{u},0,\infty)| < \varepsilon.$$

*b) If the integral $\mathcal{J}(x,u,-\infty,0)$ converges and $\lim_{t\to-\infty} Ex(t) = 0$, then there exists some $(\overline{x},\overline{u}) \in \mathfrak{B}_{(E,A,B)}(x_0)$ with compact support and*

$$|\mathcal{J}(x,u,-\infty,0) - \mathcal{J}(\overline{x},\overline{u},-\infty,0)| < \varepsilon.$$

*c) If the integral $\mathcal{J}(x,u,-\infty,\infty)$ converges and $\lim_{t\to-\infty} Ex(t) = \lim_{t\to\infty} Ex(t) = 0$, then there exists some $(\overline{x},\overline{u}) \in \mathfrak{B}_{(E,A,B)}(x_0)$ with compact support and*

$$|\mathcal{J}(x,u,-\infty,\infty) - \mathcal{J}(\overline{x},\overline{u},-\infty,\infty)| < \varepsilon.$$

*Proof.* Statement b) follows from a) by turning to the backward system $(-E, A, B) \in \Sigma_{n,m}$ and, further, c) follows from a concatenation of the trajectories obtained in a) and b). Therefore, it suffices to prove statement a).
*Step 1: We show that there exists some constant $c > 0$ such that for all $x_0 \in \mathbb{R}^n$ there exists some $(x, u) \in \mathfrak{B}_{(E,A,B)}(x_0)$ with $(x, u)|_{[1,\infty)} \equiv 0$ and*

$$|\mathcal{J}(x, u, 0, 1)| \leq c \cdot \|Ex_0\|_2^2 : \tag{3.93}$$

Let $\{z_1, \dots, z_r\}$ be a basis of $\operatorname{im} E$ and let $x_{01}, \dots, x_{0r}$ with $Ex_{01} = z_1, \dots, Ex_{0r} = z_r$ be given. By [BR13, Lem. 2.3], there exists a trajectory $(x_k, u_k) \in \mathfrak{B}_{(E,A,B)}(x_{0k})$ with $(x_k, u_k)|_{[1,\infty)} \equiv 0$ for all $k \in \{1, \dots, r\}$. Consider the linear operator

$$\begin{aligned} \mathcal{F} : \qquad \qquad \operatorname{im} E &\to \mathcal{L}^2([0,\infty), \mathbb{R}^{n+m}), \\ \lambda_1 z_1 + \dots + \lambda_r z_r &\mapsto \lambda_1 (x_1, u_1) + \dots + \lambda_r (x_r, u_r). \end{aligned}$$

Then $\mathcal{F}(Ex_0) \in \mathfrak{B}_{[E,A,B]}(x_0)$ with $\mathcal{F}(Ex_0)|_{[1,\infty)} = 0$. The finite-dimensionality of $\operatorname{im} E$ gives rise to the boundedness of $\mathcal{F}$. Define the constant

$$\kappa = \left\| \begin{bmatrix} Q & S \\ S^\top & R \end{bmatrix} \right\|_2 .$$

Then the Cauchy-Schwarz inequality implies that for $(x, u) = \mathcal{F}(Ex_0)$ we have

$$|\mathcal{J}(x, u, 0, 1)| \leq \kappa \cdot \|(x, u)\|^2_{\mathcal{L}^2([0,\infty),\mathbb{R}^{n+m})} \leq \kappa \cdot \|\mathcal{F}\|^2 \cdot \|Ex_0\|_2^2.$$

Thus, (3.93) is satisfied with $c = \kappa \cdot \|\mathcal{F}\|^2$.
*Step 2: We conclude the result:*
Let $c > 0$ be defined as in Step 1. Since $\lim_{t\to\infty} Ex(t) = 0$ and the integral $\mathcal{J}(x, u, 0, \infty)$ converges, there exists some $T \geq 0$ with

$$|\mathcal{J}(x, u, T-1, \infty)| < \frac{\varepsilon}{2},$$

and

$$c \cdot \|Ex(T-1)\|_2^2 < \frac{\varepsilon}{2}.$$

Using the shift-invariance of the behavior and the findings in Step 1, we obtain that there exist some $(\widetilde{x}, \widetilde{u}) \in \mathfrak{B}_{(E,A,B)}$ with $E\widetilde{x}(T-1) = Ex(T-1)$, $(\widetilde{x}, \widetilde{u})|_{[T,\infty)} \equiv 0$, and

$$|\mathcal{J}(\widetilde{x}, \widetilde{u}, T-1, T)| \leq c \cdot \|Ex(T-1)\|^2 < \frac{\varepsilon}{2}.$$

Now define

$$(\overline{x}, \overline{u}) := (x, u) \underset{T-1}{\diamond} (\widetilde{x}, \widetilde{u}).$$

Then we obtain by the triangle inequality

$$\begin{aligned}
&|\mathcal{J}(x,u,0,\infty)-\mathcal{J}(\overline{x},\overline{u},0,\infty)| \\
&= |\mathcal{J}(x,u,0,\infty)-\mathcal{J}(x,u,0,T-1)-\mathcal{J}(\widetilde{x},\widetilde{u},T-1,\infty)| \\
&< |\mathcal{J}(x,u,T-1,\infty)|+|\mathcal{J}(\widetilde{x},\widetilde{u},T-1,T)| \\
&< \varepsilon.
\end{aligned}$$

□

Now we can formulate the main result. Note that some of the implications have already been shown for ODE systems in [Wil71, Thm. 2].

**Theorem 3.8.9.** *Let $(E,A,B)\in\Sigma_{n,m}$ with the system space $\mathcal{V}_{\mathrm{sys}}$ be strongly controllable, let $Q=Q^{\mathsf{T}}\in\mathbb{R}^{n\times n}$, $S\in\mathbb{R}^{n\times m}$, and $R=R^{\mathsf{T}}\in\mathbb{R}^{m\times m}$ be given. Let the cost functional $\mathcal{J}$ be defined as in* (3.74)*, and let the functions $V^+$, $V^-:\operatorname{im}E\to\mathbb{R}$ be defined as in* (3.75) *and* (3.76)*. Then the following statements are equivalent:*

*a) For all $T\geq 0$ and $(x,u)\in\mathfrak{B}_{(E,A,B)}(0)$ with $Ex(T)=0$ it holds that $\mathcal{J}(x,u,0,T)\geq 0$.*

*b) For all trajectories $(x,u)\in\mathfrak{B}_{(E,A,B)}$ with $\lim_{t\to\infty}Ex(t)=\lim_{t\to-\infty}Ex(t)=0$ and $\mathcal{J}(x,u,-\infty,\infty)<\infty$ it holds that $\mathcal{J}(x,u,-\infty,\infty)\geq 0$.*

*c) For all $x_0\in\mathbb{R}^n$ we have $V^+(Ex_0)\in\mathbb{R}$.*

*d) For all $x_0\in\mathbb{R}^n$ we have $V^-(Ex_0)\in\mathbb{R}$.*

*e) There exists some functional $V:\operatorname{im}E\to\mathbb{R}$ that satisfies the dissipation inequality* (3.77).

*f) For all $\omega\in\mathbb{R}$ with $\mathrm{i}\omega\notin\Lambda(E,A)$, the Popov function* (3.2) *fulfills $\Phi(\mathrm{i}\omega)\geq 0$.*

*g) There exists a $P\in\mathbb{R}^{n\times n}$ that satisfies the descriptor KYP inequality* (3.22).

*h) The descriptor Lur'e equation* (3.35) *has a stabilizing solution.*

*i) The descriptor Lur'e equation* (3.35) *has an anti-stabilizing solution.*

*Moreover, if the above conditions are fulfilled, then for all $x_0\in\mathbb{R}^n$ it holds that*

$$-\infty<V^-(Ex_0)\leq V^+(Ex_0)<\infty. \tag{3.94}$$

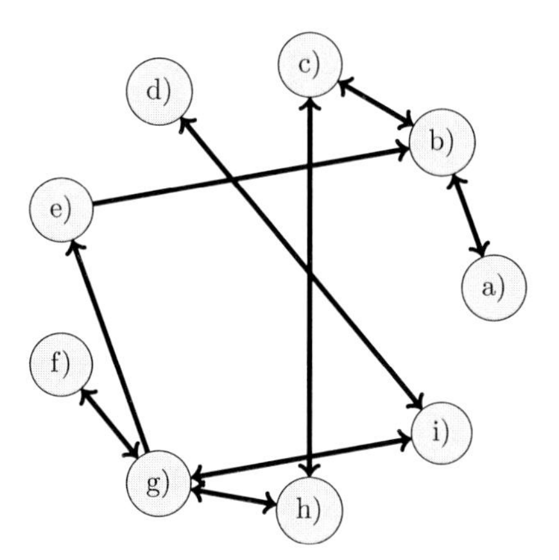


Figure 3.1: Structure of the proof of Theorem 3.8.9

*Proof.* The proof is carried out by showing the following implications as in Figure 3.1.

The equivalences "g) $\Leftrightarrow$ h)" and "g) $\Leftrightarrow$ i)" have been shown in Theorem 3.5.3. The statement "f) $\Leftrightarrow$ g)" has been proven in Theorem 3.3.1. Further, the equivalences "c) $\Leftrightarrow$ h)" and "d) $\Leftrightarrow$ i)" are subject of Theorem 3.8.3 and Theorem 3.8.7, respectively.

The implication "g) $\Rightarrow$ e)" follows from the fact that $V(Ex_0) = x_0^\top E^\top P x_0$ is a dissipation function according to (3.77).

Moreover, the implication "e) $\Rightarrow$ b)" follows from

$$\mathcal{J}(x,u,-\infty,\infty) = \lim_{t\to\infty} \mathcal{J}(x,u,-t,t) \geq \lim_{t\to\infty} \left(V(Ex(t)) - V(Ex(-t))\right) = 0.$$

For the proof of "b) $\Rightarrow$ a)" assume that there exist a $T \geq 0$ and a trajectory $(x,u) \in \mathfrak{B}_{(E,A,B)}(0)$ with $Ex(T) = 0$ such that $\mathcal{J}(x,u,0,T) < 0$. The concatenation

$$(\overline{x},\overline{u}) := (0,0)|_{(-\infty,0]} \underset{0}{\diamondsuit} (x,u) \underset{T}{\diamondsuit} (0,0)|_{[T,\infty)} \in \mathfrak{B}_{(E,A,B)}$$

yields $\mathcal{J}(\overline{x},\overline{u},-\infty,\infty) < 0$ which is a contradiction.

Now we show that a) implies b): Assume that b) does not hold true. Then there exist some $\varepsilon > 0$ and some $(x,u) \in \mathfrak{B}_{(E,A,B)}$ with

$$\lim_{t\to\infty} Ex(t) = \lim_{t\to-\infty} Ex(t) = 0$$

and

$$\mathcal{J}(x,u,-\infty,\infty) < -\varepsilon.$$

By Lemma 3.8.8, there exist some $T \geq 0$ and a $(\overline{x}, \overline{u}) \in \mathfrak{B}_{(E,A,B)}$ with support contained in $[-T, T]$, and

$$|\mathcal{J}(x, u, -\infty, \infty) - \mathcal{J}(\overline{x}, \overline{u}, -\infty, \infty)| < \frac{\varepsilon}{2}.$$

This gives

$$\mathcal{J}(\overline{x}, \overline{u}, -T, T) = \mathcal{J}(\overline{x}, \overline{u}, -\infty, \infty) < -\frac{\varepsilon}{2} < 0.$$

Since $Ex(-T) = Ex(T) = 0$, this is a contradiction to a).

Finally we show that b) implies c): Let $x_0 \in \mathbb{R}^n$ be given. By strong controllability of $(E, A, B)$, there exists a $(\widetilde{x}, \widetilde{u}) \in \mathfrak{B}_{(E,A,B)}(x_0)$ such that $\lim_{t\to-\infty} E\widetilde{x}(t) = 0$ and $\mathcal{J}(\widetilde{x}, \widetilde{u}, -\infty, 0)$ converges. Let $(x, u) \in \mathfrak{B}_{(E,A,B)}(x_0)$ be such that $\lim_{t\to\infty} Ex(t) = 0$ and $\mathcal{J}(x, u, 0, \infty)$ converges. Define

$$(\overline{x}, \overline{u}) := (\widetilde{x}, \widetilde{u}) \underset{0}{\Diamond} (x, u) \in \mathfrak{B}_{(E,A,B)}.$$

Then we obtain

$$0 \leq \mathcal{J}(\overline{x}, \overline{u}, -\infty, \infty) = \mathcal{J}(\widetilde{x}, \widetilde{u}, -\infty, 0) + \mathcal{J}(x, u, 0, \infty),$$

and thus

$$-\mathcal{J}(\widetilde{x}, \widetilde{u}, -\infty, 0) \leq \mathcal{J}(x, u, 0, \infty) \quad \forall (x, u) \in \mathfrak{B}_{(E,A,B)}(x_0).$$

Thus we obtain

$$-\infty < -\mathcal{J}(\widetilde{x}, \widetilde{u}, -\infty, 0) \leq V^+(Ex_0).$$

The inequality $V^+(Ex_0) < \infty$ follows by strong controllability of $(E, A, B)$. The inequality (3.94) follows, since by b), we have

$$0 \leq \mathcal{J}(x, u, -\infty, \infty) = \mathcal{J}(x, u, -\infty, 0) + \mathcal{J}(x, u, 0, \infty) \leq -V^-(Ex_0) + V^+(Ex_0)$$

for all $(x, u) \in \mathfrak{B}_{(E,A,B)}(x_0)$ □

*Remark* 3.8.10. The requirement in Theorem 3.3.1 that $\operatorname{rank}_{\mathbb{R}(s)} \Phi(s) = m$ (equivalently $s\mathcal{E} - \mathcal{A}$ is regular) in the case where $(E, A, B)$ is only strongly stabilizable or strongly anti-stabilizable seems to be, at least from a linear algebraic point of view, artificial for solvability of the descriptor KYP inequality. Theorem 3.8.3 and Theorem 3.8.7 however give rise to an interpretation of the solvability of the descriptor KYP inequality in terms of feasibility of linear-quadratic infimization problems. Indeed, it might happen that the infimization problem is not feasible even if the Popov function is positive semidefinite on $i\mathbb{R}$ and the system is strongly stabilizable. For instance, consider the strongly stabilizable system $(E, A, B) \in \Sigma_{1,1}$ with

$$E = 1, \quad A = -1, \quad B = Q = R = 0, \quad S = 1,$$

and the Popov function $\Phi(s) = 0$. Then the even pencil

$$s\mathcal{E} - \mathcal{A} := \begin{bmatrix} 0 & -s-1 & 0 \\ s-1 & 0 & 1 \\ 0 & 1 & 0 \end{bmatrix} \in \mathbb{R}[s]^{3\times 3}$$

is singular. Moreover, the associated descriptor KYP inequality

$$\begin{bmatrix} A^\top P + P^\top A + Q & P^\top B + S \\ B^\top P + S^\top & R \end{bmatrix} = \begin{bmatrix} -2P & 1 \\ 1 & 0 \end{bmatrix} \geq 0$$

has no solution. The cost functional

$$\mathcal{J}(x, u, 0, \infty) = \int_0^\infty 2x(\tau)^\top u(\tau) \mathrm{d}\tau$$

can indeed be made arbitrarily large and negative, since there exists a free control variable that does not affect the system dynamics but the cost functional.

In the following we briefly consider the case where even equality holds true in the dissipation inequality (3.77). In view of (3.80) and (3.92), this means that $K$ and $L$ have zero rows, which means that, by Theorem 3.6.3, the Popov function vanishes. We present some equivalent statements for the controllable case in the following theorem.

**Theorem 3.8.11.** *Let $(E, A, B) \in \Sigma_{n,m}$ with the system space $\mathcal{V}_{\mathrm{sys}}$ be strongly controllable, let $Q = Q^\top \in \mathbb{R}^{n\times n}$, $S \in \mathbb{R}^{n\times m}$, and $R = R^\top \in \mathbb{R}^{m\times m}$ be given. Let the cost functional $\mathcal{J}$ be defined as in (3.74), and let the functions $V^+$, $V^- : \operatorname{im} E \to \mathbb{R}$ be defined as in (3.75) and (3.76). Then the following statements are equivalent:*

*a) For all $T \geq 0$ and $(x, u) \in \mathfrak{B}_{(E,A,B)}(0)$ with $Ex(T) = 0$ it holds that $\mathcal{J}(x, u, 0, T) = 0$.*

*b) For all trajectories $(x, u) \in \mathfrak{B}_{(E,A,B)}$ with $\lim_{t\to\infty} Ex(t) = \lim_{t\to-\infty} Ex(t) = 0$ and $\mathcal{J}(x, u, -\infty, \infty) < \infty$ it holds that $\mathcal{J}(x, u, -\infty, \infty) = 0$.*

*c) There exists some functional $V : \operatorname{im} E \to \mathbb{R}$ such that*

$$\mathcal{J}(x, u, t_0, t_1) + V(Ex(t_1)) = V(Ex(t_0)) \quad \forall (x, u) \in \mathfrak{B}_{(E,A,B)}, \text{ and } t_0,\ t_1 \in \mathbb{R} \text{ with } t_0 \leq t_1. \tag{3.95}$$

*d) The Popov function (3.2) fulfills $\Phi(s) = 0$.*

*e) There exists some $P \in \mathbb{R}^{n\times n}$ with*

$$\begin{bmatrix} A^\top P + P^\top A + Q & P^\top B + S \\ B^\top P + S^\top & R \end{bmatrix} =_{\mathcal{V}_{\mathrm{sys}}} 0, \quad E^\top P = P^\top E.$$

*f) There exists a stabilizing solution $(X, K, L)$ of the descriptor Lur'e equation* (3.35). *This solution has the property $K \in \mathbb{R}^{0 \times n}$ and $L \in \mathbb{R}^{0 \times m}$.*

*g) There exists an anti-stabilizing solution $(X, K, L)$ of the descriptor Lur'e equation* (3.35). *This solution has the property $K \in \mathbb{R}^{0 \times n}$ and $L \in \mathbb{R}^{0 \times m}$.*

*Moreover, if the above conditions are fulfilled, then for all $x_0 \in \mathbb{R}^n$ it holds that*

$$-\infty < V^-(Ex_0) = V^+(Ex_0) < \infty. \tag{3.96}$$

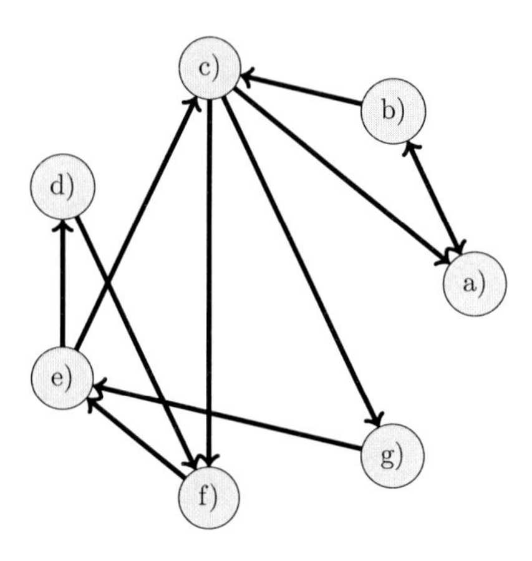


Figure 3.2: Structure of the proof of Theorem 3.8.11

*Proof.* The proof is carried out by showing the following implications as in Figure 3.2.

The proof that a) and b) are equivalent is analogous to the proof of "a) $\Leftrightarrow$ b)" in Theorem 3.8.9.

The implication "c) $\Rightarrow$ a)" is trivial.

We show that b) implies c): Assume that for $(x, u) \in \mathfrak{B}_{(E,A,B)}$ with $\lim_{t\to\infty} Ex(t) = \lim_{t\to-\infty} Ex(t) = 0$ it holds $\mathcal{J}(x, u, -\infty, \infty) = 0$. From Theorem 3.8.9 b) and e) we obtain that there exists a function $V : \operatorname{im} E \to \mathbb{R}$ that fulfills the dissipation inequality. In particular, it holds that

$$\mathcal{J}(x, u, t_0, t_1) + V(Ex(t_1)) \geq V(Ex(t_0)). \tag{3.97}$$

On the other hand we have

$$\begin{aligned} \mathcal{J}(x, u, -\infty, t_0) + V(Ex(t_0)) &\geq 0, \\ \mathcal{J}(x, u, t_1, \infty) &\geq V(Ex(t_1)). \end{aligned} \tag{3.98}$$

Since by assumption

$$\mathcal{J}(x,u,-\infty,t_0)+\mathcal{J}(x,u,t_1,\infty)=-\mathcal{J}(x,u,t_0,t_1),$$

we obtain

$$-\mathcal{J}(x,u,t_0,t_1)+V(Ex(t_0))\geq V(Ex(t_1))$$

by adding the two inequalities in (3.98). Together with (3.97) this yields the claim.

Now we prove that c) implies f): It follows from c) that

$$V(Ex_0)=\mathcal{J}(x,u,0,\infty)\quad \forall (x,u)\in\mathfrak{B}_{(E,A,B)}(x_0)\text{ with }\lim_{t\to\infty}Ex(t)=0.$$

and therefore, $V^+=V$. By Theorem 3.8.3, there exists a stabilizing solution $(X,K,L)$ of the descriptor Lur'e equation with $K\in\mathbb{K}^{q\times n}$ and $L\in\mathbb{K}^{q\times m}$. It remains to show that $q=0$:

Therefore, assume that $q>0$. Then, by the assumption of strong controllability, we can use Lemma 3.8.2 to infer that there exists some $x_0\in\mathbb{R}^n$ and a *smooth* trajectory $(x,u)\in\mathfrak{B}_{(E,A,B)}(x_0)$ with $Kx_0+Lu(0)\neq 0$, $\int_0^\infty\|Kx(\tau)+Lu(\tau)\|_2^2\mathrm{d}\tau<\infty$ and $\lim_{t\to\infty}Ex(t)=0$. The first two statements imply that

$$0<\int_0^\infty\|Kx(\tau)+Lu(\tau)\|_2^2\mathrm{d}\tau<\infty.$$

We can further conclude from Theorem 3.8.3 and $V^+=V$ that

$$V^+(Ex_0)=\mathcal{J}(x,u,0,\infty)=V^+(Ex_0)+\int_0^\infty\|Kx(\tau)+Lu(\tau)\|_2^2\mathrm{d}\tau>V^+(Ex_0),$$

which is a contradiction.

The proof that c) implies g) is completely analogous to the proof of "c) $\Rightarrow$ f)".

The implications "f) $\Rightarrow$ e)" and "g) $\Rightarrow$ e)" are trivial.

Further, by Theorem 3.6.3, we see that e) implies d).

Moreover, e) implies c) since with $V(Ex_0)=x_0^\top X^\top Px_0$ we have

$$\begin{pmatrix}x(t)\\u(t)\end{pmatrix}^\top\begin{bmatrix}Q&S\\S^\top&R\end{bmatrix}\begin{pmatrix}x(t)\\u(t)\end{pmatrix}=-\nabla V(Ex(t))(Ax(t)+Bu(t))=-\nabla V(Ex(t))E\dot{x}(t)$$
$$\forall (x,u)\in\mathfrak{B}_{(E,A,B)}\text{ and almost all }t\in\mathbb{R},$$

see also (3.78).

Finally, we show that d) implies f): If $(E,A,B)$ is strongly controllable and $\Phi(s)=0$, the existence of a stabilizing solution follows from Theorem 3.5.3. By Theorem 3.6.4, we further have that $K\in\mathbb{R}^{q\times n}$ and $L\in\mathbb{R}^{q\times m}$ for

$$q=\operatorname{rank}_{\mathbb{R}(s)}\Phi(s)=0.$$

Equation (3.96) follows immediately from b) since for all $(x,u) \in \mathfrak{B}_{(E,A,B)}(x_0)$ we have

$$0 = \mathcal{J}(x,u,-\infty,\infty) = \mathcal{J}(x,u,-\infty,0) + \mathcal{J}(x,u,0,\infty) = -V^-(Ex_0) + V^+(Ex_0).$$

□

*Remark* 3.8.12.

a) We can conclude from (3.96) and Theorem 3.5.4 that, if one of the assertions a)–g) and the assumptions in Theorem 3.8.11 hold true, then the descriptor KYP inequality has exactly one solution in the sense that $E^\top P$ is unique.

b) If one of the assertions a)–g) and the assumptions in Theorem 3.8.11 hold true, it holds that

$$\begin{bmatrix} 0 & -sE+A & B \\ sE^\top + A^\top & Q & S \\ B^\top & S^\top & R \end{bmatrix} \begin{bmatrix} X & 0 \\ I_n & 0 \\ 0 & I_m \end{bmatrix} = \begin{bmatrix} I_n \\ -X^\top + H^\top \Sigma M T_\infty^\top \\ J^\top \Sigma M T_\infty^\top \end{bmatrix} \begin{bmatrix} -sE+A & B \end{bmatrix}.$$

We can further conclude from strong controllability of $(E,A,B)$ that the KCF of the augmented pencil $\begin{bmatrix} -sE+A & B \end{bmatrix} \in \mathbb{R}[s]^{n\times n+m}$ consists of $m$ blocks of type K3, and $n-r$ blocks of type K2. Hence, by Theorem 3.4.2 and Remark 3.4.3, the statements a)–g) are equivalent to the EKCF of $s\mathcal{E}-\mathcal{A}$ being of the following structure:

i) There exist neither blocks of type E1 nor blocks of type E2.

ii) There exist exactly $2(n-r)$ blocks of type E3. These are all of size $1\times 1$. The number of blocks with positive and negative sign-characteristic is equal.

iii) There exist exactly $m$ blocks of type E4.

### 3.8.2 Optimal Control with Free Terminal Condition

Now we turn to the linear-quadratic optimal control problem with free terminal conditions. Therefore, we consider the following two optimal value functions [Wil71]:

a) $V_{\mathrm{f}}^+(Ex_0) = \inf\left\{\mathcal{J}(x,u,0,\infty) : (x,u) \in \mathfrak{B}_{(E,A,B)}(x_0)\right\},$ (3.99)

b) $V_{\mathrm{n}}^+(Ex_0) = \inf\left\{\mathcal{J}(x,u,0,T) : (x,u) \in \mathfrak{B}_{(E,A,B)}(x_0) \text{ and } T \geq 0\right\}.$ (3.100)

Clearly, $V_{\mathrm{n}}^+(Ex_0) \leq 0$, since $\mathcal{J}(x,u,0,0) = 0$ holds for all $(x,u) \in \mathfrak{B}_{(E,A,B)}(x_0)$.

We will prove that, for strongly controllable systems, the finiteness of $V_{\mathrm{n}}^+$ or $V_{\mathrm{f}}^+$ is equivalent to the existence of a nonpositive solution of the descriptor KYP inequality. We also refer to [Wil71, Thm. 1] where some of the following implications are shown for ODE systems.

**Theorem 3.8.13.** *Let $(E, A, B) \in \Sigma_{n,m}$ with the system space $\mathcal{V}_{\mathrm{sys}}$ be strongly controllable, and let $Q = Q^{\mathsf{T}} \in \mathbb{R}^{n\times n}$, $S \in \mathbb{R}^{n\times m}$, and $R = R^{\mathsf{T}} \in \mathbb{R}^{m\times m}$ be given. Further, let the functions $V^-$, $V_{\mathrm{f}}^+$, and $V_{\mathrm{n}}^+$ be defined as in (3.76), (3.99), and (3.100), respectively. Then the following statements are equivalent:*

*a) For all $T \geq 0$ and $(x, u) \in \mathfrak{B}_{(E,A,B)}(0)$ it holds that $\mathcal{J}(x, u, 0, T) \geq 0$.*

*b) For all $(x, u) \in \mathfrak{B}_{(E,A,B)}$ with $\lim_{t\to-\infty} Ex(t) = 0$ and $\mathcal{J}(x, u, -\infty, 0) < \infty$ it holds that $\mathcal{J}(x, u, -\infty, 0) \geq 0$.*

*c) For all $x_0 \in \mathbb{R}^n$ we have $-\infty < V^-(Ex_0) \leq 0$.*

*d) For all $x_0 \in \mathbb{R}^n$ we have $-\infty < V_{\mathrm{n}}^+(Ex_0) \leq 0$.*

*e) For all $x_0 \in \mathbb{R}^n$ we have $V_{\mathrm{f}}^+(Ex_0) \in \mathbb{R}$.*

*f) There exists some functional $V : \operatorname{im} E \to (-\infty, 0]$ that satisfies the dissipation inequality (3.77).*

*g) The descriptor KYP inequality (3.22) has a nonpositive solution $P \in \mathbb{R}^{n\times n}$.*

*h) There exists an anti-stabilizing solution $(X, K, L)$ to the descriptor Lur'e equation (3.35) which satisfies $E^{\mathsf{T}}X \leq 0$.*

*Furthermore, if the above conditions are fulfilled, then for all $x_0 \in \mathbb{K}^n$, it holds that*

$$-\infty < V^-(Ex_0) \leq V_{\mathrm{n}}^+(Ex_0) \leq V_{\mathrm{f}}^+(Ex_0) \leq V^+(Ex_0) < \infty.$$

*Proof.* The proof is carried out by showing the following implications as in Figure 3.3.

The proof of "a) $\Rightarrow$ b)" is obtained by contradiction: Assume that b) is not satisfied. Then there exists an $\varepsilon > 0$ and a trajectory $(x, u) \in \mathfrak{B}_{(E,A,B)}(x_0)$ with $\lim_{t\to-\infty} Ex(t) = 0$ such that

$$\mathcal{J}(x, u, -\infty, 0) < -\varepsilon.$$

From Lemma 3.8.8 we obtain that there exists a $T > 0$ and some $(\overline{x}, \overline{u}) \in \mathfrak{B}_{(E,A,B)}(x_0)$ with support in $[-T, 0]$ such that

$$|\mathcal{J}(\overline{x}, \overline{u}, -\infty, 0) - \mathcal{J}(x, u, -\infty, 0)| < \frac{\varepsilon}{2}.$$

By shift-invariance of the behavior this gives

$$\mathcal{J}(\overline{x}, \overline{u}, -\infty, 0) = \mathcal{J}(\overline{x}, \overline{u}, -T, 0) = \mathcal{J}(\overline{x}(\cdot - T), \overline{u}(\cdot - T), 0, T) < -\frac{\varepsilon}{2},$$

which is a contradiction to a).

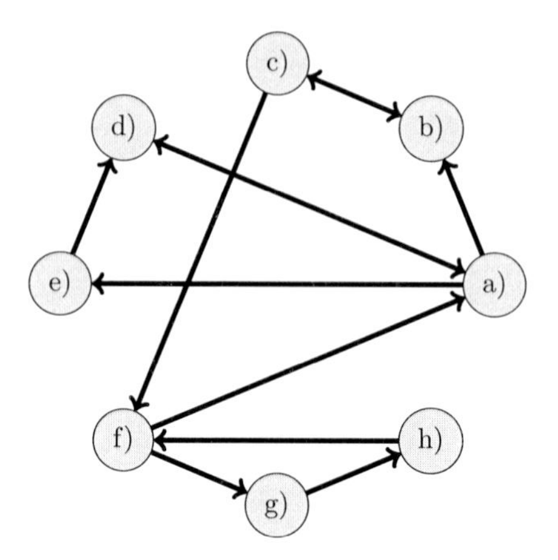


Figure 3.3: Structure of the proof of Theorem 3.8.13

The equivalence between b) and c) is trivial.

The implication "c) $\Rightarrow$ f)" follows from the fact that the dissipation inequality is fulfilled by $V^-$, see Theorem 3.8.3.

The implication "f) $\Rightarrow$ a)" follows from

$$\mathcal{J}(x,u,t_0,t_1) + V(Ex(t_1)) \geq V(Ex(t_0)),$$

and setting $t_0 = 0$, $Ex(t_0) = 0$ and by using the normalization $V(0) = 0$.

Now we prove "a) $\Rightarrow$ d), e)": Let $(x,u) \in \mathfrak{B}_{(E,A,B)}$ with $\lim_{t\to-\infty} Ex(t) = 0$, $Ex(0) = Ex_0$, and $Ex(T) = Ex_T$ for $T \geq 0$ be given. From "a) $\Rightarrow$ c)" it follows that

$$-\mathcal{J}(x,u,-\infty,0) \leq \mathcal{J}(x,u,0,T).$$

Therefore, the inequalities $-\infty < V^-(Ex_0) \leq V_{\mathrm{n}}^+(Ex_0) \leq V_{\mathrm{f}}^+(Ex_0)$ hold which directly yields d) and e).

The statement "e) $\Rightarrow$ d)" trivially follows from definition.

Next we prove the implication "d) $\Rightarrow$ a)": Assume that $(x,u) \in \mathfrak{B}_{(E,A,B)}(0)$ and $T \geq 0$ with $\mathcal{J}(x,u,0,T) < 0$. Then, by linearity of the behavior, we have $(k\cdot x, k\cdot u) \in \mathfrak{B}_{(E,A,B)}(0)$ for all $k \in \mathbb{N}$, and we obtain

$$\lim_{k\to\infty} \mathcal{J}(k\cdot x, k\cdot u, 0, T) = \lim_{k\to\infty} k^2 \cdot \mathcal{J}(x,u,0,T) = -\infty.$$

This is a contradiction to d).

Now we prove "f) $\Rightarrow$ g)": Since f) is equivalent to c) and by Theorem 3.8.7, $V^-$ is a quadratic functional, the claim directly follows since $V^-(Ex_0) = x_0^\top E^\top P x_0 \leq 0$, where $P$ is a solution of the descriptor KYP inequality (3.22).

The implication "g) $\Rightarrow$ h)" is shown as follows: If the descriptor KYP inequality has a solution $P$ with $E^\top P \leq 0$, then by Theorem 3.5.3, there exists an anti-stabilizing solution $(X, K, L)$ of the descriptor Lur'e equation (3.35). By Theorem 3.5.4 this solution fulfills $E^\top X \leq E^\top P \leq 0$.

To complete the proof, we show "h) $\Rightarrow$ f)": Obviously, if the descriptor Lur'e equation has an anti-stabilizing solution triple $(X, K, L)$ with $E^\top X \leq 0$, then the desired functional is given by $V^-(Ex_0) = x_0^\top E^\top X x_0$.

The last inequality $V_{\mathrm{f}}^+(Ex_0) \leq V^+(Ex_0) < \infty$ is trivially fulfilled by definition of the functionals. Finiteness of $V^+$ follows from strong controllability. □

*Remark* 3.8.14.

a) We obtain from Theorem 3.8.13 and Theorem 3.7.1 that the finiteness of the functionals $V_{\mathrm{n}}^+$ and $V_{\mathrm{f}}^+$ implies that the modified Popov function (3.62) is positive semidefinite in $\overline{\mathbb{C}^+} \setminus \Lambda(E, A)$.

b) If Theorem 3.8.13 holds true and $\Psi(\mathrm{i}\omega) = 0$ for all $\omega \in \mathbb{R}$ with $\mathrm{i}\omega \notin \Lambda(E, A)$, then Theorem 3.8.11 implies that the descriptor KYP inequality has exactly one solution. This solution is nonpositive by Theorem 3.8.13.

c) In [Wil71], it has been claimed that the functionals $V_{\mathrm{n}}^+$ and $V_{\mathrm{f}}^+$ are quadratic and that they can be expressed by a solution of the descriptor KYP inequality. A proof that $V_{\mathrm{f}}^+$ is a quadratic functional for ODE systems is given in [Gee89] and an extension of this result to DAE systems seems possible. On the other hand, it is doubtful that $V_{\mathrm{n}}^+$ is quadratic, since the time $T$ for which $\mathcal{J}(x, u, 0, T)$ is minimized, is variable.

## 3.9 Dissipative and Cyclo-Dissipative Systems

### 3.9.1 Systems with General Quadratic Supply Rates

The results of the previous sections will now be applied to the analysis of dissipativity of differential-algebraic systems. Dissipative systems are provided with an energy functional depending on inputs and outputs. Loosely speaking, dissipativity means that the system only consumes and does not produce energy [Wil72a, Wil72b, Brü11b, Brü11a].

**Definition 3.9.1** (Supply rate, dissipative system)**.** Let $Q_s = Q_s^\top \in \mathbb{R}^{p\times p}$, $S_s \in \mathbb{R}^{p\times m}$, and $R_s = R_s^\top \in \mathbb{R}^{m\times m}$ be given. The mapping

$$\begin{aligned} s:\ & \mathbb{R}^m \times \mathbb{R}^p \to \mathbb{R} \\ & (u,y) \mapsto \begin{pmatrix} y \\ u \end{pmatrix}^\top \begin{bmatrix} Q_s & S_s \\ S_s^\top & R_s \end{bmatrix} \begin{pmatrix} y \\ u \end{pmatrix} \end{aligned} \tag{3.101}$$

is called *supply rate*. The differential-algebraic system $(E,A,B,C,D) \in \Sigma_{n,m,p}$ is called

a) *dissipative* (with respect to $s(\cdot,\cdot)$), if for all $T \geq 0$ and all $(x,u,y) \in \mathfrak{B}_{(E,A,B,C,D)}(0)$ it holds that
$$\int_0^T s(u(\tau), y(\tau))\mathrm{d}\tau \geq 0;$$

b) *cyclo-dissipative* (with respect to $s(\cdot,\cdot)$), if for all $T \geq 0$ and all $(x,u,y) \in \mathfrak{B}_{(E,A,B,C,D)}(0)$ with $Ex(T) = 0$ it holds that
$$\int_0^T s(u(\tau), y(\tau))\mathrm{d}\tau \geq 0;$$

c) *lossless dissipative* (with respect to $s(\cdot,\cdot)$), if it is dissipative and, furthermore, for all $T \geq 0$ and all $(x,u,y) \in \mathfrak{B}_{(E,A,B,C,D)}(0)$ it holds that
$$\int_0^T s(u(\tau), y(\tau))\mathrm{d}\tau = 0;$$

d) *lossless cyclo-dissipative* (with respect to $s(\cdot,\cdot)$), if it is cyclo-dissipative and, furthermore, for all $T \geq 0$ and all $(x,u,y) \in \mathfrak{B}_{(E,A,B,C,D)}(0)$ with $Ex(T) = 0$ it holds that
$$\int_0^T s(u(\tau), y(\tau))\mathrm{d}\tau = 0.$$

We will characterize dissipativity and cyclo-dissipativity by means of a linear matrix inequality which is intimately related to the descriptor KYP inequality.

**Definition 3.9.2** (Descriptor dissipativity inequality/equation)**.** Let the linear system $(E,A,B,C,D) \in \Sigma_{n,m,p}$ with the system space $\mathcal{V}_{\mathrm{sys}}$ as in (3.5) be given and assume that $Q_s = Q_s^\top \in \mathbb{R}^{p\times p}$, $S_s \in \mathbb{R}^{p\times m}$, and $R_s = R_s^\top \in \mathbb{R}^{m\times m}$. Let the supply rate $s(\cdot,\cdot)$ be defined as in (3.101). Then we say that $P \in \mathbb{R}^{n\times n}$ solves the

a) *descriptor dissipativity inequality*, if

$$\begin{aligned} \begin{bmatrix} A^\top P + P^\top A - C^\top Q_s C & P^\top B - C^\top Q_s D - C^\top S_s \\ B^\top P - D^\top Q_s C - S_s^\top C & -D^\top Q_s D - S_s^\top D - D^\top S_s - R_s \end{bmatrix} &\leq_{\mathcal{V}_{\mathrm{sys}}} 0, \\ E^\top P &= P^\top E; \end{aligned} \tag{3.102}$$

b) *descriptor dissipativity equation*, if

$$\begin{bmatrix} A^\mathsf{T}P + P^\mathsf{T}A - C^\mathsf{T}Q_sC & P^\mathsf{T}B - C^\mathsf{T}Q_sD - C^\mathsf{T}S_s \\ B^\mathsf{T}P - D^\mathsf{T}Q_sC - S_s^\mathsf{T}C & -D^\mathsf{T}Q_sD - S_s^\mathsf{T}D - D^\mathsf{T}S_s - R_s \end{bmatrix} =_{\mathcal{V}_{\mathrm{sys}}} 0,$$
$$E^\mathsf{T}P = P^\mathsf{T}E. \quad (3.103)$$

*Remark* 3.9.3 (Descriptor dissipativity inequality/equation and (modified) Popov functions).

a) By defining $Q := C^\mathsf{T}Q_sC$, $S := C^\mathsf{T}Q_sD + C^\mathsf{T}S_s$, $R := D^\mathsf{T}Q_sD + S_s^\mathsf{T}D + D^\mathsf{T}S_s + R_s$, and replacing $P$ by $-P$, we can make use of Theorem 3.8.9 to see that for strongly controllable systems, cyclo-dissipativity is equivalent to the solvability of the descriptor dissipativity inequality.

   This however, is equivalent to the transfer function $G(s) \in \mathbb{R}(s)^{p\times m}$ of the system $(E, A, B, C, D) \in \Sigma_{n,m,p}$ fulfilling

$$\begin{bmatrix} G(\mathrm{i}\omega) \\ I_m \end{bmatrix}^\mathsf{H} \begin{bmatrix} Q_s & S_s \\ S_s^\mathsf{T} & R_s \end{bmatrix} \begin{bmatrix} G(\mathrm{i}\omega) \\ I_m \end{bmatrix} \geq 0 \quad \forall \omega \in \mathbb{R} \text{ with } \mathrm{i}\omega \notin \Lambda(E, A). \quad (3.104)$$

   Further, the existence of a solution of the descriptor dissipativity equation is equivalent to

$$\begin{bmatrix} G(\mathrm{i}\omega) \\ I_m \end{bmatrix}^\mathsf{H} \begin{bmatrix} Q_s & S_s \\ S_s^\mathsf{T} & R_s \end{bmatrix} \begin{bmatrix} G(\mathrm{i}\omega) \\ I_m \end{bmatrix} = 0 \quad \forall \omega \in \mathbb{R} \text{ with } \mathrm{i}\omega \notin \Lambda(E, A). \quad (3.105)$$

b) On the other hand, we can employ Theorem 3.8.13 to see that for strongly controllable systems, dissipativity is equivalent to the solvability of the descriptor dissipativity inequality with $E^\mathsf{T}P \geq 0$. Since the descriptor dissipation inequality is a special descriptor KYP inequality, it can be treated by Theorem 3.7.1, Theorem 3.7.4, and Proposition 3.7.6.

   In particular, we obtain from Theorem 3.7.1 that solvability of the descriptor dissipativity inequality with $E^\mathsf{T}P \geq 0$ implies that the transfer function $G(s) \in \mathbb{R}(s)^{p\times m}$ of the system $(E, A, B, C, D) \in \Sigma_{n,m,p}$ fulfills

$$\begin{bmatrix} G(\lambda) \\ I_m \end{bmatrix}^\mathsf{H} \begin{bmatrix} Q_s & S_s \\ S_s^\mathsf{T} & R_s \end{bmatrix} \begin{bmatrix} G(\lambda) \\ I_m \end{bmatrix} \geq 0 \quad \forall \lambda \in \overline{\mathbb{C}^+} \setminus \Lambda(E, A). \quad (3.106)$$

   Further, the existence of a nonnegative solution of the descriptor dissipativity equation gives rise to (3.106) with the additional property (3.105).

Following from the above remark we can directly conclude conditions for dissipativity and cyclo-dissipativity in terms of an associated (modified) Popov function, formulated in the following corollary.

**Corollary 3.9.4.** *Let* $(E, A, B, C, D) \in \Sigma_{n,m,p}$ *with the system space* $\mathcal{V}_{\mathrm{sys}}$ *as in* (3.5) *and the transfer function* $G(s) \in \mathbb{R}(s)^{p \times m}$ *be given and assume that* $Q_s = Q_s^{\top} \in \mathbb{R}^{p \times p}$, $S_s \in \mathbb{R}^{p \times m}$, *and* $R_s = R_s^{\top} \in \mathbb{R}^{m \times m}$. *Then the following statements hold true:*

*a) The system* $(E, A, B, C, D)$ *is*

*i) cyclo-dissipative with respect to the supply rate* $s(\cdot, \cdot)$ *if and only if* (3.104) *is satisfied;*

*ii) lossless cyclo-dissipative with respect to the supply rate* $s(\cdot, \cdot)$ *if and only if* (3.105) *is satisfied.*

*b) Assume that* $K_1 \in \mathbb{R}^{m \times p}$, $K_2 \in \mathbb{R}^{p_2 \times p}$, $L_1 \in \mathbb{R}^{m \times m}$, *and* $L_2 \in \mathbb{R}^{p_2 \times m}$ *are given such that*

$$\begin{bmatrix} Q_s & S_s \\ S_s^{\top} & R_s \end{bmatrix} =_{\widetilde{\mathcal{V}}_{\mathrm{sys}}} \begin{bmatrix} K_1^{\top} K_1 & K_1^{\top} L_1 \\ L_1^{\top} K_1 & L_1^{\top} L_1 \end{bmatrix} - \begin{bmatrix} K_2^{\top} K_2 & K_2^{\top} L_2 \\ L_2^{\top} K_2 & L_2^{\top} L_2 \end{bmatrix},$$

*with*

$$\widetilde{\mathcal{V}}_{\mathrm{sys}} := \left\{ \begin{pmatrix} y \\ u \end{pmatrix} \in \mathbb{R}^{p+m} : y = Cx + Du \text{ with } \begin{pmatrix} x \\ u \end{pmatrix} \in \mathcal{V}_{\mathrm{sys}} \right\}$$

*and* $K_1 G(s) + L_1 \in \mathrm{Gl}_m(\mathbb{R}(s))$. *Then the system* $(E, A, B, C, D)$ *is*

*i) dissipative with respect to the supply rate* $s(\cdot, \cdot)$ *if and only if* (3.106) *holds;*

*ii) lossless dissipative with respect to the supply rate* $s(\cdot, \cdot)$ *if and only if* (3.106) *and* (3.105) *hold.*

*Proof.* The proof is based on the fact that a system $(E, A, B, C, D) \in \Sigma_{n,m,p}$ can be transformed into *Kalman decomposition for controllability* [BT14]. That is, there exist $W, T \in \mathrm{Gl}_n(\mathbb{R})$, such that

$$(WET, WAT, WB, CT, D) = \left( \begin{bmatrix} E_{11} & E_{12} \\ 0 & E_{22} \end{bmatrix}, \begin{bmatrix} A_{11} & A_{12} \\ 0 & A_{22} \end{bmatrix}, \begin{bmatrix} B_1 \\ 0 \end{bmatrix}, \begin{bmatrix} C_1 & C_2 \end{bmatrix}, D \right),$$

where $(E_{11}, A_{11}, B_1, C_1, D) \in \Sigma_{n_1,m,p}$ is completely controllable (and hence, strongly controllable). The transfer functions of $(E, A, B, C, D)$ and $(E_{11}, A_{11}, B_1, C_1, D)$ coincide. Furthermore, by a conforming partition of the state, we obtain that

$$(x, u, y) \in \mathfrak{B}_{(E,A,B,C,D)}(0) \iff (x_1, u, y) \in \mathfrak{B}_{(E_{11},A_{11},B_1,C_1,D)}(0),$$
$$\text{where } x(t) = T^{-1} \begin{pmatrix} x_1(t) \\ 0 \end{pmatrix}.$$

Thus we can restrict our considerations to the completely controllable subsystem of $(E, A, B, C, D)$.

Statement a) now follows from Remark 3.9.3 a), whereas Statement b) can be concluded from Remark 3.9.3 b) together with Theorem 3.7.4 and Theorem 3.8.11. □

In the following we give a corollary that summarizes the relationship between cyclo-dissipativity and the spectrum of two associated even matrix pencils. For a similar result see also [Brü11b, Cor. 4.2]. For the formulation of our result we need the sign-sum function for a Hermitian matrix $H \in \mathbb{C}^{k\times k}$. If $\mathrm{In}(H) = (\pi_+, \pi_0, \pi_-)$, then the sign-sum function is defined by

$$\eta(H) = \pi_+ + \pi_0 - \pi_-. \tag{3.107}$$

**Corollary 3.9.5.** *Assume that* $(E, A, B, C, D) \in \Sigma_{n,m,p}$ *be given and let* $Q_s = Q_s^\mathsf{T} \in \mathbb{R}^{p\times p}$, $S_s \in \mathbb{R}^{p\times m}$, *and* $R_s = R_s^\mathsf{T} \in \mathbb{R}^{m\times m}$ *be given. Let the supply rate* $s(\cdot,\cdot)$ *be defined as in* (3.101). *Furthermore, define* $Q := C^\mathsf{T} Q_s C$, $S := C^\mathsf{T} Q_s D + C^\mathsf{T} S_s$, $R := D^\mathsf{T} Q_s D + S_s^\mathsf{T} D + D^\mathsf{T} S_s + R_s$. *Consider the two even matrix pencils*

$$s\mathcal{E}_1 - \mathcal{A}_1 := \begin{bmatrix} 0 & -sE + A & B \\ sE^\mathsf{T} + A^\mathsf{T} & Q & S \\ B^\mathsf{T} & S^\mathsf{T} & R \end{bmatrix} \in \mathbb{R}[s]^{2n+m\times 2n+m},$$

*and*

$$s\mathcal{E}_2 - \mathcal{A}_2 := \begin{bmatrix} 0 & 0 & 0 & -sE + A & B \\ 0 & 0 & -I_p & C & D \\ 0 & -I_p & Q_s & 0 & S_s \\ sE^\mathsf{T} + A^\mathsf{T} & C^\mathsf{T} & 0 & 0 & 0 \\ B^\mathsf{T} & D^\mathsf{T} & S_s^\mathsf{T} & 0 & R_s \end{bmatrix} \in \mathbb{R}[s]^{2(n+p)+m\times 2(n+p)+m},$$

*Then the following statements are equivalent:*

*a) The system* $(E, A, B, C, D) \in \Sigma_{n,m,p}$ *is cyclo-dissipative with respect to* $s(\cdot,\cdot)$.

*b) The relation* (3.104) *holds.*

*c) It holds* $\eta(\mathrm{i}\omega\mathcal{E}_1 - \mathcal{A}_1) = m$ *for all* $\omega \in \mathbb{R}$ *with* $\mathrm{i}\omega \notin \Lambda(E, A)$.

*d) It holds* $\eta(\mathrm{i}\omega\mathcal{E}_2 - \mathcal{A}_2) = m$ *for all* $\omega \in \mathbb{R}$ *with* $\mathrm{i}\omega \notin \Lambda(E, A)$.

*If furthermore, the system* $(E, A, B)$ *has no uncontrollable modes on the imaginary axis, then the statements a)–d) are moreover equivalent to:*

*e) The EKCF of the pencil* $s\mathcal{E}_1 - \mathcal{A}_1$ *fulfills the statements b) and c) of Theorem 3.4.2.*

*f) The EKCF of the pencil* $s\mathcal{E}_2 - \mathcal{A}_2$ *fulfills the statements b) and c) of Theorem 3.4.2 with* $n$ *replaced by* $n + p$.

*Proof.* The equivalence of a) and b) has already been shown in Corollary 3.9.4. The equivalence of b) and c) follows from

$$\begin{aligned}\Phi(s) :&= \begin{bmatrix} G(s) \\ I_m \end{bmatrix}^{\sim} \begin{bmatrix} Q_s & S_s \\ S_s^{\mathsf{T}} & R_s \end{bmatrix} \begin{bmatrix} G(s) \\ I_m \end{bmatrix} \\ &= \begin{bmatrix} (sE-A)^{-1}B \\ I_m \end{bmatrix}^{\sim} \begin{bmatrix} Q & S \\ S^{\mathsf{T}} & R \end{bmatrix} \begin{bmatrix} (sE-A)^{-1}B \\ I_m \end{bmatrix} \in \mathbb{R}(s)^{m\times m},\end{aligned}$$

and an application of Lemma 3.4.1 with the fact that

$$\operatorname{In}\left(\begin{bmatrix} 0 & -\mathrm{i}\omega E + A \\ \mathrm{i}\omega E^{\mathsf{T}} + A^{\mathsf{T}} & Q \end{bmatrix}\right) = (n,0,n) \quad \forall \omega \in \mathbb{R} \text{ with } \mathrm{i}\omega \notin \Lambda(E,A).$$

Now we show that b) is equivalent to d): It holds

$$\begin{aligned}\Phi(s) :&= \begin{bmatrix} G(s) \\ I_m \end{bmatrix}^{\sim} \begin{bmatrix} Q_s & S_s \\ S_s^{\mathsf{T}} & R_s \end{bmatrix} \begin{bmatrix} G(s) \\ I_m \end{bmatrix} \\ &= \begin{bmatrix} C(sE-A)^{-1}B + D \\ (sE-A)^{-1}B \\ I_m \end{bmatrix}^{\sim} \begin{bmatrix} Q_s & 0 & S_s \\ 0 & 0 & 0 \\ S_s^{\mathsf{T}} & 0 & R_s \end{bmatrix} \begin{bmatrix} C(sE-A)^{-1}B + D \\ (sE-A)^{-1}B \\ I_m \end{bmatrix} \\ &= \begin{bmatrix} (sE_{\mathrm{e}}-A_{\mathrm{e}})^{-1}B_{\mathrm{e}} \\ I_m \end{bmatrix}^{\sim} \begin{bmatrix} Q_{\mathrm{e}} & S_{\mathrm{e}} \\ S_{\mathrm{e}}^{\mathsf{T}} & R_{\mathrm{e}} \end{bmatrix} \begin{bmatrix} (sE_{\mathrm{e}}-A_{\mathrm{e}})^{-1}B_{\mathrm{e}} \\ I_m \end{bmatrix}\end{aligned}$$

with

$$sE_{\mathrm{e}} - A_{\mathrm{e}} := \begin{bmatrix} 0 & sE-A \\ I_p & -C \end{bmatrix}, \quad B_{\mathrm{e}} = \begin{bmatrix} B \\ D \end{bmatrix}, \quad Q_{\mathrm{e}} = \begin{bmatrix} Q_s & 0 \\ 0 & 0 \end{bmatrix}, \quad S_{\mathrm{e}} = \begin{bmatrix} S_s \\ 0 \end{bmatrix}, \quad R_{\mathrm{e}} = R_s.$$

By using the same argumentation as for the proof of the equivalence of b) and c) we obtain the result.

The equivalence of the statements a)–d) and e) and f) in case that $(E, A, B)$ has no uncontrollable modes on the imaginary axis follows directly from Theorem 3.4.2 and the relation

$$\operatorname{rank}\begin{bmatrix} \lambda E_{\mathrm{e}} - A_{\mathrm{e}} & B_{\mathrm{e}} \end{bmatrix} = \operatorname{rank}\begin{bmatrix} 0 & \lambda E - A & B \\ I_p & -C & D \end{bmatrix} = p + \operatorname{rank}\begin{bmatrix} \lambda E - A & B \end{bmatrix} \quad \forall \lambda \in \mathbb{C},$$

which means that $(E, A, B) \in \Sigma_{n,m}$ has no uncontrollable modes on the imaginary axis if and only if $(E_{\mathrm{e}}, A_{\mathrm{e}}, B_{\mathrm{e}}) \in \Sigma_{n+p,m}$ has no uncontrollable modes on the imaginary axis. □

In the sequel we turn to energy considerations for differential-algebraic systems. Therefore, we need the following definitions.

**Definition 3.9.6** ((Virtual) available storage and required supply). [HM80] Let the system $(E, A, B, C, D) \in \Sigma_{n,m,m}$ be given, and for $Q_s = Q_s^\mathsf{T} \in \mathbb{R}^{p\times p}$, $S_s \in \mathbb{R}^{p\times m}$ and $R_s = R_s^\mathsf{T} \in \mathbb{R}^{m\times m}$, let $s(\cdot,\cdot)$ be a supply rate as in (3.101). Then

a) the *available storage* is given by $V_{\mathrm{as}} : \operatorname{im} E \to \mathbb{R}$ with

$$V_{\mathrm{as}}(Ex_0) = \sup \left\{ -\int_0^T s(u(\tau), y(\tau))\mathrm{d}\tau : \right. \\ \left. T \geq 0 \text{ and } (x, u, y) \in \mathfrak{B}_{(E,A,B,C,D)}(x_0) \right\}; \quad (3.108)$$

b) the *virtual available storage* is given by $V_{\mathrm{vas}} : \operatorname{im} E \to \mathbb{R}$ with

$$V_{\mathrm{vas}}(Ex_0) = \sup \left\{ -\int_0^\infty s(u(\tau), y(\tau))\mathrm{d}\tau : \right. \\ \left. (x, u, y) \in \mathfrak{B}_{(E,A,B,C,D)}(x_0) \text{ with } \lim_{t\to\infty} Ex(t) = 0 \right\}; \quad (3.109)$$

c) the *required supply* is given by $V_{\mathrm{rs}} : \operatorname{im} E \to \mathbb{R}$ with

$$V_{\mathrm{rs}}(Ex_0) = \inf \left\{ \int_{-\infty}^0 s(u(\tau), y(\tau))\mathrm{d}\tau : \right. \\ \left. (x, u, y) \in \mathfrak{B}_{(E,A,B,C,D)}(x_0) \text{ with } \lim_{t\to-\infty} Ex(t) = 0 \right\}. \quad (3.110)$$

*Remark* 3.9.7. The integral over the supply rate has the physical interpretation of the energy supplied to the system. Therefore, the available storage is the maximum amount of energy which can be extracted from the system initialized with $Ex(0) = Ex_0$. The virtual available storage is the maximum energy that can be extracted from stabilizing trajectories, i.e., those that fulfill $\lim_{t\to\infty} Ex(t) = 0$. On the other hand, the required supply is the minimum amount of energy that has to be put into the system to steer it from zero state to a final state with $Ex(0) = Ex_0$. Clearly, this amount is infinite, if $Ex_0$ cannot be reached from zero. Note that "lossless" means that no energy is dissipated for any input that controls from zero state.

With the above findings, it makes sense to define the following two terms.

**Definition 3.9.8** ((Virtual) storage function). [HM80] A function $V : \operatorname{im} E \to \mathbb{R}$ is called

a) a *storage function* if it is continuous, $0 \leq V(Ex_0) < \infty$ for all $x_0 \in \mathbb{R}^n$, $V(0) = 0$, and

$$\int_{t_0}^{t_1} s(u(\tau), y(\tau))\mathrm{d}\tau + V(Ex(t_0)) \geq V(Ex(t_1))$$
$$\forall (x, u, y) \in \mathfrak{B}_{(E,A,B,C,D)},\ t_0,\ t_1 \in \mathbb{R} \text{ with } t_0 \leq t_1; \quad (3.111)$$

b) a *virtual storage function* if it is continuous, $-\infty < V(Ex_0) < \infty$ for all $x_0 \in \mathbb{R}^n$, $V(0) = 0$, and (3.111) holds.

In the following we will give some corollaries that state equivalent characterizations of dissipativity and cyclo-dissipativity. This will also show that under certain assumptions, $V_{\mathrm{as}}$, $V_{\mathrm{vas}}$, and $V_{\mathrm{rs}}$ are (virtual) storage functions. These follow by a combination of Theorem 3.5.3, Theorem 3.5.4, Theorem 3.8.7, Theorem 3.8.13, Remark 3.9.3, and Corollary 3.9.4.

**Corollary 3.9.9.** *Let $(E, A, B, C, D) \in \Sigma_{n,m,p}$ with the system space $\mathcal{V}_{\mathrm{sys}}$ as in (3.5) be given and assume that $Q_s = Q_s^\top \in \mathbb{R}^{p\times p}$, $S_s \in \mathbb{R}^{p\times m}$, and $R_s = R_s^\top \in \mathbb{R}^{m\times m}$. Let the supply rate $s(\cdot,\cdot)$ be defined as in (3.101) and let the functions $V_{\mathrm{as}}$, $V_{\mathrm{vas}}$, and $V_{\mathrm{rs}}$ be defined as in (3.108), (3.109), and (3.110), respectively. Then the following statements are equivalent:*

*a) The system $(E, A, B, C, D)$ is strongly anti-stabilizable and there exists some $P \in \mathbb{R}^{n\times n}$ with $E^\top P \geq 0$ solving the descriptor dissipativity inequality (3.102).*

*b) The required supply fulfills $0 \leq V_{\mathrm{rs}}(Ex_0) < \infty$ for all $x_0 \in \mathbb{R}^n$.*

*If (one of) the conditions a), b) is fulfilled, then the following statements hold true:*

*i) The system $(E, A, B, C, D)$ is dissipative with respect to the supply rate $s(\cdot,\cdot)$.*

*ii) There exists some $P_+ \in \mathbb{R}^{n\times n}$ fulfilling (3.102) with $E^\top P_+ \geq 0$, such that for all $x_0 \in \mathbb{R}^n$ it holds that*

$$V_{\mathrm{rs}}(Ex_0) = x_0^\top E^\top P_+ x_0 \geq 0. \quad (3.112)$$

*iii) With $P_+$ as in (3.112) we have: For all solutions $Y \in \mathbb{R}^{n\times n}$ of the descriptor dissipativity inequality (3.102) it holds that $E^\top Y \leq E^\top P_+$.*

*iv) The transfer function of $(E, A, B, C, D)$ fulfills (3.106).*

*If (one of) the conditions a), b) is fulfilled and, moreover, $(E, A, B, C, D)$ is strongly stabilizable, then there exists some $P_- \in \mathbb{R}^{n\times n}$ fulfilling (3.102) such that for all $x_0 \in \mathbb{R}^n$ we obtain*

$$V_{\mathrm{vas}}(Ex_0) = x_0^\top E^\top P_- x_0. \quad (3.113)$$

*With $P_-$ as in (3.113) it holds: For all $Y \in \mathbb{R}^{n\times n}$ solving (3.102) we have $E^\mathsf{T} Y \geq E^\mathsf{T} P_-$. Moreover, the relation*

$$0 \leq V_{\mathrm{as}}(Ex_0) < \infty \quad \forall x_0 \in \mathbb{R}^n$$

*is fulfilled.*

**Corollary 3.9.10.** *Let $(E, A, B, C, D) \in \Sigma_{n,m,p}$ with the system space $\mathcal{V}_{\mathrm{sys}}$ as in (3.5) be given and assume that $Q_s = Q_s^\mathsf{T} \in \mathbb{R}^{p\times p}$, $S_s \in \mathbb{R}^{p\times m}$, and $R_s = R_s^\mathsf{T} \in \mathbb{R}^{m\times m}$. Let the supply rate $s(\cdot,\cdot)$ be defined as in (3.101) and let the functions $V_{\mathrm{vas}}$ and $V_{\mathrm{rs}}$ be defined as in (3.109) and (3.110), respectively.. Then the following statements are equivalent:*

*a) The system $(E, A, B, C, D)$ is strongly anti-stabilizable and there exists some $P \in \mathbb{R}^{n\times n}$ solving the descriptor dissipativity inequality (3.102).*

*b) The required supply fulfills $-\infty < V_{\mathrm{rs}}(Ex_0) < \infty$ for all $x_0 \in \mathbb{R}^n$.*

*If (one of) the conditions a), b) is fulfilled, then the following statements hold true:*

*i) $(E, A, B, C, D)$ is cyclo-dissipative with respect to the supply rate $s(\cdot,\cdot)$.*

*ii) There exists some $P_+ \in \mathbb{R}^{n\times n}$ fulfilling (3.102) such that for all $x_0 \in \mathbb{R}^n$ it holds that*

$$V_{\mathrm{rs}}(Ex_0) = x_0^\mathsf{T} E^\mathsf{T} P_+ x_0. \tag{3.114}$$

*iii) With $P_+$ as in (3.114) holds: For all solutions $Y \in \mathbb{R}^{n\times n}$ of the descriptor dissipativity inequality (3.102) it holds that $E^\mathsf{T} Y \leq E^\mathsf{T} P_+$.*

*iv) The transfer function of $(E, A, B, C, D)$ fulfills (3.104).*

*If (one of) the conditions a), b) is fulfilled and, moreover, $(E, A, B, C, D)$ is strongly stabilizable, then there exists some $P \in \mathbb{R}^{n\times n}$ fulfilling (3.102), such that for all $x_0 \in \mathbb{R}^n$ we obtain*

$$V_{\mathrm{vas}}(Ex_0) = x_0^\mathsf{T} E^\mathsf{T} P_- x_0. \tag{3.115}$$

*With $P_-$ as in (3.115) it holds: For all $Y \in \mathbb{R}^{n\times n}$ solving (3.102) we have $E^\mathsf{T} Y \geq E^\mathsf{T} P_-$.*

*Remark* 3.9.11.

a) The above two corollaries show that if $V_{\mathrm{as}}$ and $V_{\mathrm{rs}}$ exist, they are both storage functions in case of a dissipative system. On the other hand, if the functionals $V_{\mathrm{vas}}$ and $V_{\mathrm{rs}}$ exist, they are virtual storage functions in case of a cyclo-dissipative system.

b) Define

$$\mathcal{V}_{\text{sys},0} := \{x \in \mathbb{R}^n : Ax \in \operatorname{im} E\}.$$

If the index of $sE - A \in \mathbb{R}[s]^{n\times n}$ is at most one, $C^\top Q_s C \leq_{\mathcal{V}_{\text{sys},0}} 0$ and if $V(Ex_0) = x_0^\top E^\top P x_0$ is a quadratic storage function with $E^\top P >_{\mathcal{V}_{\text{sys},0}} 0$, it is simultaneously a *Lyapunov function*, that is $V(Ex_0) > 0$ for all $x_0 \in \mathcal{V}_{\text{sys},0}$, $V(0) = 0$, and $\frac{\mathrm{d}}{\mathrm{d}t}V(Ex(t)) \leq 0$ for all $x \in \mathfrak{B}_{(E,A)}$ [TSH01, p. 55]. The latter statement indeed follows from

$$\begin{aligned}\frac{\mathrm{d}}{\mathrm{d}t}V(Ex(t)) &= \dot{x}(t)^\top E^\top P x(t) + x(t)^\top P^\top E\dot{x}(t)\\ &= x(t)^\top (A^\top P + P^\top A)x(t)\\ &\leq x(t)^\top C^\top Q_s C x(t) \leq 0.\end{aligned}$$

Therefore, from the theory of Lyapunov functions [Sty02], every dissipative system that satisfies the above properties is automatically stable.

In the following two corollaries we consider the respective equivalences for the lossless notions of dissipativity and cyclo-dissipativity. These results follow from the previous two corollaries and an application of Theorem 3.8.11.

**Corollary 3.9.12.** *Let $(E, A, B, C, D) \in \Sigma_{n,m,p}$ with the system space $\mathcal{V}_{\text{sys}}$ as in* (3.5) *be strongly controllable and let $Q_s = Q_s^\top \in \mathbb{R}^{p\times p}$, $S_s \in \mathbb{R}^{p\times m}$, and $R_s = R_s^\top \in \mathbb{R}^{m\times m}$ be given. Let the supply rate $s(\cdot,\cdot)$ be defined as in* (3.101) *and let the functions $V_{\text{as}}$, $V_{\text{vas}}$, and $V_{\text{rs}}$ be defined as in* (3.108), (3.109), *and* (3.110), *respectively. Then the following statements are equivalent:*

*a) There exists a solution $P \in \mathbb{R}^{n\times n}$ of the descriptor dissipativity equation* (3.103) *with $E^\top P \geq 0$.*

*b) $(E, A, B, C, D)$ is lossless dissipative with respect to the supply rate $s(\cdot,\cdot)$.*

*If (one of) the conditions a), b) is fulfilled, then the following statements hold true:*

*i) With $P \in \mathbb{R}^{n\times n}$ as in a) it holds that*

$$V_{\text{vas}}(Ex_0) = V_{\text{as}}(Ex_0) = V_{\text{rs}}(Ex_0) = x_0^\top E^\top P x_0 \geq 0 \quad \forall x_0 \in \mathbb{R}^n.$$

*ii) The transfer function of $(E, A, B, C, D)$ fulfills* (3.106) *and* (3.105).

**Corollary 3.9.13.** *Let $(E, A, B, C, D) \in \Sigma_{n,m,p}$ with the system space $\mathcal{V}_{\text{sys}}$ as in* (3.5) *be strongly controllable and let $Q_s = Q_s^\top \in \mathbb{R}^{p\times p}$, $S_s \in \mathbb{R}^{p\times m}$, and $R_s = R_s^\top \in \mathbb{R}^{m\times m}$ be given. Let the supply rate $s(\cdot,\cdot)$ be defined as in* (3.101) *and let the functions $V_{\text{as}}$, $V_{\text{vas}}$, and $V_{\text{rs}}$ be defined as in* (3.108), (3.109), *and* (3.110), *respectively. Then the following statements are equivalent:*

*a) There exists a solution* $P \in \mathbb{R}^{n \times n}$ *of the descriptor dissipativity equation* (3.103).

*b)* $(E, A, B, C, D)$ *is lossless cyclo-dissipative with respect to the supply rate* $s(\cdot,\cdot)$.

*If (one of) the conditions a), b) is fulfilled, then the following statements hold true:*

*i) With* $P \in \mathbb{R}^{n \times n}$ *as in a) it holds that*

$$V_{\text{vas}}(Ex_0) = V_{\text{as}}(Ex_0) = V_{\text{rs}}(Ex_0) = x_0^\top E^\top P x_0 \quad \forall x_0 \in \mathbb{R}^n.$$

*ii) The transfer function of* $(E, A, B, C, D)$ *fulfills* (3.105).

Next we analyze whether the condition (3.106) on the transfer function is also sufficient for dissipativity. The following result is an immediate consequence of Theorem 3.7.4 (see also Remark 3.9.3).

**Corollary 3.9.14.** *Let the system* $(E, A, B, C, D) \in \Sigma_{n,m,p}$ *with the system space* $\mathcal{V}_{\text{sys}}$ *as in* (3.5) *and the transfer function* $G(s) \in \mathbb{R}(s)^{p \times m}$ *be given. Further assume that* $Q_s = Q_s^\top \in \mathbb{R}^{p \times p}$, $S_s \in \mathbb{R}^{p \times m}$, $R_s = R_s^\top \in \mathbb{R}^{m \times m}$. *Let the supply rate* $s(\cdot,\cdot)$ *be defined as in* (3.101). *Assume that* (3.106) *holds true. Further, suppose that at least one of the following two assumptions holds true:*

*a)* $(E, A, B, C, D)$ *is strongly stabilizable and there exists some* $\omega \in \mathbb{R}$ *with* $\mathrm{i}\omega \notin \Lambda(E, A)$ *such that*

$$\begin{bmatrix} G(\mathrm{i}\omega) \\ I_m \end{bmatrix}^{\mathsf{H}} \begin{bmatrix} Q_s & S_s \\ S_s^\top & R_s \end{bmatrix} \begin{bmatrix} G(\mathrm{i}\omega) \\ I_m \end{bmatrix} > 0.$$

*b)* $(E, A, B, C, D)$ *is strongly controllable.*

*Assume that there exist matrices* $K_1 \in \mathbb{R}^{m \times p}$, $K_2 \in \mathbb{R}^{p_2 \times p}$, $L_1 \in \mathbb{R}^{m \times m}$, $L_2 \in \mathbb{R}^{p_2 \times m}$ *with*

$$\begin{bmatrix} Q_s & S_s \\ S_s^\top & R_s \end{bmatrix} =_{\widetilde{\mathcal{V}}_{\text{sys}}} \begin{bmatrix} K_1^\top K_1 & K_1^\top L_1 \\ L_1^\top K_1 & L_1^\top L_1 \end{bmatrix} - \begin{bmatrix} K_2^\top K_2 & K_2^\top L_2 \\ L_2^\top K_2 & L_2^\top L_2 \end{bmatrix} \tag{3.116}$$

*and*

$$\widetilde{\mathcal{V}}_{\text{sys}} := \left\{ \begin{pmatrix} y \\ u \end{pmatrix} \in \mathbb{R}^{p+m} : y = Cx + Du \text{ with } \begin{pmatrix} x \\ u \end{pmatrix} \in \mathcal{V}_{\text{sys}} \right\}$$

*such that* $K_1 G(s) + L_1 \in \mathrm{Gl}_m(\mathbb{R}(s))$. *Then there exists a* $P \in \mathbb{R}^{n \times n}$ *with* $E^\top P \geq 0$ *that solves the descriptor dissipativity inequality* (3.102). *If, furthermore, the system*

$$\left( \begin{bmatrix} E & 0 \\ 0 & 0 \end{bmatrix}, \begin{bmatrix} A & B \\ K_1 C & K_1 D + L_1 \end{bmatrix}, \begin{bmatrix} 0 \\ -I_m \end{bmatrix}, \begin{bmatrix} K_2 C & K_2 D + L_2 \end{bmatrix} \right) \in \Sigma_{n+m,m,p_2} \tag{3.117}$$

*is strongly detectable, then all* $P \in \mathbb{R}^{n \times n}$ *solving the descriptor dissipativity inequality* (3.102) *fulfill* $E^\top P \geq 0$.

*Remark* 3.9.15. If the condition (3.117) is fulfilled and thus all solutions of the descriptor dissipativity inequality (3.102) are nonnegative, then even the virtual available storage is a (proper) storage function.

### 3.9.2 Contractive and Passive Systems

Important special cases of dissipative systems are so-called contractive and passive systems. These classes play a fundamental role for instance in the analysis and design of electrical circuits as well as mechanical systems [AV73, Smi02, Rei10, RS10a, RS10b, Rei14]. We will see in the following that, in the case where dissipativity corresponds to contractivity or passivity, the results from the above subsection simplify drastically.

**Definition 3.9.16** (Contractivity, passivity)**.** [AV73] A system $(E, A, B, C, D) \in \Sigma_{n,m,p}$ is called

a) *contractive*, if it is dissipative with respect to the supply rate $s(u, y) = \|u\|_2^2 - \|y\|_2^2$, i. e., $Q_s = -I_p$, $S_s = 0_{p\times m}$, and $R_s = I_m$.

b) *passive*, if it is dissipative with respect to the supply rate $s(u, y) = 2u^\mathsf{T} y$, i. e., $p = m$, $Q_s = R_s = 0_{m\times m}$, and $S_s = I_m$.

*Remark* 3.9.17.

a) In case of a contractive system, the differential-algebraic dissipativity inequality reads

$$\begin{bmatrix} A^\mathsf{T} P + P^\mathsf{T} A + C^\mathsf{T} C & P^\mathsf{T} B + C^\mathsf{T} D \\ B^\mathsf{T} P + D^\mathsf{T} C & D^\mathsf{T} D - I_m \end{bmatrix} \leq_{\mathcal{V}_{\mathrm{sys}}} 0, \quad E^\mathsf{T} P = P^\mathsf{T} E. \tag{3.118}$$

For systems governed by ordinary differential equations, solvability of this inequality with $E^\mathsf{T} P \geq 0$ is covered by the *bounded real lemma* [And66, AV73]. Note that equation (3.106) reduces to

$$I_m - G^\mathsf{H}(\lambda) G(\lambda) \geq 0 \quad \forall \lambda \in \overline{\mathbb{C}^+} \setminus \Lambda(E, A). \tag{3.119}$$

This property is called *bounded realness*. In particular, due to the special structure of the modified Popov function, (3.119) even holds for all $\lambda \in \overline{\mathbb{C}^+}$. This however is equivalent to $G \in \mathcal{RH}_\infty^{p\times m}$ with $\|G\|_{\mathcal{H}_\infty} \leq 1$. Furthermore, equation (3.105) is equivalent to $G(s)$ being inner, see Definition 2.2.17.

b) In case of a passive system, the associated descriptor dissipativity inequality is

$$\begin{bmatrix} A^\mathsf{T} P + P^\mathsf{T} A & P^\mathsf{T} B - C^\mathsf{T} \\ B^\mathsf{T} P - C & -D^\mathsf{T} - D \end{bmatrix} \leq_{\mathcal{V}_{\mathrm{sys}}} 0, \quad E^\mathsf{T} P = P^\mathsf{T} E. \tag{3.120}$$

In the case of ordinary differential equations, the solvability of this equation with $E^\mathsf{T} P \geq 0$ is subject of the *positive real lemma* [AV73]. The equation (3.106) reduces to

$$G^\mathsf{H}(\lambda) + G(\lambda) \geq 0 \quad \forall \lambda \in \overline{\mathbb{C}^+} \setminus \Lambda(E, A). \tag{3.121}$$

This property is called *positive realness* [AV73]. Furthermore, for every positive real function $G(s) \in \mathbb{R}(s)^{m\times m}$, it holds that $\mathfrak{P}(G) \subset \overline{\mathbb{C}^-}$. To see this, we make use of the pole-residue representation

$$G(s) = \sum_{k=1}^{\ell} \frac{R_k}{(s-\lambda_k)^{p_k}} + G_{\text{poly}}(s)$$

with the poles $\lambda_k$ of multiplicity $p_k$ and the associated residues $R_k \in \mathbb{C}^{m\times m}$ for $k = 1, \ldots, \ell$. Assume that there exists a pole $\lambda_j$ with positive real part. Then for $\lambda \in \mathbb{C}^+ \setminus \{\lambda_j\}$ close to $\lambda_j$ we have the estimate

$$\begin{aligned} G(\lambda) + G^{\mathsf{H}}(\lambda) &\approx \frac{R_j}{(\lambda-\lambda_j)^{p_j}} + \frac{R_j^{\mathsf{H}}}{(\overline{\lambda}-\overline{\lambda}_j)^{p_j}} \\ &= \frac{R_j(\overline{\lambda}-\overline{\lambda}_j)^{p_j} + R_j^{\mathsf{H}}(\lambda-\lambda_j)^{p_j}}{\left((\lambda-\lambda_j)(\overline{\lambda}-\overline{\lambda}_j)\right)^{p_j}} \end{aligned} \tag{3.122}$$

The denominator of (3.122) is always positive, however, the numerator changes definiteness in a neighborhood of $\lambda_j$. This is a contradiction to positive realness. Moreover, it can be shown that positive real functions can have poles on the imaginary axis with multiplicity at most one [AV73].

In the lossless case, the transfer function is positive real with, moreover,

$$G^{\mathsf{H}}(\mathrm{i}\omega) + G(\mathrm{i}\omega) = 0 \quad \forall \omega \in \mathbb{R} \text{ with } \mathrm{i}\omega \notin \Lambda(E,A). \tag{3.123}$$

We call the latter property *lossless positive realness*.

Now we first state a new version of the (lossless) bounded real lemma for DAEs.

**Theorem 3.9.18** (Bounded real lemma for DAEs)**.** *Let $(E,A,B,C,D) \in \Sigma_{n,m,p}$ with the system space $\mathcal{V}_{\text{sys}}$ as in (3.5) and the transfer function $G(s) \in \mathbb{R}(s)^{p\times m}$ be given.*

*a) If there exists some $P \in \mathbb{R}^{n\times n}$ with $E^{\mathsf{T}}P \geq 0$ fulfilling (3.118), then $G(s)$ is bounded real, or equivalently, $G \in \mathcal{RH}_\infty^{p\times m}$ and $\|G\|_{\mathcal{H}_\infty} \leq 1$.*

*b) Assume that $G(s)$ is bounded real, and at least one of the following properties holds true:*

*i) $(E,A,B,C,D)$ is strongly stabilizable and there exists some $\omega \in \mathbb{R}$ with $\mathrm{i}\omega \notin \Lambda(E,A)$ such that $\|G(\mathrm{i}\omega)\|_2 \neq 1$;*

*ii) $(E,A,B,C,D)$ is strongly controllable.*

*Then there exists some $P \in \mathbb{R}^{n\times n}$ with $E^{\mathsf{T}}P \geq 0$ fulfilling (3.118).*

*c) If, in addition to the assumptions in b), $(E, A, B, C, D)$ is strongly detectable, then all $P \in \mathbb{R}^{n \times n}$ satisfying (3.118) fulfill $E^\mathsf{T} P \geq 0$.*

*Proof.* The reason for statement a) has already been mentioned in Remark 3.9.3 b).

Now we prove b): The special structure of the supply rate (i. e., $Q_s = -I_p$, $S_s = 0_{p \times m}$ and $R_s = I_m$) implies the following facts:

1) It holds $\|G\|_{\mathcal{H}_\infty} \leq 1$ if and only if $I_m - G^\mathsf{H}(\lambda)G(\lambda) \geq 0$ for all $\lambda \in \overline{\mathbb{C}^+}$.

2) For $\omega \in \mathbb{R}$ with $\mathrm{i}\omega \notin \Lambda(E, A)$ we have $\|G(\mathrm{i}\omega)\|_2 < 1$ if and only if the relation $I_m - G^\mathsf{H}(\mathrm{i}\omega)G(\mathrm{i}\omega) > 0$ is satisfied.

3) Equation (3.116) holds true for $K_1 = 0_{m \times p}$, $K_2 = I_p$, $L_1 = I_m$, and $L_2 = 0_{p \times m}$.

Since $L_1 + K_1 G(s) = I_m$ is invertible, b) now follows immediately from Corollary 3.9.14.

To prove Statement c), it suffices by Corollary 3.9.14 to check that strong detectability of $(E, A, B, C, D)$ implies that the system (3.117) is strongly detectable. This however is a direct consequence of

$$\operatorname{rank} \begin{bmatrix} -\lambda E + A & B \\ K_1 C & K_1 D + L_1 \\ K_2 C & K_2 D + L_2 \end{bmatrix} = \operatorname{rank} \begin{bmatrix} -\lambda E + A & B \\ 0 & I_m \\ C & D \end{bmatrix} = m + \operatorname{rank} \begin{bmatrix} -\lambda E + A \\ C \end{bmatrix} \quad \forall \lambda \in \mathbb{C}.$$

□

**Theorem 3.9.19** (Lossless bounded real lemma for DAEs)**.** *Let $(E, A, B, C, D) \in \Sigma_{n,m,p}$ with the system space $\mathcal{V}_{\mathrm{sys}}$ as in (3.5) and the transfer function $G(s) \in \mathbb{R}(s)^{p \times m}$ be given.*

*a) If there exists some $P \in \mathbb{R}^{n \times n}$ with*

$$\begin{bmatrix} A^\mathsf{T} P + P^\mathsf{T} A + C^\mathsf{T} C & P^\mathsf{T} B + C^\mathsf{T} D \\ B^\mathsf{T} P + D^\mathsf{T} C & D^\mathsf{T} D - I_m \end{bmatrix} =_{\mathcal{V}_{\mathrm{sys}}} 0, \quad E^\mathsf{T} P \geq 0, \tag{3.124}$$

*then $G(s)$ is inner.*

*b) If $G(s)$ is inner and $(E, A, B, C, D) \in \Sigma_{n,m,p}$ is strongly controllable, then there exists some $P \in \mathbb{R}^{n \times n}$ fulfilling (3.124).*

*Proof.* The reason for statement a) is presented in Remark 3.9.3 b).

To prove b), assume that $G(s)$ is inner. Then Theorem 3.8.11 implies that there exists some $P \in \mathbb{R}^{n \times n}$ with $E^\mathsf{T} P = P^\mathsf{T} E$ and

$$\begin{bmatrix} A^\mathsf{T} P + P^\mathsf{T} A + C^\mathsf{T} C & P^\mathsf{T} B + C^\mathsf{T} D \\ B^\mathsf{T} P + D^\mathsf{T} C & D^\mathsf{T} D - I_m \end{bmatrix} =_{\mathcal{V}_{\mathrm{sys}}} 0.$$

Theorem 3.9.18 c) now implies that $E^\mathsf{T} P \geq 0$, hence the result is proven. □

In the following two theorems we finally present new differential-algebraic formulations of the positive real lemma and the lossless positive real lemma.

**Theorem 3.9.20** (Positive real lemma for DAEs). *Let $(E, A, B, C, D) \in \Sigma_{n,m,p}$ with the system space $\mathcal{V}_{\rm sys}$ as in (3.5) and the transfer function $G(s) \in \mathbb{R}(s)^{p\times m}$ be given.*

*a) If there exists some $P \in \mathbb{R}^{n\times n}$ fulfilling (3.120), then $G(s)$ is positive real, i. e., $G(\lambda) + G^{\mathsf{H}}(\lambda) \geq 0$ for all $\lambda \in \overline{\mathbb{C}^+}$.*

*b) Assume that $G(s)$ is positive real, and least one of the following properties holds true:*

*i) $(E, A, B, C, D)$ is strongly stabilizable and there exists some $\omega \in \mathbb{R}$ with $\mathrm{i}\omega \notin \Lambda(E, A)$ and $G^{\mathsf{H}}(\mathrm{i}\omega) + G(\mathrm{i}\omega) > 0$;*

*ii) $(E, A, B, C, D)$ is strongly controllable.*

*Then there exists some $P \in \mathbb{R}^{n\times n}$ fulfilling (3.120).*

*c) If, in addition to the assumptions in b), $(E, A, B, C, D) \in \Sigma_{n,m,p}$ is strongly detectable, then all $P \in \mathbb{R}^{n\times n}$ satisfying (3.120) fulfill $E^{\mathsf{T}} P \geq 0$.*

*Proof.* The reason for statement a) has already been mentioned in Remark 3.9.3 b).

Now we prove b): The special structure of the supply rate (i. e., $Q_s = R_s = 0_{m\times m}$, $S_s = I_m$) implies that (3.116) holds true for $K_1 = L_1 = K_2 = \frac{1}{\sqrt{2}} I_m$ and $L_2 = -\frac{1}{\sqrt{2}} I_m$. Then we obtain $L_1 + K_1 G(s) = \frac{1}{\sqrt{2}}(I_m + G(s))$. For $\lambda \in \overline{\mathbb{C}^+}$ and $u \in \mathbb{C}^m \setminus \{0\}$, we obtain

$$u^{\mathsf{H}}(I_m + G(\lambda))u = \|u\|_2^2 + \frac{1}{2} u^{\mathsf{H}}\big(G^{\mathsf{H}}(\lambda) + G(\lambda)\big)u \geq \|u\|_2^2 > 0,$$

hence $L_1 + K_1 G(s) = \frac{1}{\sqrt{2}}(I_m + G(s))$ is invertible. Assertion b) now follows from Corollary 3.9.14.

To prove Statement c) it suffices by Corollary 3.9.14 to check that strong detectability of $(E, A, B, C, D)$ implies that the system (3.117) is strongly detectable. However, this is a consequence of

$$\begin{aligned} \operatorname{rank} \begin{bmatrix} -\lambda E + A & B \\ K_1 C & K_1 D + L_1 \\ K_2 C & K_2 D + L_2 \end{bmatrix} &= \operatorname{rank} \begin{bmatrix} -\lambda E + A & B \\ \frac{1}{\sqrt{2}} C & \frac{1}{\sqrt{2}}(D + I_m) \\ \frac{1}{\sqrt{2}} C & \frac{1}{\sqrt{2}}(D - I_m) \end{bmatrix} \\ &= m + \operatorname{rank} \begin{bmatrix} -\lambda E + A \\ C \end{bmatrix} \quad \forall \lambda \in \mathbb{C}. \quad (3.125) \end{aligned}$$

□

**Theorem 3.9.21** (Lossless positive real lemma for DAEs). *Let $(E, A, B, C, D) \in \Sigma_{n,m,p}$ with the system space $\mathcal{V}_{\mathrm{sys}}$ as in* (3.5) *and the transfer function $G(s) \in \mathbb{R}(s)^{p\times m}$ be given.*

*a) If there exists some $P \in \mathbb{R}^{n\times n}$ with*

$$\begin{bmatrix} A^{\mathsf{T}}P + P^{\mathsf{T}}A & P^{\mathsf{T}}B - C^{\mathsf{T}} \\ B^{\mathsf{T}}P - C & -D^{\mathsf{T}} - D \end{bmatrix} =_{\mathcal{V}_{\mathrm{sys}}} 0, \quad E^{\mathsf{T}}P \geq 0, \tag{3.126}$$

*then $G(s)$ is lossless positive real, i. e., positive real with $G(\mathrm{i}\omega) + G^{\mathsf{H}}(\mathrm{i}\omega) = 0$ for all $\omega \in \mathbb{R}$ with $\mathrm{i}\omega \notin \Lambda(E, A)$.*

*b) If $G(s)$ is lossless positive real and $(E, A, B, C, D)$ is strongly controllable, then there exists some $P \in \mathbb{R}^{n\times n}$ fulfilling* (3.126)

*Proof.* This result can be inferred from Theorem 3.8.11 and Theorem 3.9.20 by using an argumentation analogous to that in the proof of Theorem 3.9.19. □

*Remark* 3.9.22 (Passivity, contractivity, positive realness, bounded realness).

a) If $(E, A, B, C, D) \in \Sigma_{n,m,p}$ is strongly controllable, then by Theorem 3.9.18 its transfer function is bounded real if and only if (3.118) has a solution $P \in \mathbb{R}^{n\times n}$ with $E^{\mathsf{T}}P \geq 0$. Corollary 3.9.9 implies that (3.118) has such a a solution if and only if $(E, A, B, C, D)$ is contractive.

b) Analogous argumentations to b) can be made to see that a strongly controllable system $(E, A, B, C, D) \in \Sigma_{n,m,p}$ is lossless contractive (passive, lossless passive) if and only if $G(s)$ is inner (positive real, lossless positive real).

c) We can conclude from Corollary 3.9.4 that an arbitrary system $(E, A, B, C, D) \in \Sigma_{n,m,p}$ is

   i) contractive if and only if $G \in \mathcal{RH}_\infty^{p\times m}$ and $\|G\|_{\mathcal{H}_\infty} \leq 1$;

   ii) lossless contractive if and only if $G(s)$ is inner;

   iii) passive if and only if $G(s)$ is positive real;

   iv) lossless passive if and only if $G(s)$ is lossless positive real.

## 3.10 Normalized Coprime Factorizations

In this part we show that the theory of descriptor Lur'e equations can be used to construct normalized coprime factorizations of transfer functions. These play an important role in $\mathcal{H}_\infty$-controller design [MG89, BB90a], system identification [dHSdCB95], the computation of the gap metric [Geo88], and model order reduction [Mey88, MRS11].

**Definition 3.10.1** (Coprime factorizations). Let $G(s) \in \mathbb{R}(s)^{p\times m}$ be given.

a) A *right coprime factorization* of $G(s) \in \mathbb{R}(s)^{p\times m}$ is given by

$$\begin{bmatrix} M_\mathrm{r} \\ N_\mathrm{r} \end{bmatrix} \in \mathcal{RH}_\infty^{m+p\times m}$$

with $M_\mathrm{r}(s) \in \mathrm{Gl}_m(\mathbb{R}(s))$ and $N_\mathrm{r}(s) \in \mathbb{R}(s)^{p\times m}$ such that $G(s) = N_\mathrm{r}(s)M_\mathrm{r}^{-1}(s)$, and there exist $Y_\mathrm{r} \in \mathcal{RH}_\infty^{m\times m}$ and $Z_\mathrm{r} \in \mathcal{RH}_\infty^{m\times p}$ that satisfy the *right Bézout identity*

$$Y_\mathrm{r}(s)M_\mathrm{r}(s) + Z_\mathrm{r}(s)N_\mathrm{r}(s) = I_m.$$

b) A *left coprime factorization* of $G(s) \in \mathbb{R}(s)^{p\times m}$ is given by

$$\begin{bmatrix} M_\mathrm{l} & N_\mathrm{l} \end{bmatrix} \in \mathcal{RH}_\infty^{p\times p+m}$$

with $M_\mathrm{l}(s) \in \mathrm{Gl}_p(\mathbb{R}(s))$ and $N_\mathrm{l}(s) \in \mathbb{R}(s)^{p\times m}$ such that $G(s) = M_\mathrm{l}^{-1}(s)N_\mathrm{l}(s)$, and there exist $Y_\mathrm{l} \in \mathcal{RH}_\infty^{p\times p}$ and $Z_\mathrm{l} \in \mathcal{RH}_\infty^{m\times p}$ that satisfy the *left Bézout identity*

$$M_\mathrm{l}(s)Y_\mathrm{l}(s) + N_\mathrm{l}(s)Z_\mathrm{l}(s) = I_p.$$

c) A *doubly coprime factorization* of $G(s) \in \mathbb{R}(s)^{p\times m}$ is given by a pair

$$\begin{bmatrix} M_\mathrm{r} \\ N_\mathrm{r} \end{bmatrix} \in \mathcal{RH}_\infty^{m+p\times m},$$
$$\begin{bmatrix} M_\mathrm{l} & N_\mathrm{l} \end{bmatrix} \in \mathcal{RH}_\infty^{p\times p+m},$$

such that $G(s) = N_\mathrm{r}(s)M_\mathrm{r}^{-1}(s) = M_\mathrm{l}^{-1}(s)N_\mathrm{l}(s)$ and there exist $Y_\mathrm{r} \in \mathcal{RH}_\infty^{m\times m}$, $Z_\mathrm{r} \in \mathcal{RH}_\infty^{m\times p}$, $Y_\mathrm{l} \in \mathcal{RH}_\infty^{p\times p}$, and $Z_\mathrm{l} \in \mathcal{RH}_\infty^{m\times p}$ that satisfy the *double Bézout identity*

$$\begin{bmatrix} Y_\mathrm{r}(s) & Z_\mathrm{r}(s) \\ -N_\mathrm{l}(s) & M_\mathrm{l}(s) \end{bmatrix} \begin{bmatrix} M_\mathrm{r}(s) & -Z_\mathrm{l}(s) \\ N_\mathrm{r}(s) & Y_\mathrm{l}(s) \end{bmatrix} = I_{m+p}.$$

d) A right coprime factorization is called *normalized*, if additionally

$$M_\mathrm{r}^\sim(s)M_\mathrm{r}(s) + N_\mathrm{r}^\sim(s)N_\mathrm{r}(s) = I_m \tag{3.127}$$

holds true.

e) A left coprime factorization is called *normalized*, if additionally

$$M_\mathrm{l}(s)M_\mathrm{l}^\sim(s) + N_\mathrm{l}(s)N_\mathrm{l}^\sim(s) = I_p \tag{3.128}$$

holds true.

f) A doubly coprime factorization is called *normalized*, if additionally

$$\begin{bmatrix} M_{\mathrm{r}}^{\sim}(s) & N_{\mathrm{r}}^{\sim}(s) \\ -N_{\mathrm{l}}(s) & M_{\mathrm{l}}(s) \end{bmatrix} \begin{bmatrix} M_{\mathrm{r}}(s) & -N_{\mathrm{l}}^{\sim}(s) \\ N_{\mathrm{r}}(s) & M_{\mathrm{l}}^{\sim}(s) \end{bmatrix} = I_{m+p}$$

holds true.

*Remark* 3.10.2. Property (3.127) is equivalent to

$$\begin{bmatrix} M_{\mathrm{r}}(s) \\ N_{\mathrm{r}}(s) \end{bmatrix}$$

being inner. Moreover, relation (3.128) is equivalent to

$$\begin{bmatrix} M_{\mathrm{l}}(s) & N_{\mathrm{l}}(s) \end{bmatrix}$$

begin co-inner, equivalently $\begin{bmatrix} M_{\mathrm{l}}^{\mathsf{H}}(\overline{s}) \\ N_{\mathrm{l}}^{\mathsf{H}}(\overline{s}) \end{bmatrix}$ is inner.

We will construct normalized coprime factorizations on the basis of the following descriptor Lur'e equations:

$$\begin{bmatrix} A^{\mathsf{T}} X_{\mathrm{r}} + X_{\mathrm{r}}^{\mathsf{T}} A + C^{\mathsf{T}} C & X_{\mathrm{r}}^{\mathsf{T}} B + C^{\mathsf{T}} D \\ B^{\mathsf{T}} X_{\mathrm{r}} + D^{\mathsf{T}} C & I_m + D^{\mathsf{T}} D \end{bmatrix} =_{\mathcal{V}_{\mathrm{sys,r}}} \begin{bmatrix} K_{\mathrm{r}}^{\mathsf{T}} \\ L_{\mathrm{r}}^{\mathsf{T}} \end{bmatrix} \begin{bmatrix} K_{\mathrm{r}} & L_{\mathrm{r}} \end{bmatrix}, \quad E^{\mathsf{T}} X_{\mathrm{r}} = X_{\mathrm{r}}^{\mathsf{T}} E, \tag{3.129}$$

where

$$\mathcal{V}_{\mathrm{sys,r}} := \left\{ \begin{pmatrix} x \\ u \end{pmatrix} \in \mathbb{R}^{n+m} : Ax + Bu \in \operatorname{im} E \right\},$$

and

$$\begin{bmatrix} A X_{\mathrm{l}} + X_{\mathrm{l}}^{\mathsf{T}} A^{\mathsf{T}} + B B^{\mathsf{T}} & X_{\mathrm{l}}^{\mathsf{T}} C^{\mathsf{T}} + B D^{\mathsf{T}} \\ C X_{\mathrm{l}} + D B^{\mathsf{T}} & I_p + D D^{\mathsf{T}} \end{bmatrix} =_{\mathcal{V}_{\mathrm{sys,l}}} \begin{bmatrix} K_{\mathrm{l}}^{\mathsf{T}} \\ L_{\mathrm{l}}^{\mathsf{T}} \end{bmatrix} \begin{bmatrix} K_{\mathrm{l}} & L_{\mathrm{l}} \end{bmatrix}, \quad E X_{\mathrm{l}} = X_{\mathrm{l}}^{\mathsf{T}} E^{\mathsf{T}}, \tag{3.130}$$

where

$$\mathcal{V}_{\mathrm{sys,l}} := \left\{ \begin{pmatrix} x \\ u \end{pmatrix} \in \mathbb{R}^{n+m} : A^{\mathsf{T}} x + C^{\mathsf{T}} u \in \operatorname{im} E^{\mathsf{T}} \right\}.$$

Before the normalized coprime factorizations will be constructed, we will analyze the eigenstructure of the even matrix pencils associated to (3.129) and (3.130). We will in particular show that the EKCFs of these matrix pencils do not contain any "critical blocks".

**Lemma 3.10.3.** *The following two statements hold true:*

a) *Assume that* $(E, A, B, C, D) \in \Sigma_{n,m,p}$ *is strongly stabilizable and let* $r = \operatorname{rank} E$. *Then the EKCF of the pencil*

$$s\mathcal{E}_{\mathrm{r}} - \mathcal{A}_{\mathrm{r}} := \begin{bmatrix} 0 & -sE + A & B \\ sE^\top + A^\top & C^\top C & C^\top D \\ B^\top & D^\top C & I_m + D^\top D \end{bmatrix} \in \mathbb{R}[s]^{2n+m\times 2n+m}$$

*has the following properties:*

i) *There do neither exist any blocks of type E2 nor blocks of type E4.*

ii) *There exist* $m + 2(n-r)$ *blocks of type E3. These are of size* $1 \times 1$; $n - r$ *of these blocks have negative sign-characteristic; the remaining* $m + n - r$ *blocks have positive sign-characteristic.*

b) *Assume that* $(E, A, B, C, D) \in \Sigma_{n,m,p}$ *is strongly detectable and let* $r = \operatorname{rank} E$. *Then the EKCF of the pencil*

$$s\mathcal{E}_{\mathrm{l}} - \mathcal{A}_{\mathrm{l}} = \begin{bmatrix} 0 & -sE^\top + A^\top & C^\top \\ sE + A & BB^\top & BD^\top \\ C & DB^\top & I_p + DD^\top \end{bmatrix} \in \mathbb{R}[s]^{2n+p\times 2n+p}$$

*has the following properties:*

i) *There do neither exist any blocks of type E2 nor blocks of type E4.*

ii) *There exist* $p + 2(n-r)$ *blocks of type E3. These are of size* $1 \times 1$; $n - r$ *of these blocks have negative sign-characteristic; the remaining* $p + n - r$ *blocks have positive sign-characteristic.*

*Proof.* We will only show statement a), since b) follows from a) by turning to the dual system $\left(E^\top, A^\top, C^\top, B^\top, D^\top\right) \in \Sigma_{n,p,m}$.

First we prove statement i). This assertion will follow from the fact that $\mathrm{i}\omega\mathcal{E}_{\mathrm{r}} - \mathcal{A}_{\mathrm{r}} \in \mathrm{Gl}_{2n+m}(\mathbb{C})$ for all $\omega \in \mathbb{R}$. The latter statement will be proven in the following:

Assume that $\omega \in \mathbb{R}$ and let $\mu$, $x \in \mathbb{C}^n$ and $u \in \mathbb{C}^m$ be such that

$$\begin{bmatrix} 0 & -\mathrm{i}\omega E + A & B \\ \mathrm{i}\omega E^\top + A^\top & C^\top C & C^\top D \\ B^\top & D^\top C & I_m + D^\top D \end{bmatrix} \begin{pmatrix} \mu \\ x \\ u \end{pmatrix} = 0. \tag{3.131}$$

In particular, we have $(-\mathrm{i}\omega E + A)x + Bu = 0$. A multiplication of (3.131) from the left with $\begin{pmatrix} 0_{1\times n} & x^{\mathsf{H}} & u^{\mathsf{H}} \end{pmatrix}$ yields

$$0 = \|Cx + Du\|_2^2 + \|u\|_2^2 + ((-\mathrm{i}\omega E + A)x + Bu)^{\mathsf{H}}\mu = \|Cx + Du\|_2^2 + \|u\|_2^2.$$

We can now conclude that $u = 0$ and $Cx = 0$. Then (3.131) gives rise to

$$\begin{bmatrix} \mathrm{i}\omega E^\top + A^\top \\ B^\top \end{bmatrix} \mu = 0.$$

By the assumption of strong stabilizability, we can use Proposition 2.2.6 to see that $\mu = 0$. Altogether, we have $\ker(\mathrm{i}\omega\mathcal{E}_\mathrm{r} - \mathcal{A}_\mathrm{r}) = \{0\}$, hence $\mathrm{i}\omega\mathcal{E}_\mathrm{r} - \mathcal{A}_\mathrm{r} \in \mathrm{Gl}_{2n+m}(\mathbb{C})$.

Now we show ii). The Popov function associated to the descriptor Lur'e equation (3.129) reads

$$\Phi(s) = I_m + G^{\sim}(s)G(s). \tag{3.132}$$

In particular, we have $\Phi(\mathrm{i}\omega) > 0$ for all $\omega \in \mathbb{R}$ with $\mathrm{i}\omega \notin \Lambda(E, A)$. Using Theorem 3.4.2, it suffices to show that all blocks of type E3 in the EKCF of $s\mathcal{E}_\mathrm{r} - \mathcal{A}_\mathrm{r}$ are of size $1 \times 1$. This is proven in the following:

Since $(E, A, B)$ is impulse controllable, there exist $W, T \in \mathrm{Gl}_n(\mathbb{R})$ and $F \in \mathbb{R}^{m \times n}$ such that

$$sE_F - A_F = W(sE - (A + BF))T = \begin{bmatrix} sI_r - A_{11} & 0 \\ 0 & -I_{n-r} \end{bmatrix}, \quad B_F = WB = \begin{bmatrix} B_1 \\ B_2 \end{bmatrix},$$
$$C_F = CT = \begin{bmatrix} C_1 & C_2 \end{bmatrix}, \quad D_F = D.$$

Following the argumentation of Remark 3.4.3, it remains to show that the EKCF of the pencil

$$s\widehat{\mathcal{E}}_\mathrm{r} - \widehat{\mathcal{A}}_\mathrm{r} :=$$
$$\begin{bmatrix} 0 & -sI_r + A_{11} & B_1 \\ sI_r + A_{11}^\top & C_1^\top C_1 & -C_1^\top C_2 B_2 + C_1^\top D \\ B_1^\top & -B_2^\top C_2^\top C_1 + D^\top C_1 & B_2^\top C_2^\top C_2 B_2 - B_2^\top C_2^\top D - D^\top C_2 B_2 + I_m + D^\top D \end{bmatrix}$$

has exactly $m$ blocks of type E3 of size $1 \times 1$. However, this fact follows from $B_2^\top C_2^\top C_2 B_2 - B_2^\top C_2^\top D - D^\top C_2 B_2 + I_m + D^\top D = I_m + (D - C_2B_2)^\top (D - C_2B_2) > 0$, which implies that the pencil $s\widehat{\mathcal{E}}_\mathrm{r} - \widehat{\mathcal{A}}_\mathrm{r}$ is regular and of index at most one.

□

**Lemma 3.10.4.** *The following two statements hold true:*

*a) Assume that $(E, A, B, C, D) \in \Sigma_{n,m,p}$ is strongly stabilizable. Then a stabilizing solution $(X_\mathrm{r}, K_\mathrm{r}, L_\mathrm{r})$ of the descriptor Lur'e equation (3.129) exists. Further, the Rosenbrock pencil $R_\mathrm{r}(s)$ of $(E, A, B, K_\mathrm{r}, L_\mathrm{r}) \in \Sigma_{n,m,m}$ is regular with $R_\mathrm{r}^{-1} \in \mathcal{RH}_\infty^{n+m \times n+m}$.*

*b) Assume that $(E, A, B, C, D) \in \Sigma_{n,m,p}$ is strongly detectable. Then a stabilizing solution $(X_\mathrm{l}, K_\mathrm{l}, L_\mathrm{l})$ of the descriptor Lur'e equation (3.130) exists. Further, the Rosenbrock pencil $R_\mathrm{l}(s)$ of $(E^\top, A^\top, C^\top, K_\mathrm{l}, L_\mathrm{l}) \in \Sigma_{n,p,p}$ is regular with $R_\mathrm{l}^{-1} \in \mathcal{RH}_\infty^{n+p \times n+p}$.*

*Proof.* It suffices to prove statement a) since b) follows from a) by turning to the dual system $(E^\top, A^\top, C^\top, B^\top, D^\top) \in \Sigma_{n,p,m}$. Assume that the system $(E, A, B, C, D)$ is strongly stabilizable. Furthermore, $P = 0$ solves the descriptor KYP inequality associated to (3.129). The existence of a stabilizing solution $(X_\mathrm{r}, K_\mathrm{r}, L_\mathrm{r})$ then follows from Theorem 3.5.3.

By (3.52) and the fact that, by Lemma 3.10.3, the associated even matrix pencil $s\mathcal{E}_\mathrm{r} - \mathcal{A}_\mathrm{r}$ is regular, does not have any imaginary eigenvalues and its index is bounded from above by one, we obtain that $R_\mathrm{r}(s)$ is as well regular, it does not have any eigenvalues on $\mathrm{i}\mathbb{R}$, and its index does not exceed one. This means that $\lambda \mapsto R_\mathrm{r}(\lambda)^{-1}$ is bounded on $\mathrm{i}\mathbb{R}$. On the other hand, since $(X_\mathrm{r}, K_\mathrm{r}, L_\mathrm{r})$ is a stabilizing solution, $R_\mathrm{r}^{-1}(s)$ further does not have any poles in $\mathbb{C}^+$. Altogether, we obtain that $R_\mathrm{r}^{-1} \in \mathcal{RH}_\infty^{n+m\times n+m}$. □

**Theorem 3.10.5.** *Let $(E, A, B, C, D) \in \Sigma_{n,m,p}$ be given and let $G(s) \in \mathbb{R}(s)^{p\times m}$ be its transfer function. Then the following statements hold true:*

*a) Assume that $(E, A, B, C, D)$ is strongly stabilizable. Let $(X_\mathrm{r}, K_\mathrm{r}, L_\mathrm{r})$ be a stabilizing solution of the descriptor Lur'e equation* (3.129). *Then a normalized right coprime factorization $\begin{bmatrix} M_\mathrm{r}(s) \\ N_\mathrm{r}(s) \end{bmatrix} \in \mathbb{R}(s)^{m+p\times m}$ of $G(s)$ is realized by*

$$(E_\mathrm{r}, A_\mathrm{r}, B_\mathrm{r}, C_\mathrm{r}) = \left( \begin{bmatrix} E & 0 \\ 0 & 0 \end{bmatrix}, \begin{bmatrix} A & B \\ K_\mathrm{r} & L_\mathrm{r} \end{bmatrix}, \begin{bmatrix} 0 \\ -I_m \end{bmatrix}, \begin{bmatrix} 0 & I_m \\ C & D \end{bmatrix} \right) \in \Sigma_{n+m,m,p+m}. \tag{3.133}$$

*b) Assume that $(E, A, B, C, D)$ is strongly detectable. Let $(X_\mathrm{l}, K_\mathrm{l}, L_\mathrm{l})$ be a stabilizing solution of the descriptor Lur'e equation* (3.130)*. Then a normalized left coprime factorization $\begin{bmatrix} M_\mathrm{l}(s) & N_\mathrm{l}(s) \end{bmatrix} \in \mathbb{R}(s)^{p\times p+m}$ of $G(s)$ is realized by*

$$(E_\mathrm{l}, A_\mathrm{l}, B_\mathrm{l}, C_\mathrm{l}) = \left( \begin{bmatrix} E & 0 \\ 0 & 0 \end{bmatrix}, \begin{bmatrix} A & K_\mathrm{l}^\mathsf{T} \\ C & L_\mathrm{l}^\mathsf{T} \end{bmatrix}, \begin{bmatrix} 0 & B \\ I_p & D \end{bmatrix}, \begin{bmatrix} 0 & -I_p \end{bmatrix} \right) \in \Sigma_{n+p,p+m,p}. \tag{3.134}$$

*c) Assume that $(E, A, B, C, D)$ is strongly stabilizable and strongly detectable. Then a normalized doubly coprime factorization is given by the normalized left and right coprime factorizations in* (3.134) *and* (3.133).

*Proof.* First we prove statement a): Lemma 3.10.4 implies that $M_\mathrm{r} \in \mathcal{RH}_\infty^{m\times m}$ and $N_\mathrm{r} \in \mathcal{RH}_\infty^{p\times m}$. By Lemma 3.7.3 we obtain

$$M_\mathrm{r}(s) = \begin{bmatrix} 0 & I_m \end{bmatrix} \begin{bmatrix} sE - A & -B \\ -K_\mathrm{r} & -L_\mathrm{r} \end{bmatrix}^{-1} \begin{bmatrix} 0 \\ -I_m \end{bmatrix} = \left(L_\mathrm{r} + K_\mathrm{r}(sE - A)^{-1}B\right)^{-1},$$

and

$$N_\mathrm{r}(s) = \begin{bmatrix} C & D \end{bmatrix} \begin{bmatrix} sE - A & -B \\ -K_\mathrm{r} & -L_\mathrm{r} \end{bmatrix}^{-1} \begin{bmatrix} 0 \\ -I_m \end{bmatrix} = G(s) \cdot \left(L_\mathrm{r} + K_\mathrm{r}(sE - A)^{-1}B\right)^{-1}.$$

Thus we obtain

$$N_{\mathrm{r}}(s)M_{\mathrm{r}}^{-1}(s) = G(s).$$

Since by Lemma 3.10.4, $sE_{\mathrm{r}} - A_{\mathrm{r}}$ is regular and of index at most one, we have $L_{\mathrm{r}} \in \mathrm{Gl}_m(\mathbb{R})$. By using the Schur complement, we obtain another realization for $\begin{bmatrix} M_{\mathrm{r}}(s) \\ N_{\mathrm{r}}(s) \end{bmatrix}$ given by

$$\begin{aligned}(\widetilde{E}_{\mathrm{r}}, \widetilde{A}_{\mathrm{r}}, \widetilde{B}_{\mathrm{r}}, \widetilde{C}_{\mathrm{r}}, \widetilde{D}_{\mathrm{r}}) \\ := \left(E, A - BL_{\mathrm{r}}^{-1}K_{\mathrm{r}}, BL_{\mathrm{r}}^{-1}, \begin{bmatrix} -L_{\mathrm{r}}^{-1}K_{\mathrm{r}} \\ C - DL_{\mathrm{r}}^{-1}K_{\mathrm{r}} \end{bmatrix}, \begin{bmatrix} L_{\mathrm{r}}^{-1} \\ DL_{\mathrm{r}}^{-1} \end{bmatrix}\right) \in \Sigma_{n,m,m+p}.\end{aligned}$$

Since $s\widetilde{E}_{\mathrm{r}} - \widetilde{A}_{\mathrm{r}}$ is regular and of index at most one and all its eigenvalues are in $\mathbb{C}^-$, we see from [WB89] (see also [Var98, Lem. 1]) that $\begin{bmatrix} M_{\mathrm{r}}(s) \\ N_{\mathrm{r}}(s) \end{bmatrix}$ is indeed a right coprime factorization.

It remains to show that $\begin{bmatrix} M_{\mathrm{r}}(s) \\ N_{\mathrm{r}}(s) \end{bmatrix}$ is inner, i. e., there exists a solution to the descriptor dissipativity equality

$$\begin{bmatrix} A_{\mathrm{r}}^\top P_{\mathrm{r}} + P_{\mathrm{r}}^\top A_{\mathrm{r}} + C_{\mathrm{r}}^\top C_{\mathrm{r}} & P_{\mathrm{r}}^\top B_{\mathrm{r}} \\ B_{\mathrm{r}}^\top P_{\mathrm{r}} & -I_m \end{bmatrix} =_{\widetilde{\mathcal{V}}_{\mathrm{sys,r}}} 0, \quad E_{\mathrm{r}}^\top P_{\mathrm{r}} \geq 0$$

with the subspace

$$\widetilde{\mathcal{V}}_{\mathrm{sys,r}} := \left\{ \begin{pmatrix} x_1 \\ x_2 \\ u \end{pmatrix} \in \mathbb{R}^{n+2m} : \begin{pmatrix} x_1 \\ x_2 \end{pmatrix} \in \mathcal{V}_{\mathrm{sys,r}} \text{ and } K_{\mathrm{r}}x_1 + L_{\mathrm{r}}x_2 = u \right\}.$$

Indeed, for

$$\begin{pmatrix} x_1 \\ x_2 \\ u \end{pmatrix} \in \widetilde{\mathcal{V}}_{\mathrm{sys,r}} \quad \text{and} \quad P_{\mathrm{r}} = \begin{bmatrix} X_{\mathrm{r}} & 0 \\ 0 & 0 \end{bmatrix}$$

we obtain

$$\begin{pmatrix} x_1 \\ x_2 \\ u \end{pmatrix}^{\mathsf{T}} \begin{bmatrix} \begin{bmatrix} A & B \\ K_{\mathrm{r}} & L_{\mathrm{r}} \end{bmatrix}^{\mathsf{T}} \begin{bmatrix} X_{\mathrm{r}} & 0 \\ 0 & 0 \end{bmatrix} + \begin{bmatrix} X_{\mathrm{r}} & 0 \\ 0 & 0 \end{bmatrix}^{\mathsf{T}} \begin{bmatrix} A & B \\ K_{\mathrm{r}} & L_{\mathrm{r}} \end{bmatrix} + \begin{bmatrix} 0 & I_m \\ C & D \end{bmatrix}^{\mathsf{T}} \begin{bmatrix} 0 & I_m \\ C & D \end{bmatrix} & \begin{bmatrix} X_{\mathrm{r}} & 0 \\ 0 & 0 \end{bmatrix}^{\mathsf{T}} \begin{bmatrix} 0 \\ -I_m \end{bmatrix} \\ \begin{bmatrix} 0 \\ -I_m \end{bmatrix}^{\mathsf{T}} \begin{bmatrix} X_{\mathrm{r}} & 0 \\ 0 & 0 \end{bmatrix} & -I_m \end{bmatrix} \begin{pmatrix} x_1 \\ x_2 \\ u \end{pmatrix}$$

$$= \begin{pmatrix} x_1 \\ x_2 \\ K_{\mathrm{r}}x_1 + L_{\mathrm{r}}x_2 \end{pmatrix}^{\mathsf{T}} \begin{bmatrix} A^{\mathsf{T}}X_{\mathrm{r}} + X_{\mathrm{r}}^{\mathsf{T}}A + C^{\mathsf{T}}C & X_{\mathrm{r}}^{\mathsf{T}}B + C^{\mathsf{T}}D & 0 \\ B^{\mathsf{T}}X_{\mathrm{r}} + D^{\mathsf{T}}C & I_m + D^{\mathsf{T}}D & 0 \\ 0 & 0 & -I_m \end{bmatrix} \cdot \begin{pmatrix} x_1 \\ x_2 \\ K_{\mathrm{r}}x_1 + L_{\mathrm{r}}x_2 \end{pmatrix}$$

$$= \begin{pmatrix} x_1 \\ x_2 \end{pmatrix}^{\mathsf{T}} \left( \begin{bmatrix} A^{\mathsf{T}}X_{\mathrm{r}} + X_{\mathrm{r}}^{\mathsf{T}}A + C^{\mathsf{T}}C & X_{\mathrm{r}}^{\mathsf{T}}B + C^{\mathsf{T}}D \\ B^{\mathsf{T}}X_{\mathrm{r}} + D^{\mathsf{T}}C & I_m + D^{\mathsf{T}}D \end{bmatrix} - \begin{bmatrix} K_{\mathrm{r}}^{\mathsf{T}} \\ L_{\mathrm{r}}^{\mathsf{T}} \end{bmatrix} \begin{bmatrix} K_{\mathrm{r}} & L_{\mathrm{r}} \end{bmatrix} \right) \begin{pmatrix} x_1 \\ x_2 \end{pmatrix}$$

$$= 0,$$

and

$$E_{\mathrm{r}}^{\mathsf{T}}P_{\mathrm{r}} = \begin{bmatrix} E^{\mathsf{T}}X_{\mathrm{r}} & 0 \\ 0 & 0 \end{bmatrix} \geq 0$$

which shows the claim.

Statement b) is proven in an analogous fashion. Lemma 3.10.4 implies that $M_{\mathrm{l}} \in \mathcal{RH}_{\infty}^{p\times p}$ and $N_{\mathrm{l}} \in \mathcal{RH}_{\infty}^{p\times m}$. By Lemma 3.7.3 we obtain

$$M_{\mathrm{l}}(s) = \begin{bmatrix} 0 & -I_p \end{bmatrix} \begin{bmatrix} sE - A & -K_{\mathrm{l}}^{\mathsf{T}} \\ -C & -L_{\mathrm{l}}^{\mathsf{T}} \end{bmatrix}^{-1} \begin{bmatrix} 0 \\ I_p \end{bmatrix} = \left( L_{\mathrm{l}}^{\mathsf{T}} + C(sE - A)^{-1}K_{\mathrm{l}}^{\mathsf{T}} \right)^{-1},$$

and

$$N_{\mathrm{l}}(s) = \begin{bmatrix} 0 & -I_p \end{bmatrix} \begin{bmatrix} sE - A & -K_{\mathrm{l}}^{\mathsf{T}} \\ -C & -L_{\mathrm{l}}^{\mathsf{T}} \end{bmatrix}^{-1} \begin{bmatrix} B \\ D \end{bmatrix} = \left( L_{\mathrm{l}}^{\mathsf{T}} + C(sE - A)^{-1}K_{\mathrm{l}}^{\mathsf{T}} \right)^{-1} \cdot G(s).$$

Thus we obtain

$$M_{\mathrm{l}}^{-1}(s)N_{\mathrm{l}}(s) = G(s).$$

From [WB89] it follows that $\begin{bmatrix} M_{\mathrm{l}}(s) & N_{\mathrm{l}}(s) \end{bmatrix}$ is a left coprime factorization.

Now we show that $\begin{bmatrix} M_{\mathrm{l}}(s) & N_{\mathrm{l}}(s) \end{bmatrix}$ is co-inner, i. e.,

$$\begin{bmatrix} M_{\mathrm{l}}^{\mathsf{H}}(\overline{s}) \\ N_{\mathrm{l}}^{\mathsf{H}}(\overline{s}) \end{bmatrix} = \begin{bmatrix} 0 & I_p \\ B^{\mathsf{T}} & D^{\mathsf{T}} \end{bmatrix} \begin{bmatrix} sE^{\mathsf{T}} - A^{\mathsf{T}} & -C^{\mathsf{T}} \\ -K_{\mathrm{l}} & -L_{\mathrm{l}} \end{bmatrix}^{-1} \begin{bmatrix} 0 \\ -I_p \end{bmatrix}$$

is inner. This rational function, however, has a similar structure as $\begin{bmatrix} M_\mathrm{r}(s) \\ N_\mathrm{r}(s) \end{bmatrix}$ (in fact the matrices $E$, $A$, $B$, $C$, and $D$ are replaced by those of the dual system). Therefore, we can use analogous arguments as in the proof of a) to show that $\begin{bmatrix} M_\mathrm{l}^\mathsf{H}(\overline{s}) \\ N_\mathrm{l}^\mathsf{H}(\overline{s}) \end{bmatrix}$ is inner.

Finally we show statement c). The doubly coprimeness follows from [WB89]. By construction it holds that

$$\begin{aligned} M_\mathrm{r}^\sim(s)M_\mathrm{r}(s) + N_\mathrm{r}^\sim(s)N_\mathrm{r}(s) &= I_m, \\ M_\mathrm{l}(s)M_\mathrm{l}^\sim(s) + N_\mathrm{l}(s)N_\mathrm{l}^\sim(s) &= I_p. \end{aligned}$$

Moreover, by construction we have

$$\begin{aligned} -N_\mathrm{l}(s)M_\mathrm{r}(s) + M_\mathrm{l}(s)N_\mathrm{r}(s) &= 0, \\ -M_\mathrm{r}^\sim(s)N_\mathrm{l}^\sim(s) + N_\mathrm{r}^\sim(s)M_\mathrm{l}^\sim(s) &= 0. \end{aligned}$$

This shows the result and finalizes the proof. □

**Example 3.10.6.** To illustrate the result of Theorem 3.10.5, we consider the matrices

$$E = \begin{bmatrix} 1 & 0 \\ 0 & 0 \end{bmatrix}, \quad A = \begin{bmatrix} 1 & 0 \\ 0 & -1 \end{bmatrix}, \quad B = \begin{bmatrix} 1 \\ 0 \end{bmatrix}, \quad C = \begin{bmatrix} 1 & 0 \end{bmatrix}, \quad D = 0.$$

The system $(E, A, B, C, D) \in \Sigma_{2,1,1}$ is strongly stabilizable with transfer function

$$G(s) = \begin{bmatrix} 1 & 0 \end{bmatrix} \begin{bmatrix} s-1 & 0 \\ 0 & 1 \end{bmatrix}^{-1} \begin{bmatrix} 1 \\ 0 \end{bmatrix} = \frac{1}{s-1}.$$

From $E^\mathsf{T} X_\mathrm{r} = X_\mathrm{r}^\mathsf{T} E$ we obtain

$$X_\mathrm{r} = \begin{bmatrix} x_{11} & 0 \\ x_{21} & x_{22} \end{bmatrix},$$

and the descriptor Lur'e equation (3.129) is

$$\begin{bmatrix} 2x_{11}+1 & -x_{21} & x_{11} \\ -x_{21} & -2x_{22} & 0 \\ x_{11} & 0 & 1 \end{bmatrix} =_{\mathcal{V}_{\mathrm{sys,r}}} \begin{bmatrix} k_1 \\ k_2 \\ \ell \end{bmatrix} \begin{bmatrix} k_1 & k_2 & \ell \end{bmatrix}, \tag{3.135}$$

where $K_\mathrm{r} = \begin{bmatrix} k_1 & k_2 \end{bmatrix}$ and $L_\mathrm{r} = \ell$. Then a straight-forward calculation shows that a stabilizing and maximal solution of (3.135) is given by

$$(X_\mathrm{r}, K_\mathrm{r}, L_\mathrm{r}) = \left( \begin{bmatrix} 1+\sqrt{2} & 0 \\ 0 & 0 \end{bmatrix}, \begin{bmatrix} 1+\sqrt{2} & 0 \end{bmatrix}, 1 \right).$$

Then Theorem 3.10.5 yields

$$M_{\mathrm{r}}(s) = \begin{bmatrix} 0 & 0 & 1 \end{bmatrix} \begin{bmatrix} s-1 & 0 & -1 \\ 0 & 1 & 0 \\ -1-\sqrt{2} & 0 & -1 \end{bmatrix}^{-1} \begin{bmatrix} 0 \\ 0 \\ -1 \end{bmatrix} = \frac{s-1}{s+\sqrt{2}} \in \mathcal{RH}_\infty^{1\times 1},$$

$$N_{\mathrm{r}}(s) = \begin{bmatrix} 1 & 0 & 0 \end{bmatrix} \begin{bmatrix} s-1 & 0 & -1 \\ 0 & 1 & 0 \\ -1-\sqrt{2} & 0 & -1 \end{bmatrix}^{-1} \begin{bmatrix} 0 \\ 0 \\ -1 \end{bmatrix} = \frac{1}{s+\sqrt{2}} \in \mathcal{RH}_\infty^{1\times 1}.$$

Obviously, we have $G(s) = N_{\mathrm{r}}(s)M_{\mathrm{r}}^{-1}(s)$. Moreover, a simple calculation shows $M_{\mathrm{r}}^{\sim}(s)M_{\mathrm{r}}(s) + N_{\mathrm{r}}(s)^{\sim}N_{\mathrm{r}}(s) = 1$, i.e., we have indeed calculated a right normalized coprime factorization. By a completely analogous computation, we further obtain $M_{\mathrm{l}}(s) = M_{\mathrm{r}}(s)$ and $N_{\mathrm{l}}(s) = N_{\mathrm{r}}(s)$ as the corresponding left normalized coprime factors.

*Remark* 3.10.7 (Normalized coprime factorizations).

a) If $E = I_n$, then the descriptor Lur'e equations (3.129) and (3.130) reduce to algebraic Riccati equations

$$A^{\mathsf{T}}X_{\mathrm{r}} + X_{\mathrm{r}}A + C^{\mathsf{T}}C - \left(X_{\mathrm{r}}B + C^{\mathsf{T}}D\right)\left(I_m + D^{\mathsf{T}}D\right)^{-1}\left(X_{\mathrm{r}}B + C^{\mathsf{T}}D\right)^{\mathsf{T}} = 0, \qquad X_{\mathrm{r}} = X_{\mathrm{r}}^{\mathsf{T}},$$

$$AX_{\mathrm{l}} + X_{\mathrm{l}}A^{\mathsf{T}} + BB^{\mathsf{T}} - \left(X_{\mathrm{l}}C^{\mathsf{T}} + BD^{\mathsf{T}}\right)\left(I_p + DD^{\mathsf{T}}\right)^{-1}\left(X_{\mathrm{l}}C^{\mathsf{T}} + BD^{\mathsf{T}}\right)^{\mathsf{T}} = 0, \qquad X_{\mathrm{l}} = X_{\mathrm{l}}^{\mathsf{T}}.$$

The matrices $K_{\mathrm{r}}$ and $L_{\mathrm{r}}$ can be chosen as $L_{\mathrm{r}} \in \mathrm{Gl}_m(\mathbb{R})$ with $L_{\mathrm{r}} = \left(I_m + D^{\mathsf{T}}D\right)^{1/2}$ and $K_{\mathrm{r}} = L_{\mathrm{r}}^{-\mathsf{T}}\left(B^{\mathsf{T}}X_{\mathrm{r}} + D^{\mathsf{T}}C\right)$. In this case we can see that the realization of $\begin{bmatrix} M_{\mathrm{r}}(s) \\ N_{\mathrm{r}}(s) \end{bmatrix}$ in Theorem 3.10.5 reduces to

$$\left(I_n, A - BL_{\mathrm{r}}^{-1}K_{\mathrm{r}}, BL_{\mathrm{r}}^{-1}, \begin{bmatrix} -L_{\mathrm{r}}^{-1}K_{\mathrm{r}} \\ C - DL_{\mathrm{r}}^{-1}K_{\mathrm{r}} \end{bmatrix}, \begin{bmatrix} L_{\mathrm{r}}^{-1} \\ DL_{\mathrm{r}}^{-1} \end{bmatrix}\right) \in \Sigma_{n,m,m+p}.$$

Similarly, the matrices $K_{\mathrm{l}}$ and $L_{\mathrm{l}}$ can be chosen such that $L_{\mathrm{l}} \in \mathrm{Gl}_p(\mathbb{R})$ with $L_{\mathrm{l}} = \left(I_p + DD^{\mathsf{T}}\right)^{1/2}$ and $K_{\mathrm{l}} = L_{\mathrm{l}}^{-\mathsf{T}}\left(CX_{\mathrm{l}} + DB^{\mathsf{T}}\right)$. In this case the realization of $\begin{bmatrix} M_{\mathrm{l}}(s) & N_{\mathrm{l}}(s) \end{bmatrix}$ in Theorem 3.10.5 reduces to

$$\left(I_n, A - K_{\mathrm{l}}^{\mathsf{T}}L_{\mathrm{l}}^{-\mathsf{T}}C, \begin{bmatrix} -K_{\mathrm{l}}^{\mathsf{T}}L_{\mathrm{l}}^{-\mathsf{T}} & B - K_{\mathrm{l}}^{\mathsf{T}}L_{\mathrm{l}}^{-\mathsf{T}}D \end{bmatrix}, L_{\mathrm{l}}^{-\mathsf{T}}C, \begin{bmatrix} L_{\mathrm{l}}^{-\mathsf{T}} & L_{\mathrm{l}}^{-\mathsf{T}}D \end{bmatrix}\right) \in \Sigma_{n,p+m,p}.$$

These are also the realizations that have been obtained in [ZDG96, Sect. 13.8].

b) Algorithms for computing normalized coprime factorizations via (generalized) algebraic Riccati equations are outlined in [Var98]. These methods are based on an orthogonal factorization of the Rosenbrock pencil

$$\begin{bmatrix} -sE+A & B \\ C & D \end{bmatrix} \in \mathbb{R}[s]^{n+p\times n+m}$$

to extract a Rosenbrock subpencil

$$\begin{bmatrix} -s\widehat{E}+\widehat{A} & \widehat{B} \\ \widehat{C} & \widehat{D} \end{bmatrix} \in \mathbb{R}[s]^{r+p\times r+m}$$

with nonsingular $\widehat{E} \in \mathbb{R}^{r\times r}$, which corresponds to an inherent generalized state-space system. The generalized ARE for computing the right normalized coprime factorization is then given by

$$\begin{aligned} &\widehat{A}^{\mathsf{T}}\widehat{X}_{\mathrm{r}}\widehat{E} + \widehat{E}^{\mathsf{T}}\widehat{X}_{\mathrm{r}}\widehat{A} + \widehat{C}^{\mathsf{T}}\widehat{C} \\ &\quad - \big(\widehat{E}^{\mathsf{T}}\widehat{X}_{\mathrm{r}}\widehat{B} + \widehat{C}^{\mathsf{T}}\widehat{D}\big)\big(I_m + \widehat{D}^{\mathsf{T}}\widehat{D}\big)^{-1}\big(\widehat{E}^{\mathsf{T}}\widehat{X}_{\mathrm{r}}\widehat{B} + \widehat{C}^{\mathsf{T}}\widehat{D}\big)^{\mathsf{T}} = 0, \quad \widehat{X}_{\mathrm{r}} = \widehat{X}_{\mathrm{r}}^{\mathsf{T}}. \end{aligned}$$

Note that this equation can always be formed since $I_m + \widehat{D}^{\mathsf{T}}\widehat{D} > 0$.

## 3.11 Inner-Outer Factorizations

We show that the previously obtained results can be used to construct inner-outer factorizations of arbitrary rational matrices. Such factorizations play a crucial role in $\mathcal{H}_\infty$-control [Fra87, Gre88, ZDG96].

**Definition 3.11.1.** Let $G(s) \in \mathbb{R}(s)^{p\times m}$ with $q = \operatorname{rank}_{\mathbb{R}(s)} G(s)$ be given. An *inner-outer factorization* of $G(s)$ is a representation

$$G(s) = G_{\mathrm{i}}(s)G_{\mathrm{o}}(s),$$

where $G_{\mathrm{i}}(s) \in \mathbb{R}(s)^{p\times q}$ is inner, and $G_{\mathrm{o}}(s) \in \mathbb{R}(s)^{q\times m}$ is outer.

We will construct a realization of the inner and outer factors of the transfer function $G(s)$. The basis for such a construction will be the following descriptor Lur'e equation:

$$\begin{bmatrix} A^{\mathsf{T}}X + X^{\mathsf{T}}A + C^{\mathsf{T}}C & X^{\mathsf{T}}B + C^{\mathsf{T}}D \\ B^{\mathsf{T}}X + D^{\mathsf{T}}C & D^{\mathsf{T}}D \end{bmatrix} =_{\mathcal{V}_{\mathrm{sys}}} \begin{bmatrix} K^{\mathsf{T}} \\ L^{\mathsf{T}} \end{bmatrix} \begin{bmatrix} K & L \end{bmatrix}, \quad E^{\mathsf{T}}X = X^{\mathsf{T}}E, \tag{3.136}$$

where $\mathcal{V}_{\mathrm{sys}}$ is the system space of $(E, A, B)$ as in (3.5). We first prove that, if $(E, A, B, C, D) \in \Sigma_{n,m,p}$ is strongly stabilizable, then (3.136) has a positive stabilizing solution.

**Proposition 3.11.2.** *Let $(E, A, B, C, D) \in \Sigma_{n,m,p}$ with the system space $\mathcal{V}_{\text{sys}}$ as in* (3.5) *be strongly stabilizable. Then the descriptor Lur'e equation* (3.136) *has a stabilizing solution $(X, K, L)$. This solution fulfills $E^\top X \geq 0$.*

*Proof.* The existence of a stabilizing solution follows from Theorem 3.5.3. Since, furthermore, $P = 0$ solves the descriptor KYP inequality associated to (3.136), we obtain from Theorem 3.5.4 that $E^\top X \geq 0$. □

This solution will be the basis for our construction of inner-outer factorizations. The following two auxiliary results will be further used for this construction and its proof.

**Lemma 3.11.3.** *Let a system $(E, A, B, C, D) \in \Sigma_{n,m,p}$ with transfer function $G(s) \in \mathbb{R}(s)^{p\times m}$ be given. Denote $q = \operatorname{rank}_{\mathbb{R}(s)} G(s)$. Then there exists some matrix $Z \in \mathbb{R}^{m\times q}$ such that*

*a) each column vector of $Z$ consists of a canonical unit vector, and*

*b)* $\operatorname{rank}_{\mathbb{R}(s)} G(s)Z = q$.

*If, further, $q = p$ and $Z$ has the above properties, then the following statements hold true:*

*i) It holds that*

$$\begin{bmatrix} -sE + A & BZ \\ C & DZ \end{bmatrix} \in \mathrm{Gl}_{n+p}(\mathbb{R}(s)).$$

*ii) The function*

$$\mathcal{P}(s) = Z(G(s)Z)^{-1}G(s) \in \mathbb{R}(s)^{p\times p}$$

*is a projector onto* $\operatorname{im}_{\mathbb{R}(s)} G(s)$ *and along* $\ker_{\mathbb{R}(s)} G(s)$.

*Proof.* Denote the $k$-th canonical unit vector by $e_k \in \mathbb{R}^m$. Since the set of column vectors of $G(s)$ can be reduced to a basis of $\operatorname{im}_{\mathbb{R}(s)} G(s)$ and $\dim \operatorname{im}_{\mathbb{R}(s)} G(s) = q$, there exist $i_1, \ldots, i_q$ such that

$$\{G(s)e_{i_1}, \ldots, G(s)e_{i_q}\}$$

is a basis of $\operatorname{im}_{\mathbb{R}(s)} G(s)$. Then the matrix

$$Z = \begin{bmatrix} e_{i_1} & \cdots & e_{i_q} \end{bmatrix}$$

has the properties a) and b).

If, further, $q = p$, then the function $G(s)Z$ is invertible. Statement i) can now be concluded from Lemma 2.2.26, ii) follows by simple calculations. □

**Lemma 3.11.4.** *Let $G(s) \in \mathbb{R}(s)^{p\times m}$ be given. Assume that $\mathcal{P}(s) \in \mathbb{R}(s)^{m\times m}$ is a projector with $\ker_{\mathbb{R}(s)} \mathcal{P}(s) \subseteq \ker_{\mathbb{R}(s)} G(s)$. Then we have $G(s)\mathcal{P}(s) = G(s)$.*

*Proof.* Let $x(s) \in \mathbb{R}(s)^m$ be given. Since $\mathcal{P}(s)$ is a projector, we have $x(s) = x_1(s) + x_2(s)$ with $x_1(s) \in \operatorname{im}_{\mathbb{R}(s)} \mathcal{P}(s)$ and $x_2(s) \in \ker_{\mathbb{R}(s)} \mathcal{P}(s)$. Moreover, since we have $x_1(s) = \mathcal{P}(s)y_1(s)$ for some $y_1(s) \in \mathbb{R}(s)^m$, we obtain $x_1(s) = \mathcal{P}(s)x_1(s)$. This yields the claim, since it holds that

$$G(s)x(s) = G(s)x_1(s) = G(s)\mathcal{P}(s)x_1(s) = G(s)\mathcal{P}(s)x(s).$$

□

Now we formulate our main result on inner-outer factorizations.

**Theorem 3.11.5.** *Let $(E, A, B, C, D) \in \Sigma_{n,m,p}$ with the system space $\mathcal{V}_{\text{sys}}$ as in (3.5) be strongly stabilizable and let $G(s) \in \mathbb{R}(s)^{p\times m}$ be the associated transfer function. For $q = \operatorname{rank}_{\mathbb{R}(s)} G(s)$, let $Z \in \mathbb{R}^{m\times q}$ be a matrix whose column vectors are canonical unit vectors, and $\operatorname{rank}_{\mathbb{R}(s)} G(s)Z = q$. Let $(X, K, L)$ be a stabilizing solution of the descriptor Lur'e equation (3.136). Then an inner-outer factorization of $G(s)$ is given by $G(s) = G_{\text{i}}(s)G_{\text{o}}(s)$, where $G_{\text{i}}(s) \in \mathbb{R}(s)^{p\times q}$ is realized by*

$$(E_{\text{i}}, A_{\text{i}}, B_{\text{i}}, C_{\text{i}}) := \left(\begin{bmatrix} E & 0 \\ 0 & 0 \end{bmatrix}, \begin{bmatrix} A & BZ \\ K & LZ \end{bmatrix}, \begin{bmatrix} 0 \\ -I_q \end{bmatrix}, \begin{bmatrix} C & DZ \end{bmatrix}\right) \in \Sigma_{n+q,q,p},$$

*and $G_{\text{o}}(s) \in \mathbb{R}(s)^{q\times m}$ is realized by*

$$(E_{\text{o}}, A_{\text{o}}, B_{\text{o}}, C_{\text{o}}, D_{\text{o}}) := (E, A, B, K, L) \in \Sigma_{n,m,q}.$$

*Proof.* We prove the theorem in five steps.
*Step 1: We show that $G_{\text{o}}(s)$ is outer:*
This claim directly follows from Theorem 3.6.3.
*Step 2: We show that $\ker_{\mathbb{R}(s)} G(s) = \ker_{\mathbb{R}(s)} G_{\text{o}}(s)$:*
By Theorem 3.6.3 we have

$$G_{\text{o}}^{\sim}(s)G_{\text{o}}(s) = \Phi(s) = G^{\sim}(s)G(s).$$

In particular, it holds that

$$G_{\text{o}}^{\mathsf{H}}(\mathrm{i}\omega)G_{\text{o}}(\mathrm{i}\omega) = G^{\mathsf{H}}(\mathrm{i}\omega)G(\mathrm{i}\omega) \quad \forall \omega \in \mathbb{R} \text{ with } \mathrm{i}\omega \notin \Lambda(E, A). \tag{3.137}$$

First we show that $\ker_{\mathbb{R}(s)} G(s) \subseteq \ker_{\mathbb{R}(s)} G_{\text{o}}(s)$. Therefore, assume that $v(s) \in \ker_{\mathbb{R}(s)} G(s)$. Let $\Gamma \subset \mathbb{C}$ be the (finite) set of poles of $v(s) \in \mathbb{R}(s)^m$. Then, we obtain from (3.137) that for all $\omega \in \mathbb{R}$ with $\mathrm{i}\omega \notin (\Gamma \cup \Lambda(E, A))$ it holds that

$$\|G_{\text{o}}(\mathrm{i}\omega)v(\mathrm{i}\omega)\|_2^2 = \|G(\mathrm{i}\omega)v(\mathrm{i}\omega)\|_2^2 = 0.$$

Hence, $\lambda \mapsto G_{\rm o}(\lambda)v(\lambda)$ is a vector-valued rational function which vanishes on $\mathrm{i}\mathbb{R} \setminus \Gamma$. The finiteness of the set $\Gamma$ leads to $G_{\rm o}(s)v(s) = 0$, i. e., $v(s) \in \ker_{\mathbb{R}(s)} G_{\rm o}(s)$. The reverse implication $\ker_{\mathbb{R}(s)} G_{\rm o}(s) \subseteq \ker_{\mathbb{R}(s)} G(s)$ can be proven analogously.

*Step 3: We prove that $G_{\rm o}(s)Z$ is invertible:*

We obtain from Step 2 that

$$\operatorname{rank}_{\mathbb{R}(s)} G_{\rm o}(s) = \operatorname{rank}_{\mathbb{R}(s)} \Phi(s) = \operatorname{rank}_{\mathbb{R}(s)} G(s) = q.$$

The outerness of $G_{\rm o}(s)$ implies $G_{\rm o}(s) \in \mathbb{R}(s)^{q\times m}$. Further, by

$$(G_{\rm o}(\mathrm{i}\omega)Z)^{\mathsf{H}}(G_{\rm o}(\mathrm{i}\omega)Z) = (G(\mathrm{i}\omega)Z)^{\mathsf{H}}(G(\mathrm{i}\omega)Z) \quad \forall \omega \in \mathbb{R} \text{ with } \mathrm{i}\omega \notin \Lambda(E,A),$$

the assumption $\operatorname{rank}_{\mathbb{R}(s)} G(s)Z = q$ leads to

$$G_{\rm o}(s)Z \in \mathrm{Gl}_q(\mathbb{R}(s)).$$

*Step 4: We show that $G_{\rm i}(s)G_{\rm o}(s) = G(s)$:*

Using the statement in Step 1 and the fact that $G_{\rm o}(s)Z$ is realized by the system $(E, A, BZ, K, LZ)$, Lemma 3.11.3 a) leads to regularity of its Rosenbrock pencil, i. e.,

$$\begin{bmatrix} -sE + A & BZ \\ K & LZ \end{bmatrix} \in \mathrm{Gl}_{n+q}(\mathbb{R}(s)).$$

Lemma 3.7.3 a) then gives rise to

$$G_{\rm i}(s) = \begin{bmatrix} C & DZ \end{bmatrix} \begin{bmatrix} sE - A & -BZ \\ -K & -LZ \end{bmatrix}^{-1} \begin{bmatrix} 0 \\ -I_q \end{bmatrix} = G(s)Z(G_{\rm o}(s)Z)^{-1}.$$

By Lemma 3.11.3 b), $Z(G_{\rm o}(s)Z)^{-1}G_{\rm o}(s) \in \mathbb{R}(s)^{q\times q}$ is a projector along $\ker_{\mathbb{R}(s)} G(s)$. Then, by Lemma 3.11.4, we obtain

$$G_{\rm i}(s)G_{\rm o}(s) = G(s)Z \cdot \left(LZ + K(sE - A)^{-1}BZ\right)^{-1} \cdot \left(L + K(sE - A)^{-1}B\right) = G(s).$$

*Step 5: We show that $G_{\rm i}(s)$ is inner:*

We have to show that $(E_{\rm i}, A_{\rm i}, B_{\rm i}, C_{\rm i})$ fulfills a certain descriptor dissipativity equality. Define the subspace

$$\begin{aligned} \mathcal{V}_{\rm sys,i} &:= \left\{ \begin{pmatrix} x \\ u \end{pmatrix} \in \mathbb{R}^{n+2q} : A_{\rm i}x + B_{\rm i}u \in \operatorname{im} E_{\rm i} \right\} \\ &= \left\{ \begin{pmatrix} x_1 \\ x_2 \\ u \end{pmatrix} \in \mathbb{R}^{n+2q} : \begin{pmatrix} x_1 \\ Zx_2 \end{pmatrix} \in \mathcal{V}_{\rm sys} \text{ and } Kx_1 + LZx_2 = u \right\}. \end{aligned}$$

Then for

$$\begin{pmatrix} x \\ u \end{pmatrix} = \begin{pmatrix} x_1 \\ x_2 \\ u \end{pmatrix} \in \mathcal{V}_{\mathrm{sys,i}} \quad \text{and} \quad P_{\mathrm{i}} = \begin{bmatrix} X & 0 \\ 0 & 0 \end{bmatrix}$$

we obtain

$$\begin{aligned}
&\begin{pmatrix} x \\ u \end{pmatrix}^{\mathsf{T}} \begin{bmatrix} A_{\mathrm{i}}^{\mathsf{T}} P_{\mathrm{i}} + P_{\mathrm{i}}^{\mathsf{T}} A_{\mathrm{i}} + C_{\mathrm{i}}^{\mathsf{T}} C_{\mathrm{i}} & P_{\mathrm{i}}^{\mathsf{T}} B_{\mathrm{i}} \\ B_{\mathrm{i}}^{\mathsf{T}} P_{\mathrm{i}} & -I_q \end{bmatrix} \begin{pmatrix} x \\ u \end{pmatrix} \\
&= \begin{pmatrix} x_1 \\ x_2 \\ u \end{pmatrix}^{\mathsf{T}} \left[ \begin{matrix} \begin{bmatrix} A & BZ \\ K & LZ \end{bmatrix}^{\mathsf{T}} \begin{bmatrix} X & 0 \\ 0 & 0 \end{bmatrix} + \begin{bmatrix} X & 0 \\ 0 & 0 \end{bmatrix}^{\mathsf{T}} \begin{bmatrix} A & BZ \\ K & LZ \end{bmatrix} \\ + \begin{bmatrix} C & DZ \end{bmatrix}^{\mathsf{T}} \begin{bmatrix} C & DZ \end{bmatrix} \\ \begin{bmatrix} 0 \\ -I_q \end{bmatrix}^{\mathsf{T}} \begin{bmatrix} X & 0 \\ 0 & 0 \end{bmatrix} \end{matrix} \quad \begin{matrix} \begin{bmatrix} X & 0 \\ 0 & 0 \end{bmatrix}^{\mathsf{T}} \begin{bmatrix} 0 \\ -I_q \end{bmatrix} \\ \\ -I_q \end{matrix} \right] \begin{pmatrix} x_1 \\ x_2 \\ u \end{pmatrix} \\
&= \begin{pmatrix} x_1 \\ x_2 \\ Kx_1 + LZx_2 \end{pmatrix}^{\mathsf{T}} \begin{bmatrix} A^{\mathsf{T}} X + X^{\mathsf{T}} A + C^{\mathsf{T}} C & X^{\mathsf{T}} BZ + C^{\mathsf{T}} DZ & 0 \\ Z^{\mathsf{T}} B^{\mathsf{T}} X + Z^{\mathsf{T}} D^{\mathsf{T}} C & Z^{\mathsf{T}} D^{\mathsf{T}} DZ & 0 \\ 0 & 0 & -I_q \end{bmatrix} \\
&\qquad \cdot \begin{pmatrix} x_1 \\ x_2 \\ Kx_1 + LZx_2 \end{pmatrix} \\
&= \begin{pmatrix} x_1 \\ Zx_2 \end{pmatrix}^{\mathsf{T}} \left( \begin{bmatrix} A^{\mathsf{T}} X + X^{\mathsf{T}} A + C^{\mathsf{T}} C & X^{\mathsf{T}} B + C^{\mathsf{T}} D \\ B^{\mathsf{T}} X + D^{\mathsf{T}} C & D^{\mathsf{T}} D \end{bmatrix} - \begin{bmatrix} K^{\mathsf{T}} \\ L^{\mathsf{T}} \end{bmatrix} \begin{bmatrix} K & L \end{bmatrix} \right) \begin{pmatrix} x_1 \\ Zx_2 \end{pmatrix} \\
&= 0.
\end{aligned}$$

Furthermore, by Proposition 3.11.2 we have

$$E_{\mathrm{i}}^{\mathsf{T}} P_{\mathrm{i}} = \begin{bmatrix} E^{\mathsf{T}} X & 0 \\ 0 & 0 \end{bmatrix} \geq 0.$$

Then by the lossless bounded real lemma for DAEs (Theorem 3.9.19), we obtain that $G_{\mathrm{i}}(s)$ is inner. □

**Example 3.11.6.** For illustrating the result of Theorem 3.11.5 consider the system given by

$$E = \begin{bmatrix} 1 & 0 \\ 0 & 0 \end{bmatrix}, \quad A = \begin{bmatrix} 1 & 0 \\ 0 & -1 \end{bmatrix}, \quad B = \begin{bmatrix} 1 & 0 \\ 0 & 0 \end{bmatrix}, \quad C = \begin{bmatrix} 1 & 0 \\ 0 & 0 \end{bmatrix}, \quad D = 0_{2\times 2}. \tag{3.138}$$

The system $(E, A, B, C, D) \in \Sigma_{2,2,2}$ is strongly stabilizable with transfer function

$$G(s) = \begin{bmatrix} 1 & 0 \\ 0 & 0 \end{bmatrix} \begin{bmatrix} s-1 & 0 \\ 0 & 1 \end{bmatrix}^{-1} \begin{bmatrix} 1 & 0 \\ 0 & 0 \end{bmatrix} = \begin{bmatrix} \frac{1}{s-1} & 0 \\ 0 & 0 \end{bmatrix}.$$

From $E^\top X = X^\top E$ we obtain

$$X = \begin{bmatrix} x_{11} & 0 \\ x_{21} & x_{22} \end{bmatrix},$$

and the descriptor Lur'e equation (3.136) reduces to

$$\begin{bmatrix} 2x_{11}+1 & -x_{21} & x_{11} & 0 \\ -x_{21} & -2x_{22} & 0 & 0 \\ x_{11} & 0 & 0 & 0 \\ 0 & 0 & 0 & 0 \end{bmatrix} =_{\mathcal{V}_{\mathrm{sys}}} \begin{bmatrix} k_1 \\ k_2 \\ \ell_1 \\ \ell_2 \end{bmatrix} \begin{bmatrix} k_1 & k_2 & \ell_1 & \ell_2 \end{bmatrix}, \tag{3.139}$$

where $K = \begin{bmatrix} k_1 & k_2 \end{bmatrix}$ and $L = \begin{bmatrix} \ell_1 & \ell_2 \end{bmatrix}$. A simple calculation shows that a stabilizing and maximal solution of the descriptor Lur'e equation is given by the triple

$$(X, K, L) = \left( \begin{bmatrix} 0 & 0 \\ 0 & 0 \end{bmatrix}, \begin{bmatrix} 1 & 0 \end{bmatrix}, \begin{bmatrix} 0 & 0 \end{bmatrix} \right).$$

Then from Theorem 3.11.5 we obtain $Z = \begin{bmatrix} 1 \\ 0 \end{bmatrix}$ and

$$G_{\mathrm{i}}(s) = \begin{bmatrix} 1 & 0 & 0 \\ 0 & 0 & 0 \end{bmatrix} \begin{bmatrix} s-1 & 0 & -1 \\ 0 & 1 & 0 \\ -1 & 0 & 0 \end{bmatrix}^{-1} \begin{bmatrix} 0 \\ 0 \\ -1 \end{bmatrix} = \begin{bmatrix} 1 \\ 0 \end{bmatrix} \in \mathbb{R}(s)^{2\times 1},$$

$$G_{\mathrm{o}}(s) = \begin{bmatrix} 1 & 0 \end{bmatrix} \begin{bmatrix} s-1 & 0 \\ 0 & 1 \end{bmatrix}^{-1} \begin{bmatrix} 1 & 0 \\ 0 & 0 \end{bmatrix} = \begin{bmatrix} \frac{1}{s-1} & 0 \end{bmatrix} \in \mathbb{R}(s)^{1\times 2}.$$

We obviously have $G(s) = G_{\mathrm{i}}(s)G_{\mathrm{o}}(s)$. Moreover, it is readily verified that $G_{\mathrm{i}}(s)$ and $G_{\mathrm{o}}(s)$ are inner and outer rational functions, respectively.

*Remark* 3.11.7 (Inner-outer factorization).

a) The even matrix pencil associated to (3.136) can be factored as

$$\begin{aligned} s\mathcal{E} - \mathcal{A} &= \begin{bmatrix} 0 & -sE+A & B \\ sE^\top + A^\top & C^\top C & C^\top D \\ B^\top & D^\top C & D^\top D \end{bmatrix} \\ &= \begin{bmatrix} 0 & 0 & I_n \\ sE^\top + A^\top & C^\top & 0 \\ B^\top & D^\top & 0 \end{bmatrix} \begin{bmatrix} 0 & 0 & I_n \\ 0 & I_p & 0 \\ I_n & 0 & 0 \end{bmatrix} \begin{bmatrix} 0 & -sE+A & B \\ 0 & C & D \\ I_n & 0 & 0 \end{bmatrix}. \end{aligned}$$

Thereby, we see that, for the KCF of the Rosenbrock pencil $R(s)$ of $(E, A, B, C, D)$, each block of type K1 corresponding to a non-imaginary eigenvalue corresponds to a block of type E1 in the EKCF of $s\mathcal{E} - \mathcal{A}$ of double size corresponding to the same eigenvalue. Further, each block of type K1 of the KCF of $R(s)$ corresponding to an imaginary eigenvalue corresponds to a block of type E2 in the EKCF of $s\mathcal{E} - \mathcal{A}$ with negative sign-characteristic and double size corresponding to the same eigenvalue. Analogous statements on the relation between blocks of type K2, K3, and K4 in the KCF of $R(s)$ and blocks in the EKCF of $s\mathcal{E} - \mathcal{A}$ of type E3 and E4 can further be inferred from Theorem 3.4.2.

b) In the SISO case (that is, $m = p = 1$), an inner-outer factorization can be obtained by purely algebraic considerations. The transfer function $g(s) \in \mathbb{R}(s)$ is first factorized as

$$g(s) = \frac{d^+(s) \cdot d^-(s)}{n(s)}$$

for polynomials $d^+(s)$, $d^-(s)$, $n(s) \in \mathbb{R}[s]$ with the property that all roots of $d^+(s)$ are in $\mathbb{C}^+$ and all roots of $d^-(s)$ are in $\overline{\mathbb{C}^-}$. An inner-outer factorization is then given by $g(s) = g_{\mathrm{i}}(s)g_{\mathrm{o}}(s)$, where

$$g_{\mathrm{i}}(s) = \frac{d^+(s)}{\overline{d^+(-\overline{s})}}, \quad g_{\mathrm{o}}(s) = \frac{\overline{d^+(-\overline{s})} \cdot d^-(s)}{n(s)}.$$

This approach is called *Hurwitz reflection* [IW13].

c) If $E = I_n$ and $\operatorname{rank} D = m$, then the matrix $Z$ in Theorem 3.11.5 becomes the identity. Moreover, the descriptor Lur'e equation (3.136) can be reformulated as an algebraic Riccati equation

$$A^\top X + XA + C^\top C - (XB + C^\top D)(D^\top D)^{-1}(XB + C^\top D)^\top = 0, \quad X = X^\top.$$

The matrices $K$ and $L$ can be chosen as $L \in \mathrm{Gl}_m(\mathbb{R})$ with $L = (D^\top D)^{1/2}$ and $K = L^{-\top}(B^\top X + D^\top C)$. In this case we can see that the realization of $G_{\mathrm{i}}(s)$ reduces to

$$(I_n, A - BL^{-1}K, BL^{-1}, C - DL^{-1}K, DL^{-1}) \in \Sigma_{n,m,p}.$$

This is also the realization that has been obtained in [ZDG96, Sect. 13.7].

d) In the literature, inner-outer factorizations have been computed by employing (generalized) algebraic Riccati equations [Gre88, OV00, ZDG96]. In [OV00], the Rosenbrock pencil

$$\begin{bmatrix} -sE + A & B \\ C & D \end{bmatrix} \in \mathbb{R}[s]^{n+p \times n+m}$$

is first transformed by orthogonal transformations such that a Rosenbrock subpencil

$$\begin{bmatrix} -s\widehat{E}+\widehat{A} & \widehat{B} \\ \widehat{C} & \widehat{D} \end{bmatrix} \in \mathbb{R}[s]^{r+q\times r+q}$$

with nonsingular $\widehat{E}\in\mathbb{R}^{r\times r}$ and $\widehat{D}\in\mathbb{R}^{q\times q}$ can be extracted. Since $\widehat{D}^{\mathsf{T}}\widehat{D}>0$, the generalized algebraic Riccati equation

$$\begin{aligned}&\widehat{A}^{\mathsf{T}}\widehat{X}\widehat{E}+\widehat{E}^{\mathsf{T}}\widehat{X}\widehat{A}+\widehat{C}^{\mathsf{T}}\widehat{C}\\ &\qquad -\big(\widehat{E}^{\mathsf{T}}\widehat{X}\widehat{B}+\widehat{C}^{\mathsf{T}}\widehat{D}\big)\big(\widehat{D}^{\mathsf{T}}\widehat{D}\big)^{-1}\big(\widehat{E}^{\mathsf{T}}\widehat{X}\widehat{B}+\widehat{C}^{\mathsf{T}}\widehat{D}\big)^{\mathsf{T}}=0,\quad \widehat{X}=\widehat{X}^{\mathsf{T}}.\end{aligned}$$

can be formed and solved.

## 3.12 Summary and Outlook

In this chapter we have given a complete theoretical analysis of the linear-quadratic optimal control problem for differential-algebraic systems. The main assumption of this analysis is impulse controllability, no assumptions on the index are imposed. The basis of our theory is a new differential-algebraic version of the Kalman-Yakubovich-Popov lemma. Resulting from this lemma we have derived a new type of algebraic matrix equations, namely the descriptor Lur'e equation, that attains the rank-minimizing solutions of the descriptor KYP inequality. This equation extends both generalized algebraic Riccati equations and Lur'e equations. We have studied the solution theory for this equation. In particular, we have given criteria in terms of the spectrum of an associated even matrix pencil. Furthermore, we have shown how to construct solutions via its deflating subspaces. We have shown that for strongly stabilizable and strongly anti-stabilizable systems, there exist stabilizing and anti-stabilizing solutions of the descriptor Lur'e equation, respectively. These also define extremal elements in the solution set of the corresponding descriptor KYP inequality. Moreover, we have stated conditions for the existence of nonpositive solutions to the descriptor KYP inequality.

We have discussed various applications of our new theory. In particular, we have studied the linear-quadratic optimal control problem. We could show that the linear-quadratic optimal control problem (3.1) is feasible if and only if the descriptor Lur'e equation has a stabilizing solution. Moreover, we have shown that for strongly controllable systems, the descriptor Lur'e equation has a nonpositive solution if and only if a linear-quadratic optimal control problem with free terminal conditions is feasible. These results have also given rise to the analysis of dissipativity and cyclo-dissipativity of DAEs. Finally, we have shown that certain factorizations of rational matrices, namely spectral factorizations, normalized coprime factorizations, and inner-outer factorizations can be carried out be solving descriptor Lur'e equations. In particular, our theory covers the singular case which means the rational matrices under consideration do not have full rank. In this sense, our theory presents a uniform framework for solving control problems for DAEs.

Possible extensions of this theory include the consideration of systems that are not impulse controllable or systems that are over- or underdetermined. The case of non-impulse controllable systems has recently been treated in [RRV14]. The main idea consists of a transformation of the system to *feedback equivalence form*, i. e., for $(E, A, B) \in \Sigma_{n,m,p}$ there exist matrices $W, T \in \mathrm{Gl}_n(\mathbb{R})$ and $F \in \mathbb{R}^{m \times n}$, such that

$$\begin{aligned}\begin{bmatrix} sE_F - A_F & B_F \end{bmatrix} &:= W \begin{bmatrix} sE - A & B \end{bmatrix} \begin{bmatrix} T & 0 \\ -FT & I_m \end{bmatrix} \\ &= \begin{bmatrix} sI_{n_1} - A_{11} & 0 & 0 & B_1 \\ 0 & -I_{n_2} & sE_{23} & B_2 \\ 0 & 0 & sE_{33} - I_{n_3} & 0 \end{bmatrix},\end{aligned}$$

where $E_{33} \in \mathbb{R}^{n_3 \times n_3}$ is nilpotent. Then for the transformed system, the reduced system space

$$\widehat{\mathcal{V}}_{\mathrm{sys},F} := \left\{ \begin{pmatrix} x_1 \\ -B_2 u \\ 0_{n_3 \times 1} \\ u \end{pmatrix} \in \mathbb{R}^{n+m} : x_1 \in \mathbb{R}^{n_1},\ u \in \mathbb{R}^m \right\} \subset \mathcal{V}_{\mathrm{sys},F} = \begin{bmatrix} T^{-1} & 0 \\ -F & I_m \end{bmatrix} \cdot \mathcal{V}_{\mathrm{sys}}$$

is considered, where $\mathcal{V}_{\mathrm{sys}}$ is as in (3.5). Note that $\widehat{\mathcal{V}}_{\mathrm{sys},F}$ is the smallest subspace in which the solution trajectories $(x_F, u_F) \in \mathfrak{B}_{(E_F, A_F, B_F)}$ pointwisely evolve. Using the above relations, results analogous to the ones presented in this thesis can be derived with completely the same machinery. Only the construction of the solution of the descriptor Lur'e equation via the deflating subspaces of an even matrix pencil requires an index reduction of the pencil $sE - A$, see [RRV14] for details.

In case the pencil $sE - A$ is singular, one has to perform a regularization process [CKM12, BLMV14] to the system $(E, A, B)$. This procedure includes removing redundant states as well as a reinterpretation of variables in case of a nonsquare system. Reinterpretation in this context means that some states might actually be considered as inputs and vice versa.

Another open question is the following one. In Section 3.8 we have shown that the feasibility of the optimal control problem (3.1) directly implies strong stabilizability and the existence of a stabilizing solution of the associated descriptor Lur'e equation. However, such a relation is still unknown for the case of optimal control problems with free terminal conditions. This means that it is not clear whether the finiteness of the functionals $V_{\mathrm{f}}^+$ in (3.99) or $V_{\mathrm{n}}^+$ in (3.100) already implies the existence of a nonpositive solution to the descriptor KYP inequality (without the requirement of strong controllability).

Another possible direction for future research is the development of numerical methods for the solution of descriptor Lure equations. Recently, some advances have been made in this direction for the case of standard Lur'e equations [PR11, PR12]. The corresponding algorithms are based on considerations for the deflating subspaces of the even matrix pencil (3.10). Therefore, there is the hope that these methods can be generalized to descriptor Lur'e equations using the solution theory developed in this chapter.

# 4 Systems with Counterclockwise Input/Output Dynamics

## 4.1 Introduction

Dynamical systems with dissipative or cyclo-dissipative behavior play an important role in the modeling and analysis of, e. g., flexible mechanical structures or electrical circuits and are well-studied, see Chapter 3.

A less known property of dynamical systems is a counterclockwise input/output (ccw I/O) dynamics [Ang06]. For SISO systems, this refers to the fact that the input/output trajectories have a counterclockwise motion in the input/output plane. Closely related is the concept of a negative imaginary transfer function. These properties often occur in models associated to mechanical systems and electrical circuits provided that certain measurements are taken. For instance, mechanical structures with *colocated* force actuators and *position sensors* yield such systems [PL10].

Since dynamical systems of the above type are naturally modeled via differential-algebraic equations, one is interested in the theory and numerical methods which directly work on the given system realization. In this manner, the goal of this chapter is to generalize the negative imaginary theory [LP08, PL10, XPL10] to descriptor systems without performing any reductions to standard state-space systems.

Our particular focus is on deriving algebraic characterizations of the negative imaginary property of transfer functions, similarly to the cases of bounded realness and positive realness. For such systems, there exist characterizations in terms of the solvability of certain dissipativity inequalities and descriptor Lur'e equations (see Chapter 3), given by the bounded real and positive real lemma. In this fashion, there also exists a negative imaginary lemma, for standard state-space systems [PL10, XPL10], as well as for descriptor systems [MKPL12]. However, [MKPL12] does not provide necessary and sufficient conditions for negative imaginariness and the conditions on the solution of the LMI are so strong that they are almost never fulfilled. Furthermore, LMI conditions have several disadvantages: they pose certain conditions on the system such as controllability, and LMIs are very expensive to solve, i. e., infeasible for larger systems. Therefore, spectral conditions of structured matrices and pencils have been developed. There are conditions for cyclo-dissipativity in terms of Hamiltonian matrices [Wil71, Rem. 28] for the standard state-space case and even matrix pencils, given by Theorem 3.4.2. Recently, a similar approach has been used to derive a Hamiltonian matrix to check negative imaginariness of a standard state-space system given in minimal realization [MKPL11b]. This chapter generalizes this result in multiple directions. First, we formulate conditions for descriptor systems by using the more general concept of even matrix pencils without posing any very restrictive assumptions on the realization. Second, we cover all possible boundary cases in a uniform way by analyzing the eigenstructure of this pencil in a similar fashion as in Theorem 3.4.2.

The second part of this chapter deals with the question how to restore the negative imaginary property of a transfer function in case that it is lost due to errors in the modeling process. We will give a more detailed motivation and explanation in mathematical terms in Section 4.4. There is a huge list of papers concerning restoring dissipativity of dynamical systems such as [GT04, GTU06, SS07] and more recently [BS13]. All of them rely on perturbing certain eigenvalues of Hamiltonian matrices or even matrix pencils. The paper [MKPL11a] about the restoration of the negative imaginary property is also based on [GT04, GTU06]. In this chapter we will adapt this strategy for descriptor systems using the eigenvalue characterization for even pencils. One focus of our work is also the usage of structure-preserving and -exploiting algorithms to obtain the quantities that are needed for the perturbation of the eigenvalues in a more efficient, reliable, and accurate manner.

The remainder of this chapter is structured as follows. In Section 4.2 we introduce the concepts of ccw I/O dynamics for descriptor systems, negative imaginary transfer functions, and discuss some basic properties. In Section 4.3 we derive the spectral characterizations of structured matrix pencils for negative imaginariness. In Section 4.4 we suggest an algorithm which can be used to restore the negative imaginary property of a system if it has been lost by approximating the system by, e. g., reducing the model order. Finally, in Section 4.5 we summarize this chapter and point towards further possible research directions.

## 4.2 Systems with Counterclockwise Input/Output Dynamics and Negative Imaginary Transfer Functions

First, we introduce the notion of a system with counterclockwise input-output dynamics and adapt the definition to systems $(E, A, B, C, D) \in \Sigma_{n,m,p}$.

**Definition 4.2.1.** A descriptor system $(E, A, B, C, D) \in \Sigma_{n,m,p}$ has a *counterclockwise input/output dynamics* if $m = p$ and it holds that

$$\int_0^T \dot{y}(\tau)^\mathsf{T} u(\tau) \mathrm{d}\tau \geq 0 \tag{4.1}$$

for all $T \geq 0$ and all $(x, u, y) \in \mathfrak{B}_{(E,A,B,C,D)}(0)$ such that $\dot{y} \in \mathcal{L}^2_{\mathrm{loc}}([0, \infty), \mathbb{R}^p)$.

Roughly speaking, a ccw I/O dynamics can be interpreted as passivity with respect to the derivative of the output (instead of the output itself). This will be made more precise in the following lemma.

**Lemma 4.2.2.** *Consider a system $(E, A, B, C, D) \in \Sigma_{n,m,m}$ with strictly proper transfer function $G(s) \in \mathbb{R}(s)^{m\times m}$. Assume that $G(s)$ has a state-space realization $(I_r, A_{11}, B_1, C_1) \in \Sigma_{r,m,m}$. Then $(E, A, B, C, D)$ has a ccw I/O dynamics if and only if the system $(I_r, A_{11}, B_1, C_1A_{11}, C_1B_1) \in \Sigma_{r,m,m}$ is passive.*

*Proof.* We can assume w. l. o. g. that $sE - A \in \mathbb{R}[s]^{n\times n}$ is given in QWF, i. e.,

$$(E, A, B, C, D) = \left( \begin{bmatrix} I_r & 0 \\ 0 & E_{22} \end{bmatrix}, \begin{bmatrix} A_{11} & 0 \\ 0 & I_{n-r} \end{bmatrix}, \begin{bmatrix} B_1 \\ B_2 \end{bmatrix}, \begin{bmatrix} C_1 & C_2 \end{bmatrix}, D \right),$$

where $A_{11} \in \mathbb{R}^{r\times r}$, $E_{22} \in \mathbb{R}^{n-r\times n-r}$ is nilpotent with index of nilpotency $\nu$, and $B_1 \in \mathbb{R}^{r\times m}$, $B_2 \in \mathbb{R}^{n-r\times m}$, $C_1 \in \mathbb{R}^{p\times r}$, and $C_2 \in \mathbb{R}^{p\times n-r}$. Since $G(s)$ is strictly proper, it holds that

$$D - C_2B_2 = 0, \quad C_2E_{22}B_2 = \ldots = C_2E_{22}^{\nu-1}B_2 = 0.$$

Thus, the system $(E, A, B, C, D)$ has ccw I/O dynamics if and only if $(I_r, A_{11}, B_1, C_1)$ has ccw I/O dynamics.

Now assume that $(E, A, B, C, D)$ has a ccw I/O dynamics. Since $(I_r, A_{11}, B_1, C_1)$ has ccw I/O dynamics, this implies that

$$\int_0^T \widetilde{y}(\tau)^\top u(\tau)\mathrm{d}\tau \geq 0$$

for all $T \geq 0$ and all $(x, u, \widetilde{y}) \in \mathfrak{B}_{(I_r, A_{11}, B_1, C_1A_{11}, C_1B_1)}(0)$ with $\dot{\widetilde{y}} \in \mathcal{L}^2_{\mathrm{loc}}([0,\infty), \mathbb{R}^p)$. By Definition 3.9.16 b), this means that $(I_r, A_{11}, B_1, C_1A_{11}, C_1B_1)$ is passive.

On the other hand, if $(I_r, A_{11}, B_1, C_1A_{11}, C_1B_1)$ is passive, one can conclude that $(I_r, A_{11}, B_1, C_1)$ has a ccw I/O dynamics and consequently $(E, A, B, C, D)$ has a ccw I/O dynamics. □

*Remark* 4.2.3. By the definition in [Ang06], a system $(E, A, B, C, D)$ has a ccw I/O dynamics if $m = p$ and it holds that

$$\inf \left\{ \int_0^\infty \dot{y}(\tau)^\top u(\tau)\mathrm{d}\tau : (x, u, y) \in \mathfrak{B}_{(E,A,B,C,D)} \right.$$
$$\left. \text{with } \dot{y} \in \mathcal{L}^2_{\mathrm{loc}}([0,\infty), \mathbb{R}^p) \right\} > -\infty.$$

By Theorem 3.8.13 and Lemma 4.2.2, this definition is equivalent to ours if the system $(E, A, B, C, D)$ is behaviorally controllable and has a strictly proper transfer function. In [BV13] it was erroneously claimed that both definitions generally coincide. This puts also doubts on the general validity of [Ang06, Prop. III.3] and [CH10, Lem. 2.1].

Now assume that $G_{\mathrm{poly}}(s) = M_0$ is the polynomial part of $G(s)$. In [CH10] it has been shown that if $M_0 = M_0^\mathsf{T} \geq 0$, the passivity of $(I_r, A_{11}, B_1, C_1A_{11}, C_1B_1)$ also implies a ccw I/O dynamics of $(I_r, A_{11}, B_1, C_1, M_0)$ with proper transfer function $G(s) \in \mathbb{R}(s)^{m\times m}$, or equivalently a ccw I/O dynamics of a system $(E, A, B, C, D)$ with the same transfer function $G(s)$. Furthermore, it turns out that for stable systems $(E, A, B, C, D) \in \Sigma_{n,m,m}$ with proper transfer function and $M_0 = M_0^\mathsf{T} \geq 0$, a ccw I/O dynamics is equivalent to a *negative imaginary* transfer function $G(s) \in \mathbb{R}(s)^{m\times m}$.

We will intensively study the rational function

$$\Phi(s) := \mathrm{i}\left(G(s) - G^{\sim}(s)\right) \in \mathrm{i}\mathbb{R}(s)^{m\times m}. \tag{4.2}$$

This function will play a similar role as the Popov function for dissipative systems, however here $\Phi(s)$ has some different features as discussed below.

**Definition 4.2.4** ((Strict) negative imaginariness)**.** A function $G \in \mathcal{RH}_\infty^{m\times m}$ is called

a) *negative imaginary* if $\Phi(\mathrm{i}\omega) \geq 0$ for all $\omega \geq 0$;

b) *strictly negative imaginary* if $\Phi(\mathrm{i}\omega) > 0$ for all $\omega > 0$.

Note, that the above definition can be generalized to allow $\Phi(s)$ to have poles on the imaginary axis [XPL10]. However, since this chapter focuses on transfer functions $G \in \mathcal{RH}_\infty^{m\times m}$, we do not need this more general definition.

As for ccw I/O dynamics and passivity, there exists a relation between negative imaginariness and positive realness of transfer functions.

**Lemma 4.2.5.** *[XPL10, Lem. 3, Lem. 4]*

*a) A strictly proper $G \in \mathcal{RH}_\infty^{m\times m}$ is negative imaginary if and only if $sG(s) \in \mathbb{R}(s)^{m\times m}$ is positive real.*

*b) A proper $G \in \mathcal{RH}_\infty^{m\times m}$ with constant polynomial part $M_0$ is negative imaginary if and only if $M_0 = M_0^\mathsf{T}$ and $s(G(s) - M_0) \in \mathbb{R}(s)^{m\times m}$ is positive real.*

The properties above can be related to (4.1) as the derivative in the output generates an additional factor $s$ when taking Laplace transforms. Next, we show an important symmetry property of the function $\Phi(s)$.

**Lemma 4.2.6.** *[BV13, Lem. 4] Let $G \in \mathcal{RH}_\infty^{m\times m}$ and $\Phi(s) \in \mathrm{i}\mathbb{R}(s)^{m\times m}$ as in (4.2) be given. Then we have $\Lambda\left(\Phi(\mathrm{i}\omega)\right) = \Lambda\left(-\Phi(-\mathrm{i}\omega)\right)$ for all $\omega \in \mathbb{R}$.*

*Proof.* First note that $\mathrm{i}\left(G(\mathrm{i}\omega) - G^\mathsf{H}(\mathrm{i}\omega)\right) = -\mathrm{i}\left(G^\mathsf{H}(\mathrm{i}\omega) - G(\mathrm{i}\omega)\right)$ which means that $\Phi(\mathrm{i}\omega)$ is Hermitian and thus has a purely real spectrum for all real values of $\omega$. Thus

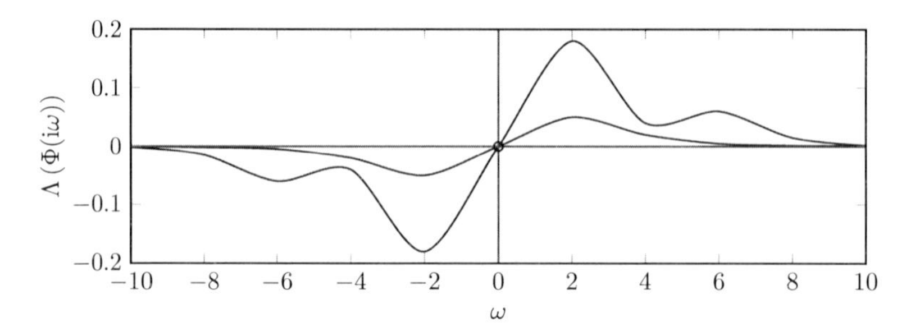


Figure 4.1: Schematic illustration of the spectrum of $\Phi(\mathrm{i}\cdot)$ for a negative imaginary transfer function

we can conclude that

$$\begin{aligned}\Lambda\left(\Phi(\mathrm{i}\omega)\right) &= \Lambda\big(\mathrm{i}\big(G(\mathrm{i}\omega) - G^{\mathsf{H}}(\mathrm{i}\omega)\big)\big) \\ &= \Lambda\Big(\big(\mathrm{i}\big(G(\mathrm{i}\omega) - G^{\mathsf{H}}(\mathrm{i}\omega)\big)\big)^{\mathsf{T}}\Big) \\ &= \Lambda\big(\mathrm{i}\big(G^{\mathsf{H}}(-\mathrm{i}\omega) - G(-\mathrm{i}\omega)\big)\big) \\ &= \Lambda\big(-\mathrm{i}\big(G(-\mathrm{i}\omega) - G^{\mathsf{H}}(-\mathrm{i}\omega)\big)\big) = \Lambda\left(-\Phi(-\mathrm{i}\omega)\right).\end{aligned}$$

□

Following from Lemma 4.2.6, the eigenvalue curves of the matrix-valued function $\Phi(\mathrm{i}\omega)$ are symmetric with respect to the origin, see also Figure 4.1.

## 4.3 Spectral Characterizations for Negative Imaginariness

In this section we derive algebraic characterizations for negative imaginariness of transfer functions in terms of spectral conditions of even matrix pencils. We formulate these conditions by using the given descriptor system realization $(E, A, B, C, D) \in \Sigma_{n,m,m}$ without transforming it to standard state-space form. This has some advantages for computational considerations as performing such a transformation might be an ill-conditioned problem and thus should be avoided whenever possible.

First we can formulate following preliminary result and give a corrected proof.

**Theorem 4.3.1.** *[BV13, Thm. 1] Let $G \in \mathcal{RH}_\infty^{m\times m}$ and $\mathrm{i}\omega_0 \notin \Lambda(E, A)$ be given. Assume that $\Phi(s) \in \mathrm{i}\mathbb{R}(s)^{m\times m}$ is as in (4.2). Then $\Phi(\mathrm{i}\omega_0)$ is singular if and only if* *the even matrix pencil*

$$s\mathcal{E} - \mathcal{A} := \begin{bmatrix} 0 & s\mathrm{i}E - \mathrm{i}A & -\mathrm{i}B \\ s\mathrm{i}E^{\mathsf{T}} + \mathrm{i}A^{\mathsf{T}} & 0 & \mathrm{i}C^{\mathsf{T}} \\ \mathrm{i}B^{\mathsf{T}} & -\mathrm{i}C & \mathrm{i}(D^{\mathsf{T}} - D) \end{bmatrix} \in \mathrm{i}\mathbb{R}[s]^{2n+m\times 2n+m} \tag{4.3}$$

*is singular or has the eigenvalue* $\mathrm{i}\omega_0$*.*

*Proof.* Let $\Phi(\mathrm{i}\omega_0)$ be a singular matrix. First, we see that

$$\begin{aligned} \Phi(\mathrm{i}\omega_0) &= \mathrm{i}C\Big(\mathrm{i}\omega_0(\mathrm{i}E) - \mathrm{i}A\Big)^{-1}\mathrm{i}B + \mathrm{i}D \\ &\quad - \mathrm{i}B^{\mathsf{T}}\Big(\mathrm{i}\omega_0(\mathrm{i}E^{\mathsf{T}}) - (-\mathrm{i}A^{\mathsf{T}})\Big)^{-1}(-\mathrm{i}C^{\mathsf{T}}) - \mathrm{i}D^{\mathsf{T}} \\ &= \begin{bmatrix}\mathrm{i}C & -\mathrm{i}B^{\mathsf{T}}\end{bmatrix}\begin{bmatrix}\mathrm{i}\omega_0(\mathrm{i}E) - \mathrm{i}A & 0 \\ 0 & \mathrm{i}\omega_0\left(\mathrm{i}E^{\mathsf{T}}\right) + \mathrm{i}A^{\mathsf{T}}\end{bmatrix}^{-1}\begin{bmatrix}\mathrm{i}B \\ -\mathrm{i}C^{\mathsf{T}}\end{bmatrix} + \mathrm{i}(D - D^{\mathsf{T}}). \end{aligned} \tag{4.4}$$

From Proposition 2.2.25 and Lemma 2.2.26 it follows that the matrix

$$\mathrm{i}\omega_0\mathcal{E} - \mathcal{A} = \begin{bmatrix} 0 & \mathrm{i}\omega_0(\mathrm{i}E) - \mathrm{i}A & -\mathrm{i}B \\ \mathrm{i}\omega_0(\mathrm{i}E^{\mathsf{T}}) + \mathrm{i}A^{\mathsf{T}} & 0 & \mathrm{i}C^{\mathsf{T}} \\ \mathrm{i}B^{\mathsf{T}} & -\mathrm{i}C & \mathrm{i}(D^{\mathsf{T}} - D) \end{bmatrix}$$

is singular, i. e., the pencil $s\mathcal{E} - \mathcal{A}$ is singular or has the eigenvalue $\mathrm{i}\omega_0$. On the other hand, if $s\mathcal{E} - \mathcal{A}$ is singular, then it follows from Lemma 2.2.26 that $\Phi(s)$ does not have full rank, and therefore, $\Phi(\mathrm{i}\omega_0)$ is singular. If $\mathrm{i}\omega_0$ is an eigenvalue of $s\mathcal{E} - \mathcal{A}$, then Proposition 2.2.25 yields that $\mathrm{i}\omega_0$ is a zero of $\Phi(s)$ since uncontrollable and unobservable modes have been excluded by assumption.

□

From Theorem 4.3.1 we conclude that $G \in \mathcal{RH}_\infty^{m\times m}$ is strictly negative imaginary if and only if $M_0 = M_0^{\mathsf{T}}$, there exists an $\omega_0 > 0$ such that $\Phi(\mathrm{i}\omega_0) > 0$, and the corresponding even matrix pencil $s\mathcal{E} - \mathcal{A}$ has no nonzero, purely imaginary eigenvalues. Graphically, this means that the eigenvalue curves of $\Phi(\mathrm{i}\omega)$ lie all above the zero level in the positive frequency range. However, there is the boundary case of eigenvalue curves that touch the zero level (and hence generate purely imaginary eigenvalues in $s\mathcal{E} - \mathcal{A}$) but do not cross it. These boundary cases are treated in the next theorem.

The following lemma is needed for the proof and is an adaption of Lemma 3.4.1.

**Lemma 4.3.2.** *[BV13, Prop. 4] Let* $s\mathcal{E} - \mathcal{A}$ *be given as in* (4.3)*. Then there exists a congruence transformation* $U(\mathrm{i}\omega) \in \mathbb{C}^{2n+m\times 2n+m}$ *for all* $\mathrm{i}\omega \notin \Lambda(E, A)$ *such that*

$$U^{\mathsf{H}}(\mathrm{i}\omega)(\mathrm{i}\omega\mathcal{E} - \mathcal{A})U(\mathrm{i}\omega) = \begin{bmatrix} 0 & \mathrm{i}\omega(\mathrm{i}E) - \mathrm{i}A & 0 \\ \mathrm{i}\omega(\mathrm{i}E^{\mathsf{T}}) + \mathrm{i}A^{\mathsf{T}} & 0 & 0 \\ 0 & 0 & -\Phi(\mathrm{i}\omega) \end{bmatrix},$$

*where* $\Phi(s) \in \mathrm{i}\mathbb{R}(s)^{m\times m}$ *is as in* (4.2) *and*

$$U(\mathrm{i}\omega) = \begin{bmatrix} I_n & 0 & -\left(\mathrm{i}\omega(\mathrm{i}E^{\mathsf{T}}) + \mathrm{i}A^{\mathsf{T}}\right)^{-1}\mathrm{i}C^{\mathsf{T}} \\ 0 & I_n & \left(\mathrm{i}\omega(\mathrm{i}E) - \mathrm{i}A\right)^{-1}\mathrm{i}B \\ 0 & 0 & I_m \end{bmatrix}.$$

**Theorem 4.3.3.** *Assume that* $\mathrm{i}\omega \notin \Lambda(E, A)$ *for all* $\omega \in \mathbb{R}$, *and let* $G \in \mathcal{RH}_\infty^{m\times m}$ *be given. Assume that* $\Phi(s) \in \mathrm{i}\mathbb{R}(s)^{m\times m}$ *is as in* (4.2) *with* $q = \operatorname{rank}_{\mathbb{C}(s)} \Phi(s)$ *and that the associated even matrix pencil* $s\mathcal{E} - \mathcal{A}$ *is as in* (4.3). *Then the following statements are equivalent:*

*a)* $G(s)$ *is negative imaginary.*

*b) The EKCF of* $s\mathcal{E} - \mathcal{A}$ *has the following structure:*

*i) Whenever there exists an even block of type E2 associated to a* $\mu = \omega_0 > 0$, *it has positive sign-characteristic and there exists an equally sized block of type E2 associated to* $\mu = -\omega_0$ *with negative sign-characteristic.*

*ii) There exist exactly* $q$ *odd blocks of type E2 corresponding to* $\mu = 0$ *with negative sign-characteristic.*

*iii) Blocks of type E3 are either of even size or otherwise, the number of odd blocks of type E3 with positive and negative sign-characteristic is equal.*

*iv) There exist exactly* $m - q$ *blocks of type E4.*

*Proof.* First we show that a) implies b): From the negative imaginariness and stability of $G(s)$ it follows that

$$\begin{aligned} \Phi(\mathrm{i}\omega) &\geq 0 \text{ for all } \omega \geq 0, \text{ and} \\ \Phi(\mathrm{i}\omega) &\leq 0 \text{ for all } \omega \leq 0, \end{aligned}$$

following from Lemma 4.2.6. Then there exists a function $a : \mathbb{R} \to \mathbb{N}$ which is zero except for a finite set of values of $\omega$ such that

- $\operatorname{In}(\mathrm{i}\omega\mathcal{E} - \mathcal{A}) = (n, m - q + a(\omega), n + q - a(\omega))$ for $\omega > 0$,
- $\operatorname{In}(-\mathcal{A}) = (n, m, n) = (n, m - q + a(0), n)$,
- $\operatorname{In}(\mathrm{i}\omega\mathcal{E} - \mathcal{A}) = (n + q - a(\omega), m - q + a(\omega), n)$ for $\omega < 0$.

Roughly speaking, the function $a(\omega)$ describes the change of inertia in the case that eigenvalue curves touch the zero level at $\omega$. Now we analyze which block structures in the EKCF of $s\mathcal{E} - \mathcal{A}$ can produce such an inertia pattern. First of all, $s\mathcal{E} - \mathcal{A}$ has at least $m - q$ zero eigenvalues for all values of $\omega$. Hence, according to the EKCF we have $m - q$ blocks of type E4. We consider the subpencil $s\mathcal{E}_1 - \mathcal{A}_1 \in \mathbb{C}[s]^{2n_1+q\times 2n_1+q}$ of $s\mathcal{E} - \mathcal{A}$ without these blocks which has the inertia pattern

- $\mathrm{In}(\mathrm{i}\omega\mathcal{E}_1 - \mathcal{A}_1) = (n_1, a(\omega), n_1 + q - a(\omega))$ for $\omega > 0$,
- $\mathrm{In}(-\mathcal{A}_1) = (n_1, q, n_1) = (n_1, a(0), n_1)$,
- $\mathrm{In}(\mathrm{i}\omega\mathcal{E}_1 - \mathcal{A}_1) = (n_1 + q - a(\omega), a(\omega), n_1)$ for $\omega < 0$.

From this structure, we can deduce that there exist $q$ odd blocks of type E2 corresponding to $\mu = 0$ with negative sign-characteristic. By again removing these from $s\mathcal{E}_1 - \mathcal{A}_1$, we obtain the subpencil $s\mathcal{E}_2 - \mathcal{A}_2 \in \mathbb{C}[s]^{2n_2 \times 2n_2}$ with

- $\mathrm{In}(\mathrm{i}\omega\mathcal{E}_2 - \mathcal{A}_2) = (n_2, a(\omega), n_2 - a(\omega))$ for $\omega > 0$,
- $\mathrm{In}(-\mathcal{A}_2) = (n_2, 0, n_2)$,
- $\mathrm{In}(\mathrm{i}\omega\mathcal{E}_2 - \mathcal{A}_2) = (n_2 - a(\omega), a(\omega), n_2)$ for $\omega < 0$.

Now, we see that the remaining blocks of type E2 are of even size. Whenever there exists such a block associated to a $\mu = \omega_0 > 0$, it has positive sign-characteristic and there exists an equally sized block of type E2 associated to $\mu = -\omega_0$ with negative sign-characteristic. When removing these blocks as well, there remains a subpencil $s\mathcal{E}_3 - \mathcal{A}_3 \in \mathbb{C}[s]^{2n_3 \times 2n_3}$ of $s\mathcal{E}_2 - \mathcal{A}_2$ with

- $\mathrm{In}(\mathrm{i}\omega\mathcal{E}_3 - \mathcal{A}_3) = (n_3, 0, n_3)$ for $\omega > 0$,
- $\mathrm{In}(-\mathcal{A}_3) = (n_3, 0, n_3)$,
- $\mathrm{In}(\mathrm{i}\omega\mathcal{E}_3 - \mathcal{A}_3) = (n_3, 0, n_3)$ for $\omega < 0$.

This shows that all blocks of type E3 are either of even size or otherwise, the number of odd blocks of type E3 with positive and negative sign-characteristic is equal. This shows the assertion.

To prove that b) implies a) one has to use the same argumentation backwards. By constructing a matrix pencil with the given blocks, one can show the properties of the inertia of the matrix pencil $s\mathcal{E} - \mathcal{A}$ as given here hold. From this fact one can conclude that $G(s)$ is negative imaginary by employing Lemma 4.3.2. □

*Remark* 4.3.4.

a) The even matrix pencil $s\mathcal{E} - \mathcal{A}$ has purely imaginary coefficients. However, from a numerical point of view it is desirable to work in real instead of complex arithmetics. This can be achieved by dividing both coefficient matrices by i. We obtain

$$s\mathcal{H} - \mathcal{S} := \begin{bmatrix} 0 & sE - A & -B \\ sE^{\mathsf{T}} + A^{\mathsf{T}} & 0 & C^{\mathsf{T}} \\ B^{\mathsf{T}} & -C & D^{\mathsf{T}} - D \end{bmatrix} \in \mathbb{R}[s]^{2n+m \times 2n+m}. \tag{4.5}$$

However, instead of an even pencil, $s\mathcal{H} - \mathcal{S}$ is an *odd matrix pencil*, see below.

b) A matrix pencil $P(s) := s\mathcal{H} - \mathcal{S} \in \mathbb{C}[s]^{n\times n}$ is called odd, if $P^{\sim}(s) = -P(s)$, i. e., $\mathcal{H} = \mathcal{H}^{\mathsf{H}}$ and $\mathcal{S} = -\mathcal{S}^{\mathsf{H}}$. Since every odd matrix pencil is equivalent to an even one (which follows by a multiplication with i), it is clear, that odd matrix pencils admit the same spectral symmetry as even ones.

## 4.4 Enforcement of Negative Imaginariness

Often, the systems that we consider are only approximations to the real system dynamics. This happens, if we, e. g., apply model order reduction [BMS05] to a large-scale system or if we approximate the system by rational interpolation via frequency response data (like vector fitting [Ant05], or interpolation via Löwner matrix pencils [LA10]). In this way it can easily happen that the negative imaginariness of the system is lost due to the modeling or approximation error. It is important to keep this property since otherwise this could lead to physically meaningless results when simulating the model. Therefore, one is interested in a post-processing procedure to restore negative imaginariness without introducing too large perturbations to the dynamical system. The method we will use here is an adaption of the concepts presented for passivity enforcement in [GT04, GTU06, SS07]. From Theorems 4.3.1 and 4.3.3 it follows that (strict) negative imaginariness is connected to the spectrum of a related imaginary even (or as shown above real odd) matrix pencil. Thus, our method is based on the computation of a perturbed descriptor system with realization $(\widetilde{E}, \widetilde{A}, \widetilde{B}, \widetilde{C}, \widetilde{D}) \in \Sigma_{n,m,m}$ and transfer function $\widetilde{G} \in \mathcal{RH}_{\infty}^{m\times m}$ which is negative imaginary and the error $\|\widetilde{G} - G\|$ is small in some system norm. The computation is performed by perturbing the nonzero, purely imaginary eigenvalues of the related matrix pencils off the imaginary axis. In our considerations, we keep the matrix pencil $sE - A$ to preserve the modes of the system. So, there is no risk of loosing stability if $\Lambda(E, A) \subset \mathbb{C}^-$. Following from the decomposition (2.12), we have to perturb $B_1$ or $C_1$ if there is a violation of negative imaginariness in the dynamic part. We will discuss in detail which matrix is the best choice for this. Furthermore, we have to modify the matrices $D$, $B_2$, or $C_2$ if the matrix $M_0$ is not symmetric. In this section we will always implicitly assume that $\operatorname{rank}_{\mathbb{C}(s)} \Phi(s) = m$ and that $\Lambda(E, A) \subset \mathbb{C}^-$ so that the system is asymptotically stable.

### 4.4.1 Some Useful Results

First, we need a basic spectral perturbation result for general matrix pencils.

**Proposition 4.4.1.** *[SS90] Let $sE - A \in \mathbb{R}[s]^{n\times n}$ be a given matrix pencil and let $v$, $w \in \mathbb{C}^n$ be right and left eigenvectors corresponding to a simple eigenvalue $\lambda = \frac{\alpha}{\beta} = \frac{w^{\mathsf{H}}Av}{w^{\mathsf{H}}Ev}$. Let $s(E + \Delta E) - (A + \Delta A)$ be a perturbed matrix pencil with* *eigenvalues* $\widetilde{\lambda} = \frac{\widetilde{\alpha}}{\widetilde{\beta}}$. *Then it holds that*

$$\frac{\widetilde{\alpha}}{\widetilde{\beta}} = \frac{\alpha + w^{\mathsf{H}}\Delta A v}{\beta + w^{\mathsf{H}}\Delta B v} + \mathcal{O}\left(\varepsilon^2\right), \tag{4.6}$$

*where* $\varepsilon = \left\| \begin{bmatrix}\Delta A & \Delta B\end{bmatrix} \right\|_2$.

Next, we want to apply this lemma to the special case of an odd matrix pencil. Let $v \in \mathbb{C}^{2n+m}$ be a right eigenvector of the odd matrix pencil $s\mathcal{H} - \mathcal{S} \in \mathbb{R}[s]^{2n+m \times 2n+m}$ corresponding to the eigenvalue $\lambda$. Then we obtain

$$0 = \lambda \mathcal{H} v - \mathcal{S} v.$$

Now, by taking the conjugate transpose of the above equation and using $\mathcal{H}^{\mathsf{T}} = \mathcal{H}$ and $\mathcal{S}^{\mathsf{T}} = -\mathcal{S}$, we obtain

$$0 = \overline{\lambda} v^{\mathsf{H}} \mathcal{H}^{\mathsf{T}} - v^{H} \mathcal{S}^{\mathsf{T}} = \overline{\lambda} v^{\mathsf{H}} \mathcal{H} + v^{H} \mathcal{S}.$$

So, when $\lambda$ is purely imaginary and hence $\lambda = -\overline{\lambda}$, we get that if $v$ is an associated right eigenvector, $v$ is also a corresponding left eigenvector. Let $\lambda = \frac{\alpha}{\beta}$ be a simple, purely imaginary eigenvalue of an odd matrix pencil $s\mathcal{H} - \mathcal{S}$. For a perturbed matrix pencil of the form $s\left(\mathcal{H} + \varepsilon \mathcal{H}'\right) - \left(\mathcal{S} + \varepsilon \mathcal{S}'\right)$, formula (4.6) can be written as

$$\frac{\widetilde{\alpha}}{\widetilde{\beta}} = \frac{\alpha + \varepsilon v^{\mathsf{H}} \mathcal{S}' v}{\beta + \varepsilon v^{\mathsf{H}} \mathcal{H}' v} + \mathcal{O}\left(\varepsilon^2\right), \tag{4.7}$$

**Theorem 4.4.2.** *[BV13, Thm. 3] Consider a transfer function* $G \in \mathcal{RH}_{\infty}^{m \times m}$ *and let* $\Phi(s) \in \mathrm{i}\mathbb{R}(s)^{m \times m}$ *be as in (4.2). Let furthermore* $v \in \mathbb{C}^{2n+m}$ *be a right eigenvector of* $s\mathcal{H} - \mathcal{S} \in \mathbb{R}[s]^{2n+m \times 2n+m}$ *as in (4.5) corresponding to a nonzero, simple, purely imaginary eigenvalue* $\mathrm{i}\omega_0$ *and let* $\nu(\omega)$ *be an eigenvalue curve of* $\Phi(\mathrm{i}\omega)$ *that crosses the level zero at* $\omega_0$*, i. e.,* $\nu(\omega_0) = 0$*. Then the slope of* $\nu(\omega)$ *is positive (negative) at* $\omega_0$ *if* $v^{\mathsf{H}} \mathcal{H} v > 0$ *(*$v^{\mathsf{H}} \mathcal{H} v < 0$*).*

*Proof.* The proof strongly follows the line of argumentation of [SS07, Thm. 2] and is motivated by the following idea. To decide whether the curve increases or decreases at the point $\omega_0$, we compute the point $\omega_0 + \delta$, where the curve crosses the level $\varepsilon$ with $\varepsilon > 0$ and then check whether $\delta$ is positive or negative. Therefore, we consider the eigenvalues of the perturbed matrix pencil

$$s\mathcal{H} - \mathcal{S}_{\varepsilon} := s\mathcal{H} - \left(\mathcal{S} + \varepsilon \mathcal{S}'\right),$$

where

$$\mathcal{S}' = \left.\frac{\mathrm{d}\mathcal{S}_{\varepsilon}}{\mathrm{d}\varepsilon}\right|_{\varepsilon=0} = \left.\frac{\mathrm{d}}{\mathrm{d}\varepsilon}\begin{bmatrix} 0 & 0 & 0 \\ 0 & 0 & 0 \\ 0 & 0 & \mathrm{i}\varepsilon I_m \end{bmatrix}\right|_{\varepsilon=0} = \begin{bmatrix} 0 & 0 & 0 \\ 0 & 0 & 0 \\ 0 & 0 & \mathrm{i} I_m \end{bmatrix}.$$

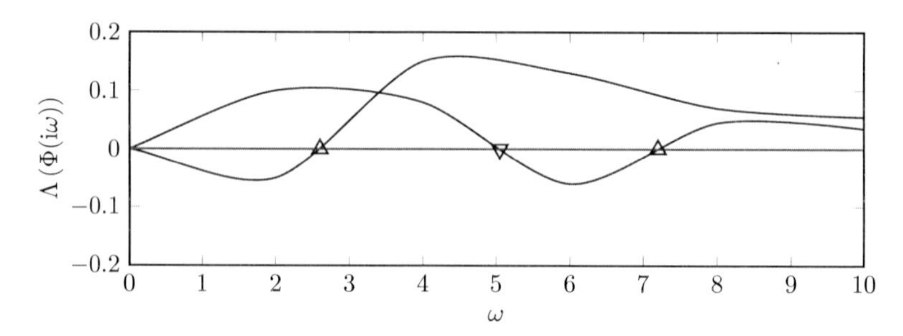


Figure 4.2: Characterization of negative imaginariness violation points and corresponding slopes

The matrix $\mathcal{S}_\varepsilon$ is obtained by analyzing at which frequencies the eigenvalue curves of $\Phi(\mathrm{i}\omega)$ cross the level $\varepsilon$, or equivalently at which frequencies the eigenvalue curves of $\Phi(\mathrm{i}\omega)-\varepsilon I_m$ cross the zero level, see (4.4). Note that we do not consider a perturbation of the matrix $\mathcal{H}$, since

$$\mathcal{H}' = \left.\frac{\mathrm{d}\mathcal{H}}{\mathrm{d}\varepsilon}\right|_{\varepsilon=0} = 0.$$

Furthermore, the matrix $\mathcal{S}_\varepsilon'$ is skew-Hermitian. Let $\mathrm{i}\omega_0$ be a finite eigenvalue of $s\mathcal{H}-\mathcal{S}$ and let $\mathrm{i}\omega_\varepsilon$ be the corresponding perturbed eigenvalue of $s\mathcal{H}-\mathcal{S}_\varepsilon$. Then, by (4.7) it follows that

$$\begin{aligned} \mathrm{i}\omega_\varepsilon &= \frac{v^{\mathsf{H}}\mathcal{S}v + \varepsilon v^{\mathsf{H}}\mathcal{S}'v}{v^{\mathsf{H}}\mathcal{H}v} + \mathcal{O}\left(\varepsilon^2\right) \\ &= \mathrm{i}\omega_0 + \varepsilon\frac{v^{\mathsf{H}}\mathcal{S}'v}{v^{\mathsf{H}}\mathcal{H}v} + \mathcal{O}\left(\varepsilon^2\right). \end{aligned} \tag{4.8}$$

In other words, we have

$$\left.\frac{\mathrm{d}\omega_\varepsilon}{\mathrm{d}\varepsilon}\right|_{\varepsilon=0} = \frac{v^{\mathsf{H}}\mathcal{S}'v}{\mathrm{i}v^{\mathsf{H}}\mathcal{H}v}.$$

Since $\nu$ and $\varepsilon$ can be interchanged, the eigenvalue curve crossing the zero level at $\omega_0$ has the slope

$$\xi := \left.\frac{\mathrm{d}\nu}{\mathrm{d}\omega_\varepsilon}\right|_{\omega_\varepsilon=\omega_0} = \frac{1}{\left.\frac{\mathrm{d}\omega_\varepsilon}{\mathrm{d}\varepsilon}\right|_{\varepsilon=0}} = \frac{\mathrm{i}v^{\mathsf{H}}\mathcal{H}v}{v^{\mathsf{H}}\mathcal{S}'v}.$$

Now, we conclude the assertion as $\mathcal{S}' = \mathrm{i}\widehat{\mathcal{S}}$ with a positive semidefinite matrix $\widehat{\mathcal{S}}$. □

In Figure 4.2, $\Phi(s)$ for a non-negative imaginary transfer function is depicted with intersection points of the eigenvalue curves with the zero level and corresponding slopes depicted by triangles. With this characterization we can now think about moving the nonzero, purely imaginary eigenvalues of $s\mathcal{H} - \mathcal{S}$ off the imaginary axis in order to enforce negative imaginariness. Therefore, we need the positive imaginary eigenvalues of $s\mathcal{H} - \mathcal{S}$ and the corresponding eigenvectors. To compute these, we reformulate the odd eigenvalue problem into a Hamiltonian/skew-Hamiltonian one and use a structure-preserving algorithm [BBMX99] to solve it. Then, we can also use a structure-exploiting technique to obtain the corresponding eigenvectors. We will describe this in detail in Subsections 4.4.7 and 4.4.8.

### 4.4.2 Choice of the New Frequencies

From Figure 4.2 we can see that it is reasonable to assume that the "size" of the violation of negative imaginariness decreases if we move the nonzero, purely imaginary eigenvalues of $s\mathcal{H} - \mathcal{S}$ with negative slope to the right and those with positive slope to the left. Let the frequencies $\omega_i$, where the eigenvalue curves cross the zero level, be ordered in increasing order, i. e., $0 < \omega_1 < \omega_2 < \ldots < \omega_k$. We choose a proportional displacement between $\omega_i$ and $\omega_{i+1}$ [SS07] and obtain

$$\widetilde{\omega}_i = \begin{cases} \omega_i + \alpha(\omega_{i+1} - \omega_i), & v_i^{\mathsf{H}} \mathcal{H} v_i < 0,\ i \neq k \neq 1, \\ (1 + 2\alpha)\omega_i & v_i^{\mathsf{H}} \mathcal{H} v_i < 0,\ i = k, \\ \omega_i - \alpha(\omega_i - \omega_{i-1}), & v_i^{\mathsf{H}} \mathcal{H} v_i > 0,\ i \neq 1 \neq k, \\ (1 - 2\alpha)\omega_i, & v_i^{\mathsf{H}} \mathcal{H} v_i > 0,\ i = 1, \end{cases} \tag{4.9}$$

where $\alpha \in (0, 0.5]$ is a tuning parameter. It seems appropriate to use $\alpha = 0.5$ since then the transfer function would be negative imaginary in just one step. However, the first order perturbation theory only holds in a small neighborhood around $\omega_i$. Therefore, taking $\alpha = 0.5$ might be ill-advised since it corresponds to a large perturbation. Instead, we suggest to use smaller values of $\alpha$ (depending on the problem) and to apply the whole method multiple times, until the negative imaginariness is enforced, see also [SS07]. We remark, that when $v_k^{\mathsf{H}} \mathcal{H} v_k < 0$, the system violates the negative imaginary property at infinity. To restore this we have to move the eigenvalue $\mathrm{i}\omega_k$ to infinity. It is not possible to do this numerically. Hence we define a threshold $\eta$ and declare all eigenvalues whose magnitudes are larger than $\eta$ as numerically infinite.

There are particular situations where the rule above might not lead to the desired result. Consider, for example, the situation depicted in Figure 4.3. Here, there are two *intersected* intervals in which negative imaginariness is violated. This is characterized by two subsequent intersection points of the eigenvalue curves with the zero level which have negative slope followed by two intersection points with positive slope. When successively applying formula (4.9), the second and third frequency point would form a double intersection point (assuming that we are able to exactly perturb these frequency points which is actually not the case). This means that

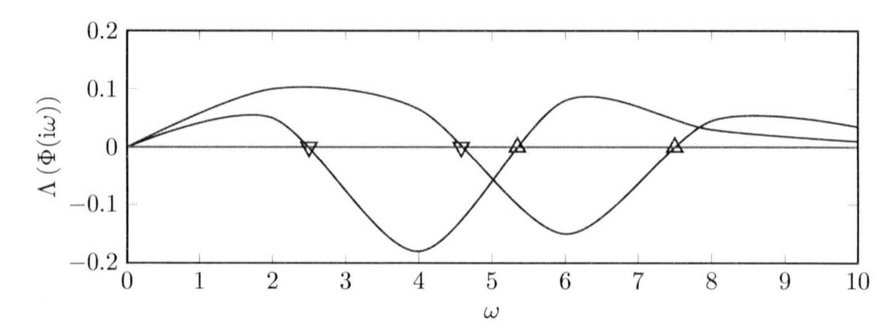


Figure 4.3: Depiction of $\Phi(\mathrm{i}\cdot)$ for a non-negative imaginary transfer function having intersected frequency intervals with negative imaginariness violations

the corresponding matrix pencil $s\mathcal{H} - \mathcal{S}$ has an algebraically double nonzero, purely imaginary eigenvalue. However, in this case we also have two linearly independent eigenvectors which means that this eigenvalue does not generate nontrivial blocks in the Kronecker canonical form. On the other hand, in the case that the frequency intervals do not intersect (like in Figure 4.2), the converged eigenvalues would form an associated block of size two in the Kronecker canonical form as there exists only one linearly independent eigenvector. So we add the following rule to the update formula (4.9):

$$\widetilde{\omega}_i = \begin{cases} \omega_i + \alpha(\omega_{i+2} - \omega_i), & v_i^{\mathsf{H}} \mathcal{H} v_i < 0,\ i \neq k-1,\ k, \\ \omega_i - \alpha(\omega_i - \omega_{i-2}), & v_i^{\mathsf{H}} \mathcal{H} v_i > 0,\ i \neq 1,\ 2, \end{cases}$$

if

$$\begin{aligned} &|\omega_{i+1} - \omega_i| < \delta \quad \text{and} \quad \left\{ \left| \frac{\mathrm{i} v_i^{\mathsf{H}} \mathcal{H} v_i}{v_i^{\mathsf{H}} \mathcal{S}' v_i} \right| > \varepsilon \quad \text{or} \quad \left| \frac{\mathrm{i} v_{i+1}^{\mathsf{H}} \mathcal{H} v_{i+1}}{v_{i+1}^{\mathsf{H}} \mathcal{S}' v_{i+1}} \right| > \varepsilon \right\}, \text{ or} \\ &|\omega_i - \omega_{i-1}| < \delta \quad \text{and} \quad \left\{ \left| \frac{\mathrm{i} v_{i-1}^{\mathsf{H}} \mathcal{H} v_{i-1}}{v_{i-1}^{\mathsf{H}} \mathcal{S}' v_{i-1}} \right| > \varepsilon \quad \text{or} \quad \left| \frac{\mathrm{i} v_i^{\mathsf{H}} \mathcal{H} v_i}{v_i^{\mathsf{H}} \mathcal{S}' v_i} \right| > \varepsilon \right\}, \end{aligned} \tag{4.10}$$

respectively, where $\delta$ and $\varepsilon$ are predefined tolerances.

### 4.4.3 Choice of the System Norm

Similarly as proposed in [GTU06, SS07], we compute the perturbation that minimizes the $\mathcal{HL}_2$-norm of the error $\mathfrak{E}(s) := \widetilde{G}(s) - G(s)$. This norm is a generalization of the $\mathcal{H}_2$-norm for non-strictly proper transfer functions, see Definition 2.2.20. Using the decomposition (2.12) into a strictly proper and polynomial part for $\mathfrak{E}(s)$, we have $\mathfrak{E}(s) = \mathfrak{E}_{\mathrm{sp}}(s) + \mathfrak{E}_{\mathrm{poly}}(s)$. Then, the $\mathcal{HL}_2$-norm is given by

$$\|\mathfrak{E}\|_{\mathcal{HL}_2} := \left( \|\mathfrak{E}_{\mathrm{sp}}\|_{\mathcal{H}_2}^2 + \frac{1}{2\pi} \int_0^{2\pi} \left\|\mathfrak{E}_{\mathrm{poly}}(\mathrm{e}^{\mathrm{i}\omega})\right\|_{\mathrm{F}}^2 \mathrm{d}\omega \right)^{1/2}. \tag{4.11}$$

Assume that the descriptor system $(E, A, B, C, D) \in \Sigma_{n,m,m}$ has been transformed such that $sE - A$ is in QWF, i. e., there exist $W$, $T \in \mathrm{Gl}_n(\mathbb{R})$ such that

$$(WET, WAT, WB, CT, D) = \left( \begin{bmatrix} I_r & 0 \\ 0 & E_{22} \end{bmatrix}, \begin{bmatrix} A_{11} & 0 \\ 0 & I_{n-r} \end{bmatrix}, \begin{bmatrix} B_1 \\ B_2 \end{bmatrix}, \begin{bmatrix} C_1 & C_2 \end{bmatrix}, D \right), \tag{4.12}$$

and suppose that $M_0 = M_0^{\mathsf{T}}$. By only perturbing $B_1$ and/or $C_1$ we ensure that the error $\mathfrak{E}(s)$ is strictly proper. Then we can drop the second term of the right-hand side of (4.11) and get

$$\|\mathfrak{E}\|_{\mathcal{HL}_2} = \|\mathfrak{E}_{\mathrm{sp}}\|_{\mathcal{HL}_2} = \|\mathfrak{E}_{\mathrm{sp}}\|_{\mathcal{H}_2}.$$

Consider the observability Gramian $\mathcal{G}_{\mathrm{o}}$ of the subsystem $(I_r, A_{11}, B_1, C_1) \in \Sigma_{r,m,m}$ which is defined as the unique, positive semidefinite solution of the Lyapunov equation [ZDG96, p. 71]

$$\mathcal{G}_{\mathrm{o}} A_{11} + A_{11}^{\mathsf{T}} \mathcal{G}_{\mathrm{o}} = -C_1^{\mathsf{T}} C_1. \tag{4.13}$$

Since $\mathcal{G}_{\mathrm{o}}$ can be written as

$$\mathcal{G}_{\mathrm{o}} = \frac{1}{2\pi} \int_{-\infty}^{\infty} (\mathrm{i}\omega I_r - A_{11})^{-\mathsf{H}} C_1^{\mathsf{T}} C_1 (\mathrm{i}\omega I_r - A_{11})^{-1} \mathrm{d}\omega,$$

and due to the relation [ZDG96, Lem. 4.6]

$$\|G\|_{\mathcal{H}_2}^2 = \mathrm{tr}\left(B^{\mathsf{T}} \mathcal{G}_{\mathrm{o}} B\right),$$

we have $\|\mathfrak{E}_{\mathrm{sp}}\|_{\mathcal{H}_2} = \|L\Delta\|_{\mathrm{F}}$ where $L$ is a lower triangular Cholesky factor of $\mathcal{G}_{\mathrm{o}}$, i. e., $\mathcal{G}_{\mathrm{o}} = L^{\mathsf{T}} L$, and $\Delta$ is a perturbation of $B_1$, i. e., $\Delta = \widetilde{B}_1 - B_1$ with $\widetilde{B}_1$ corresponding to a system with negative imaginary transfer function.

We remark that it is not necessary to compute the fully decoupled realization (4.12) to solve the Lyapunov equation (4.13) to obtain $L$. This is also not reasonable since the computation of this realization might be arbitrarily ill-conditioned and thus should be avoided. There are algorithms which compute a less condensed form of the matrix pencil $sE - A$, that is

$$W(sE - A)T = s \begin{bmatrix} E_{11} & 0 \\ 0 & E_{22} \end{bmatrix} - \begin{bmatrix} A_{11} & 0 \\ 0 & A_{22} \end{bmatrix}$$

where $W,\, T \in \mathrm{Gl}_n(\mathbb{R})$ and $sE_{11} - A_{11} \in \mathbb{R}[s]^{r\times r}$, $sE_{22} - A_{22} \in \mathbb{R}[s]^{n-r\times n-r}$ are the subpencils of $sE - A$ that correspond to its finite and infinite eigenvalues, respectively. These algorithms basically work in two steps. In Step 1, an upper triangular form with eigenvalue separation of the pencil $sE - A$ is computed, i. e.,

$$\mathcal{P}\,(sE - A)\,\mathcal{Q} = s\begin{bmatrix} E_{11} & E_{12} \\ 0 & E_{22} \end{bmatrix} - \begin{bmatrix} A_{11} & A_{12} \\ 0 & A_{22} \end{bmatrix}, \tag{4.14}$$

with orthogonal matrices $\mathcal{P},\, \mathcal{Q} \in \mathbb{R}^{n\times n}$. This can be done by the QZ algorithm with subsequent eigenvalue reordering [GL96], the GUPTRI algorithm [DK93a, DK93b], or the disk function method discussed in [BB97, SQO04]. Let $\mathcal{P}B = \begin{bmatrix} B_1 \\ B_2 \end{bmatrix}$ and $C\mathcal{Q} = \begin{bmatrix} C_1 & C_2 \end{bmatrix}$ be partitioned according to the block structure of (4.14). Now, Step 2 consists of block-diagonalizing the pencil (4.14). This can be achieved by solving the generalized Sylvester equation

$$A_{11}Y + ZA_{22} + A_{12} = 0, \quad E_{11}Y + ZE_{22} + E_{12} = 0, \tag{4.15}$$

for $Y,\, Z \in \mathbb{R}^{r\times n-r}$, see, e. g., [KD92, KW89]. Then, we define

$$\mathcal{Z} := \begin{bmatrix} I_r & Z \\ 0 & I_{n-r} \end{bmatrix}, \quad \mathcal{Y} := \begin{bmatrix} I_r & Y \\ 0 & I_{n-r} \end{bmatrix},$$

and get

$$\mathcal{Z}\left( s\begin{bmatrix} E_{11} & E_{12} \\ 0 & E_{22} \end{bmatrix} - \begin{bmatrix} A_{11} & A_{12} \\ 0 & A_{22} \end{bmatrix} \right)\mathcal{Y} = s\begin{bmatrix} E_{11} & 0 \\ 0 & E_{22} \end{bmatrix} - \begin{bmatrix} A_{11} & 0 \\ 0 & A_{22} \end{bmatrix}.$$

By updating the input and output matrices, we obtain the realization

$$\left( \begin{bmatrix} E_{11} & 0 \\ 0 & E_{22} \end{bmatrix}, \begin{bmatrix} A_{11} & 0 \\ 0 & A_{22} \end{bmatrix}, \begin{bmatrix} \widehat{B}_1 \\ B_2 \end{bmatrix}, \begin{bmatrix} C_1 & \widehat{C}_2 \end{bmatrix}, D \right) \in \Sigma_{n,m,p} \tag{4.16}$$

with $\widehat{B}_1 := B_1 + ZB_2$ and $\widehat{C}_2 := C_1Y + C_2$. To compute $\|\mathfrak{E}_{\mathrm{sp}}\|_{\mathcal{H}_2}$, we can now solve the generalized Lyapunov equation

$$E_{11}^{\mathsf{T}}\mathcal{G}_{\mathrm{o}}A_{11} + A_{11}^{\mathsf{T}}\mathcal{G}_{\mathrm{o}}E_{11} = -C_1^{\mathsf{T}}C_1. \tag{4.17}$$

instead of (4.13).

Note, that it is sufficient to perform only Step 1 since $E_{11}$, $A_{11}$, and $C_1$ are not changed while performing Step 2. However, this is only possible when we only change the matrix $B_1$ during the enforcement procedure. This would no longer hold, if we would also change $C_1$. This is the reason why we only apply perturbations to $B_1$ in this paper. Furthermore, note that we can compute $L$ directly without explicitly computing $\mathcal{G}_{\mathrm{o}}$ beforehand [BQO99, Ham82].

### 4.4.4 Enforcement Procedure

Now, as we know how to move nonzero, purely imaginary eigenvalues of odd matrix pencils $s\mathcal{H} - \mathcal{S}$ and which system norm we use to compute the optimal perturbation, we are now going to actually compute this perturbation, similarly as in [GT04, GTU06, SS07]. We follow and adapt the argumentation in [SS07] to derive our enforcement procedure. We consider the matrix pencil (4.5), where $sE - A$ is now given in the form (4.14) and $B$ and $C$ are properly updated, i. e.,

$$s\mathcal{H} - \mathcal{S} = \begin{bmatrix} 0 & 0 & sE_{11} - A_{11} & sE_{12} - A_{12} & -B_1 \\ 0 & 0 & 0 & sE_{22} - A_{22} & -B_2 \\ sE_{11}^{\mathsf{T}} + A_{11}^{\mathsf{T}} & 0 & 0 & 0 & C_1^{\mathsf{T}} \\ sE_{12}^{\mathsf{T}} + A_{12}^{\mathsf{T}} & sE_{22}^{\mathsf{T}} + A_{22}^{\mathsf{T}} & 0 & 0 & C_2^{\mathsf{T}} \\ B_1^{\mathsf{T}} & B_2^{\mathsf{T}} & -C_1 & -C_2 & D^{\mathsf{T}} - D \end{bmatrix} \tag{4.18}$$

We perturb the matrix pencil (4.18) by replacing $B_1$ with $B_1 + \Delta$. The perturbed matrix pencil $s\mathcal{H} - \widetilde{\mathcal{S}}$ can then be written as $s\mathcal{H} - \widetilde{\mathcal{S}} = s\mathcal{H} - \big(\mathcal{S} + \widehat{\mathcal{S}}\big)$ with

$$\widehat{\mathcal{S}} = \begin{bmatrix} 0 & 0 & 0 & 0 & \Delta \\ 0 & 0 & 0 & 0 & 0 \\ 0 & 0 & 0 & 0 & 0 \\ 0 & 0 & 0 & 0 & 0 \\ -\Delta^{\mathsf{T}} & 0 & 0 & 0 & 0 \end{bmatrix}. \tag{4.19}$$

Let $v_i$ be a right eigenvector of $s\mathcal{H} - \mathcal{S}$ corresponding to a simple eigenvalue $\mathrm{i}\omega_i$. Then, by (4.8) the imaginary eigenvalues $\mathrm{i}\widetilde{\omega}_i$ of $s\mathcal{H} - \widetilde{\mathcal{S}}$ and those of $s\mathcal{H} - \mathcal{S}$ are related via the first order approximation

$$\widetilde{\omega}_i - \omega_i = \frac{v_i^{\mathsf{H}} \widehat{\mathcal{S}} v_i}{\mathrm{i} v_i^{\mathsf{H}} \mathcal{H} v_i}. \tag{4.20}$$

It holds that

$$\begin{aligned} v_i^{\mathsf{H}} \widehat{\mathcal{S}} v_i &= v_{i1}^{\mathsf{H}} \Delta v_{i5} - v_{i5}^{\mathsf{H}} \Delta^{\mathsf{T}} v_{i1} \\ &= 2\mathrm{i}\, \mathrm{Im}\left(v_{i1}^{\mathsf{H}} \Delta v_{i5}\right), \end{aligned} \tag{4.21}$$

where $v_i = \begin{pmatrix} v_{i1}^{\mathsf{H}} & \dots & v_{i5}^{\mathsf{H}} \end{pmatrix}^{\mathsf{H}} \in \mathbb{C}^{2n+m}$ is partitioned according to the block structure of (4.18). Vectorizing the matrix $\Delta$ in (4.21) gives

$$v_i^{\mathsf{H}} \widehat{\mathcal{S}} v_i = 2\mathrm{i}\, \mathrm{Im}\left(v_{i5}^{\mathsf{T}} \otimes v_{i1}^{\mathsf{H}}\right) \mathrm{vec}\left(\Delta\right).$$

By inserting this into (4.20) we obtain

$$\frac{2}{v_i^{\mathsf{H}} \mathcal{H} v_i} \mathrm{Im}\left(v_{i5}^{\mathsf{T}} \otimes v_{i1}^{\mathsf{H}}\right) \mathrm{vec}\left(\Delta\right) = \widetilde{\omega}_i - \omega_i.$$

By gathering these relations for all purely imaginary eigenvalues with positive imaginary part, we get the linear system of equations

$$Z \operatorname{vec}(\Delta) = \widetilde{\omega} - \omega, \tag{4.22}$$

where $\widetilde{\omega} = \begin{pmatrix} \widetilde{\omega}_1 & \dots & \widetilde{\omega}_k \end{pmatrix}^{\mathsf{T}}$, $\omega = \begin{pmatrix} \omega_1 & \dots & \omega_k \end{pmatrix}^{\mathsf{T}}$, and the $i$-th row of $Z \in \mathbb{R}^{k \times rm}$ is given by

$$e_i^{\mathsf{T}} Z = \frac{2}{v_i^{\mathsf{H}} \mathcal{H} v_i} \operatorname{Im}\left(v_{i5}^{\mathsf{T}} \otimes v_{i1}^{\mathsf{H}}\right),$$

where $e_i \in \mathbb{R}^k$ denotes the $i$-th canonical unit vector. To compute the optimal perturbation $\Delta$, i. e., the one that satisfies (4.22) and minimizes $\|\mathfrak{E}_{\mathrm{sp}}\|_{\mathcal{H}_2}$, we have to solve the minimization problem

$$\min_{\Delta \in \mathbb{R}^{r \times m}} \|L\Delta\|_{\mathrm{F}} \text{ subject to } Z \operatorname{vec}(\Delta) = \widetilde{\omega} - \omega.$$

If $(E, A, B, C, D) \in \Sigma_{n,m,p}$ is behaviorally observable, then the observability Gramian $\mathcal{G}_{\mathrm{o}}$ is positive definite with a nonsingular Cholesky factor $L$ [ZDG96]. Therefore, we can perform the change of basis $\Delta_L := L\Delta$ and obtain the equivalent minimization problem

$$\min_{\Delta_L \in \mathbb{R}^{r \times m}} \|\Delta_L\|_{\mathrm{F}} \text{ subject to } Z_L \operatorname{vec}(\Delta_L) = \widetilde{\omega} - \omega, \tag{4.23}$$

where $Z_L = Z\left(I \otimes L^{-1}\right)$. Note that the $i$-th row of $Z_L$ can be computed as

$$e_i^{\mathsf{T}} Z_L = e_i^{\mathsf{T}} Z\left(I \otimes L^{-1}\right) = \frac{2}{v_i^{\mathsf{H}} \mathcal{H} v_i} \operatorname{Im}\left(v_{i5}^{\mathsf{T}} \otimes v_{i1}^{\mathsf{H}} L^{-1}\right). \tag{4.24}$$

This avoids the explicit construction of the matrix $I \otimes L^{-1}$. Now the minimization problem (4.23) transforms to the standard least squares problem

$$\min_{\Delta_L \in \mathbb{R}^{r \times m}} \|\operatorname{vec}(\Delta_L)\|_2 \text{ subject to } Z_L \operatorname{vec}(\Delta_L) = \widetilde{\omega} - \omega.$$

Its solution is formally given by the Moore-Penrose inverse $Z_L^+$ of $Z_L$, namely

$$\operatorname{vec}(\Delta_L) = Z_L^+ (\widetilde{\omega} - \omega).$$

Finally, the desired perturbation is given by

$$\Delta = L^{-1} \Delta_L.$$

### 4.4.5 Enforcing Symmetry of the Polynomial Part

As shown by Lemma 4.2.6, a negative imaginary transfer function $G \in \mathcal{RH}_\infty^{m\times m}$ satisfies $M_0 = M_0^\mathsf{T}$, where $M_0 = \lim_{\omega\to\infty} G(\mathrm{i}\omega)$. It might happen, that this property is also lost during the modeling process. In this section we will briefly describe how to restore the symmetry of $M_0$. First, we actually compute this matrix. This can be done by transforming the system $(E, A, B, C, D)$ into the decoupled form (4.16). Since $G(s)$ is proper, we directly obtain that $M_0 = D - \widehat{C}_2 A_{22}^{-1} B_2$. Note, that computing the form (4.16) might be ill-conditioned, as solving a generalized Sylvester equation might be. Hence, this operation should be avoided when it is clear, that $M_0$ is symmetric.

Now assume that $M_0$ is not symmetric. Then

$$M_0 - M_0^\mathsf{T} = \mathcal{F} = \mathcal{T} - \mathcal{T}^\mathsf{T},$$

where $\mathcal{T}$ is defined as the strictly upper triangular part of the skew-symmetric error matrix $\mathcal{F}$. In this way we perturb the matrix $D$ as

$$\widetilde{D} := D - \mathcal{T}.$$

The error caused by this perturbation in the $\mathcal{HL}_2$-norm of the system is given by

$$\|\mathfrak{E}\|_{\mathcal{HL}_2} = \left(\frac{1}{2\pi}\int_0^{2\pi} \left\|\mathcal{P}(\mathrm{e}^{\mathrm{i}\omega})\right\|_\mathrm{F}^2 \,\mathrm{d}\omega\right)^{1/2} = \|\mathcal{T}\|_\mathrm{F},$$

see (4.11).

### 4.4.6 The Overall Process

From the considerations above we can now state the procedures for enforcing the symmetry of $M_0$ in Algorithm 4.1 and negative imaginariness in Algorithm 4.2. Note, that when Algorithm 4.1 has been performed, the triangularization of $sE - A$ has already been done, so this step can be omitted in Algorithm 4.2.

We briefly summarize how to solve some specific subproblems with available software tools. In particular, we mention routines implemented in MATLAB and FORTRAN (within the software packages LAPACK[1] and SLICOT[2]). Algorithms which have only been implemented in FORTRAN can be called by MATLAB by using its `mex` functionality. See Table 4.1 for an overview. Note that the SLICOT routine `MB04BD` is actually designed to compute the eigenvalues of a skew-Hamiltonian/Hamiltonian matrix pencil. However, as pointed out in the next subsection, there is a close connection between skew-Hamiltonian/Hamiltonian and odd matrix pencils.

[1] `http://www.netlib.org/lapack/`
[2] `http://slicot.org/`

**Algorithm 4.1** Algorithm for enforcing symmetry of $M_0$

**Input:** Asymptotically stable descriptor system $(E, A, B, C, D) \in \Sigma_{n,m,p}$.

**Output:** Descriptor system $(E, A, B, C, \widetilde{D})$ with symmetric constant part.

1: Triangularize the matrix pencil $sE - A$, i.e., compute orthogonal $\mathcal{P} \in \mathbb{R}^{n\times n}$ and $\mathcal{Q} \in \mathbb{R}^{n\times n}$ such that

$$\mathcal{P}(sE - A)\mathcal{Q} = s\begin{bmatrix} E_{11} & E_{12} \\ 0 & E_{22} \end{bmatrix} - \begin{bmatrix} A_{11} & A_{12} \\ 0 & A_{22} \end{bmatrix}.$$

2: Set $B := \mathcal{P}B = \begin{bmatrix} B_1 \\ B_2 \end{bmatrix}$ and $C := C\mathcal{Q} = \begin{bmatrix} C_1 & C_2 \end{bmatrix}$.

3: Solve the generalized Sylvester equation

$$A_{11}Y + ZA_{22} + A_{12} = 0, \quad E_{11}Y + ZE_{22} + E_{12} = 0.$$

4: Update $C_2 := C_1 Y + C_2$.

5: Compute $M_0 := D - C_2 A_{22}^{-1} B_2$.

6: Compute the strictly upper triangular part of $M_0 - M_0^{\mathsf{T}}$, denoted by $\mathcal{T}$.

7: Set $\widetilde{D} := D - \mathcal{T}$.

### 4.4.7 Reformulation of the Odd Eigenvalue Problem

This subsection provides some details about the solution of the odd eigenvalue problem. First, we transform the matrix pencil $s\mathcal{H} - \mathcal{S}$ to a related Hamiltonian/skew-Hamiltonian pencil $s\widetilde{\mathcal{H}} - \widetilde{\mathcal{S}}$, i.e., $\widetilde{\mathcal{H}}$ is Hamiltonian and $\widetilde{\mathcal{S}}$ is skew-Hamiltonian.

Then, we can apply the structure-preserving method presented in Subsection 2.1.3 to compute the eigenvalues of $s\widetilde{\mathcal{S}} - \widetilde{\mathcal{H}}$. The related eigenvectors can also be computed in a structure-exploiting manner by using the generalized symplectic URV decomposition, see (2.4). More details on this will be given in the next subsection.

Consider now the odd matrix pencil $s\mathcal{H} - \mathcal{S} \in \mathbb{R}[s]^{2n+m\times 2n+m}$. Recall that every Hamiltonian/skew-Hamiltonian matrix pencil has even dimension. So, if $m$ is an odd number, we first increase the dimension of $s\mathcal{H} - \mathcal{S}$ by one. We define the numbers

$$r := m \bmod 2, \quad k := n + \frac{1}{2}(m + r).$$

Then, similarly as in Subsection 2.1.3, the pencil

$$s\widetilde{\mathcal{H}} - \widetilde{\mathcal{S}} := \mathcal{J}_k \begin{bmatrix} s\mathcal{H} - \mathcal{S} & 0 \\ 0 & I_r \end{bmatrix}$$

is Hamiltonian/skew-Hamiltonian. If the dimension is increased by one, an additional infinite eigenvalue is introduced. Now, we can compute the purely imaginary

**Algorithm 4.2** Algorithm for the enforcement of negative imaginariness

**Input:** Asymptotically stable and behaviorally observable descriptor system $(E, A, B, C, D)$ such that $\lim_{\omega \to \infty} \Phi(\mathrm{i}\omega) = 0$, control parameters $0 < \alpha \leq 0.5$, $\delta > 0$, $\varepsilon > 0$.

**Output:** A negative imaginary descriptor system $\left(E, A, \widetilde{B}, C, D\right)$.

1: Triangularize the matrix pencil $sE - A$, i.e., compute orthogonal $\mathcal{P} \in \mathbb{R}^{n \times n}$ and $\mathcal{Q} \in \mathbb{R}^{n \times n}$ such that

$$\mathcal{P}(sE - A)\mathcal{Q} = s \begin{bmatrix} E_{11} & E_{12} \\ 0 & E_{22} \end{bmatrix} - \begin{bmatrix} A_{11} & A_{12} \\ 0 & A_{22} \end{bmatrix}.$$

2: Set $B := \mathcal{P}B = \begin{bmatrix} B_1 \\ B_2 \end{bmatrix}$ and $C := C\mathcal{Q} = \begin{bmatrix} C_1 & C_2 \end{bmatrix}$.

3: Compute the Cholesky factor $L$ of the observability Gramian $\mathcal{G}_\mathrm{o} = L^\mathsf{T} L$ by solving the generalized Lyapunov equation (4.17).

4: Compute the purely imaginary eigenvalues of the odd matrix pencil $s\mathcal{H} - \mathcal{S}$ from (4.18) with positive imaginary part.

5: **while** $s\mathcal{H} - \mathcal{S}$ has nonzero, purely imaginary eigenvalues **do**

6: Choose new eigenvalues as in (4.9) and (4.10).

7: Solve $\min_{\Delta_L \in \mathbb{R}^{r \times m}} \|\mathrm{vec}\,(\Delta_L)\|_2$ subject to $Z_L\,\mathrm{vec}(\Delta_L) = \widetilde{\omega} - \omega$ with $Z_L$ as in (4.24).

8: Update $B_1 := B_1 + L^{-1}\Delta_L$ and update $\mathcal{S}$ accordingly.

9: Compute the positive imaginary eigenvalues and the corresponding eigenvectors of $s\mathcal{H} - \mathcal{S}$.

10: **end while**

11: Set $\widetilde{B} := \mathcal{P}^\mathsf{T} B$.

eigenvalues of $s\widetilde{\mathcal{S}} - \widetilde{\mathcal{H}}$ and the associated eigenvectors in a structure-preserving way. Due to the symmetry of $\Phi(\mathrm{i}\cdot)$ we only need the eigenvalues with positive imaginary parts. Let $\mathrm{i}\omega$ with $\omega > 0$ be an eigenvalue of $s\widetilde{\mathcal{S}} - \widetilde{\mathcal{H}}$ with eigenvector $v \in \mathbb{C}^{2k}$. Then it holds that

$$\begin{aligned} 0 &= \mathrm{i}\omega\widetilde{\mathcal{S}}v - \widetilde{\mathcal{H}}v \\ &= \widetilde{\mathcal{S}}v - \frac{1}{\mathrm{i}\omega}\widetilde{\mathcal{H}}v \\ &= \widetilde{\mathcal{S}}v + \mathrm{i}\frac{1}{\omega}\widetilde{\mathcal{H}}v \\ &= \widetilde{\mathcal{S}}\overline{v} - \mathrm{i}\frac{1}{\omega}\widetilde{\mathcal{H}}\overline{v}. \end{aligned}$$

In other words, $v$ is a right eigenvector of $s\widetilde{\mathcal{S}} - \widetilde{\mathcal{H}}$ corresponding to $\mathrm{i}\omega$ if and only if $\overline{v}$ is a right eigenvector of $s\widetilde{\mathcal{H}} - \widetilde{\mathcal{S}}$ corresponding to $\mathrm{i}\frac{1}{\omega}$. Now the theory presented for

Table 4.1: Survey of available software

| Operation | MATLAB | FORTRAN |
|---|---|---|
| Block triangularizing $sE - A$ as in (4.14) | qz, ordqz | DGGES |
| | guptri[3] | GUPTRI[3] |
| Solving generalized Sylvester equations as in (4.15) | — | SB04OD |
| Solving generalized Lyapunov equations as in (4.17) | lyapchol | SG03BD |
| Computing imaginary eigenvalues of $s\mathcal{H} - \mathcal{S}$ | — | MB04BD |

odd matrix pencils in this section directly applies to Hamiltonian/skew-Hamiltonian pencils as well. However, note that if $v$ is a right eigenvector to a purely imaginary eigenvalue $\mathrm{i}\omega$, then $\mathcal{J}_k v$ is a corresponding left eigenvector.

### 4.4.8 Computation of the Eigenvectors

To compute the eigenvectors of the skew-Hamiltonian/Hamiltonian pencil $s\widetilde{\mathcal{S}} - \widetilde{\mathcal{H}} \in \mathbb{R}[s]^{2k\times 2k}$ corresponding to the purely imaginary eigenvalues we will make use of the generalized symplectic URV decomposition, presented in Subsection 2.1.3. Since for our enforcement procedure we only need the positive imaginary eigenvalues, i. e., those with positive imaginary parts, we restrict ourselves to the computation of the eigenvectors corresponding to these eigenvalues. The following method has already been presented in [BV11, JV13]. Throughout the whole subsection we assume that all purely imaginary eigenvalues are simple.

**The Algorithm** Consider the symplectic URV decomposition of $s\widetilde{\mathcal{S}} - \widetilde{\mathcal{H}} \in \mathbb{R}[s]^{2k\times 2k}$, i. e., there exist orthogonal matrices $\mathcal{Q}_1$, $\mathcal{Q}_2 \in \mathbb{R}^{2k\times 2k}$ such that

$$\begin{aligned} \mathcal{Q}_1^\mathsf{T} \widetilde{\mathcal{S}} \mathcal{J}_k \mathcal{Q}_1 \mathcal{J}_k^\mathsf{T} &= \begin{bmatrix} S_{11} & S_{12} \\ 0 & S_{11}^\mathsf{T} \end{bmatrix}, \\ \mathcal{J}_k \mathcal{Q}_2^\mathsf{T} \mathcal{J}_k^\mathsf{T} \widetilde{\mathcal{S}} \mathcal{Q}_2 &= \begin{bmatrix} T_{11} & T_{12} \\ 0 & T_{11}^\mathsf{T} \end{bmatrix}, \\ \mathcal{Q}_1^\mathsf{T} \widetilde{\mathcal{H}} \mathcal{Q}_2 &= \begin{bmatrix} H_{11} & H_{12} \\ 0 & H_{22} \end{bmatrix}, \end{aligned} \tag{4.25}$$

where $S_{12}$ and $T_{12}$ are skew-symmetric and the formal matrix product $S_{11}^{-1} H_{11} T_{11}^{-1} H_{22}^\mathsf{T}$ is in real periodic Schur form [BGD92, HL94, Kre01]. Moreover, define the skew-

[3] http://www8.cs.umu.se/~guptri/

Hamiltonian/Hamiltonian pencil

$$s\mathcal{B}_{\widetilde{\mathcal{S}}} - \mathcal{B}_{\widetilde{\mathcal{H}}} := \mathcal{X}_k^\mathsf{T} \begin{bmatrix} s\widetilde{\mathcal{S}} - \widetilde{\mathcal{H}} & 0 \\ 0 & s\widetilde{\mathcal{S}} + \widetilde{\mathcal{H}} \end{bmatrix} \mathcal{X}_k \in \mathbb{R}[s]^{4k \times 4k} \tag{4.26}$$

as in (2.3) and (2.2). By using the decomposition (4.25) we can finally determine an orthogonal matrix $\mathcal{Q} \in \mathbb{R}^{4k\times 4k}$ such that

$$\begin{aligned} s\widehat{\mathcal{B}}_{\widetilde{\mathcal{S}}} - \widehat{\mathcal{B}}_{\widetilde{\mathcal{H}}} :&= \mathcal{J}_{2k}\mathcal{Q}^\mathsf{T}\mathcal{J}_{2k}^\mathsf{T}(s\mathcal{B}_{\widetilde{\mathcal{S}}} - \mathcal{B}_{\widetilde{\mathcal{H}}})\mathcal{Q} \\ &= s\left[\begin{array}{cc|cc} S_{11} & 0 & S_{12} & 0 \\ 0 & T_{11} & 0 & T_{12} \\ \hline 0 & 0 & S_{11}^\mathsf{T} & 0 \\ 0 & 0 & 0 & T_{11}^\mathsf{T} \end{array}\right] - \left[\begin{array}{cc|cc} 0 & H_{11} & 0 & H_{12} \\ -H_{22}^\mathsf{T} & 0 & H_{12}^\mathsf{T} & 0 \\ \hline 0 & 0 & 0 & H_{22} \\ 0 & 0 & -H_{11}^\mathsf{T} & 0 \end{array}\right] \end{aligned}$$

with $\mathcal{Q} = \mathcal{P}_k^\mathsf{T} \begin{bmatrix} \mathcal{J}_k\mathcal{Q}_1\mathcal{J}_k^\mathsf{T} & 0 \\ 0 & \mathcal{Q}_2 \end{bmatrix} \mathcal{P}_k$.

To derive an algorithm for computing the desired eigenvectors we use the following two lemmas. In the following we assume that the matrix pencil $s\begin{bmatrix} S_{11} & 0 \\ 0 & T_{11} \end{bmatrix} - \begin{bmatrix} 0 & H_{11} \\ H_{22}^\mathsf{T} & 0 \end{bmatrix}$ is regular which is also equivalent to the regularity of $s\begin{bmatrix} S_{11} & 0 \\ 0 & T_{11} \end{bmatrix} - \begin{bmatrix} 0 & H_{11} \\ -H_{22}^\mathsf{T} & 0 \end{bmatrix}$ and $s\widehat{\mathcal{B}}_{\widetilde{\mathcal{S}}} - \widehat{\mathcal{B}}_{\widetilde{\mathcal{H}}}$.

**Lemma 4.4.3.** *[JV13, Lem. 3] The vector* $\begin{pmatrix} v_1 \\ v_2 \end{pmatrix} \in \mathbb{C}^{2k}$ *with* $v_1, v_2 \in \mathbb{C}^k$ *is a right eigenvector of the matrix pencil* $s\begin{bmatrix} S_{11} & 0 \\ 0 & T_{11} \end{bmatrix} - \begin{bmatrix} 0 & H_{11} \\ H_{22}^\mathsf{T} & 0 \end{bmatrix}$ *corresponding to the eigenvalue* $\omega_0$ *if and only if* $\begin{pmatrix} -\mathrm{i}v_1 \\ v_2 \end{pmatrix}$ *is a right eigenvector of the matrix pencil* $s\begin{bmatrix} S_{11} & 0 \\ 0 & T_{11} \end{bmatrix} - \begin{bmatrix} 0 & H_{11} \\ -H_{22}^\mathsf{T} & 0 \end{bmatrix}$ *corresponding to the eigenvalue* $\mathrm{i}\omega_0$.

*Proof.* Let $\begin{pmatrix} v_1 \\ v_2 \end{pmatrix}$ be a right eigenvector of $s\begin{bmatrix} S_{11} & 0 \\ 0 & T_{11} \end{bmatrix} - \begin{bmatrix} 0 & H_{11} \\ H_{22}^\mathsf{T} & 0 \end{bmatrix}$ corresponding to the eigenvalue $\omega_0$. Then we have

$$\omega_0 S_{11} v_1 = H_{11} v_2, \quad \omega_0 T_{11} v_2 = H_{22}^\mathsf{T} v_1.$$

This is equivalent to

$$\mathrm{i}\omega_0 S_{11}(-\mathrm{i}v_1) = H_{11} v_2, \quad \mathrm{i}\omega_0 T_{11} v_2 = -H_{22}^\mathsf{T}(-\mathrm{i}v_1).$$

In other words, $\begin{pmatrix} -iv_1 \\ v_2 \end{pmatrix}$ is a right eigenvector of the matrix pencil $s\begin{bmatrix} S_{11} & 0 \\ 0 & T_{11} \end{bmatrix} - \begin{bmatrix} 0 & H_{11} \\ -H_{22}^{\mathsf{T}} & 0 \end{bmatrix}$ corresponding to the eigenvalue $i\omega_0$. The converse direction is analogous. □

**Lemma 4.4.4.** *[JV13, Lem. 4] The vector $v \in \mathbb{C}^{2k}$ is a right eigenvector of the matrix pencil $s\begin{bmatrix} S_{11} & 0 \\ 0 & T_{11} \end{bmatrix} - \begin{bmatrix} 0 & H_{11} \\ -H_{22}^{\mathsf{T}} & 0 \end{bmatrix}$ corresponding to the eigenvalue $\lambda_0$ if and only if the vector $\begin{pmatrix} v \\ 0 \end{pmatrix} \in \mathbb{C}^{4k}$ is a right eigenvector of the skew-Hamiltonian/Hamiltonian matrix pencil $s\widehat{\mathcal{B}}_{\widetilde{\mathcal{S}}} - \widehat{\mathcal{B}}_{\widetilde{\mathcal{H}}}$ corresponding to the eigenvalue $\lambda_0$.*

*Proof.* The proof is trivial. □

As an intermediate step, we compute a matrix $X$ whose columns contain the eigenvectors corresponding to the positive real eigenvalues of the pencil $s\begin{bmatrix} S_{11} & 0 \\ 0 & T_{11} \end{bmatrix} - \begin{bmatrix} 0 & H_{11} \\ H_{22}^{\mathsf{T}} & 0 \end{bmatrix}$. This is done by the following basic steps already summarized in [BV11].

*Step 1:* Reorder the positive real eigenvalues of the $k \times k$ generalized matrix product $P := S_{11}^{-1} H_{11} T_{11}^{-1} H_{22}^{\mathsf{T}}$ to the top, i. e., compute orthogonal matrices $U_i = \begin{bmatrix} U_i^{(1)} & U_i^{(2)} \end{bmatrix}$, $i = 1, \ldots, 4$, such that

$$U_2^{\mathsf{T}} S_{11} U_1 = \begin{bmatrix} S_{11}^{(11)} & S_{11}^{(12)} \\ 0 & S_{11}^{(22)} \end{bmatrix}, \quad U_2^{\mathsf{T}} H_{11} U_3 = \begin{bmatrix} H_{11}^{(11)} & H_{11}^{(12)} \\ 0 & H_{11}^{(22)} \end{bmatrix},$$

$$U_4^{\mathsf{T}} T_{11} U_3 = \begin{bmatrix} T_{11}^{(11)} & T_{11}^{(12)} \\ 0 & T_{11}^{(22)} \end{bmatrix}, \quad U_4^{\mathsf{T}} H_{22}^{\mathsf{T}} U_1 = \begin{bmatrix} H_{22}^{(11)} & H_{22}^{(12)} \\ 0 & H_{22}^{(22)} \end{bmatrix}$$

are still in upper (quasi-)triangular form, but the eigenvalues of the $q \times q$ generalized matrix product $P^{(11)} := \big(S_{11}^{(11)}\big)^{-1} H_{11}^{(11)} \big(T_{11}^{(11)}\big)^{-1} H_{22}^{(11)}$ are the positive real ones of $P$ [GKK03].

*Step 2:* Reorder the eigenvalues of $s\begin{bmatrix} S_{11}^{(11)} & 0 \\ 0 & T_{11}^{(11)} \end{bmatrix} - \begin{bmatrix} 0 & H_{11}^{(11)} \\ H_{22}^{(11)} & 0 \end{bmatrix} \in \mathbb{R}[s]^{2q \times 2q}$ by computing orthogonal matrices $V_1 = \begin{bmatrix} V_1^{(1)} & V_1^{(2)} \end{bmatrix}$, $V_2 = \begin{bmatrix} V_2^{(1)} & V_2^{(2)} \end{bmatrix} \in \mathbb{R}^{2q \times 2q}$ such that

$$V_1^T \left( s\begin{bmatrix} S_{11}^{(11)} & 0 \\ 0 & T_{11}^{(11)} \end{bmatrix} - \begin{bmatrix} 0 & H_{11}^{(11)} \\ H_{22}^{(11)} & 0 \end{bmatrix} \right) V_2 = s\begin{bmatrix} \widetilde{S}_{11} & \widetilde{S}_{12} \\ 0 & \widetilde{S}_{22} \end{bmatrix} - \begin{bmatrix} \widetilde{H}_{11} & \widetilde{H}_{12} \\ 0 & \widetilde{H}_{22} \end{bmatrix},$$

where the eigenvalues of $s\widetilde{S}_{11} - \widetilde{H}_{11} \in \mathbb{R}[s]^{q\times q}$ are positive and those of $s\widetilde{S}_{22} - \widetilde{H}_{22} \in \mathbb{R}[s]^{q\times q}$ are negative.

*Step 3:* Compute the eigenvectors of $s\widetilde{S}_{11} - \widetilde{H}_{11}$, i. e., compute a matrix $W \in \mathbb{R}^{q\times q}$ such that $\widetilde{H}_{11}W = \widetilde{S}_{11}W\mathcal{D}$, where $\mathcal{D} \in \mathbb{R}^{q\times q}$ is an appropriate diagonal matrix composed of the eigenvalues of $s\widetilde{S}_{11} - \widetilde{H}_{11}$.

*Step 4:* Collect the information contained in the relevant columns of the transformation matrices to obtain

$$X := \begin{bmatrix} X^{(1)} \\ X^{(2)} \end{bmatrix} := \begin{bmatrix} U_1^{(1)} & 0 \\ 0 & U_3^{(1)} \end{bmatrix} V_2^{(1)} W \in \mathbb{R}^{2k\times q}.$$

Now, using Lemma 4.4.3 it turns out that

$$\widetilde{X} := \begin{bmatrix} -\mathrm{i}X^{(1)} \\ X^{(2)} \end{bmatrix} \in \mathbb{C}^{2k\times q}$$

contains the eigenvectors corresponding to the positive imaginary eigenvalues of the pencil $s\begin{bmatrix} S_{11} & 0 \\ 0 & T_{11} \end{bmatrix} - \begin{bmatrix} 0 & H_{11} \\ -H_{22}^{\mathsf{T}} & 0 \end{bmatrix}$. Then, by Lemma 4.4.4, the columns of the matrix $\begin{bmatrix} \widetilde{X} \\ 0 \end{bmatrix} \in \mathbb{C}^{4k\times q}$ contain eigenvectors corresponding to the positive imaginary eigenvalues of the pencil $s\widehat{\mathcal{B}}_{\widetilde{\mathcal{S}}} - \widehat{\mathcal{B}}_{\widetilde{\mathcal{H}}}$. Note that all eigenvalues of this pencil have double algebraic, geometric, and partial multiplicities. So the matrix $\begin{bmatrix} \widetilde{X} \\ 0 \end{bmatrix}$ contains only *half* of the eigenvectors to each positive imaginary eigenvalue of $s\widehat{\mathcal{B}}_{\widetilde{\mathcal{S}}} - \widehat{\mathcal{B}}_{\widetilde{\mathcal{H}}}$. However, this is not a problem, since by later turning to the original pencil $s\widetilde{\mathcal{S}} - \widetilde{\mathcal{H}}$, we do not need the other half of the eigenvectors.

Decompose the matrices

$$\mathcal{Q}_1 =: \begin{bmatrix} \mathcal{Q}_1^{(11)} & \mathcal{Q}_1^{(12)} \\ \mathcal{Q}_1^{(21)} & \mathcal{Q}_1^{(22)} \end{bmatrix}, \quad \mathcal{Q}_1^{(ij)} \in \mathbb{R}^{k\times k}, \quad i, j = 1, 2, \text{ and}$$

$$\mathcal{Q}_2 =: \begin{bmatrix} \mathcal{Q}_2^{(11)} & \mathcal{Q}_2^{(12)} \\ \mathcal{Q}_2^{(21)} & \mathcal{Q}_2^{(22)} \end{bmatrix}, \quad \mathcal{Q}_2^{(ij)} \in \mathbb{R}^{k\times k}, \quad i, j = 1, 2.$$

Now, the corresponding eigenvectors corresponding to the positive imaginary eigenvalues of the double-sized matrix pencil $s\mathcal{B}_{\widetilde{\mathcal{S}}} - \mathcal{B}_{\widetilde{\mathcal{H}}}$ are given by

$$Y := \begin{bmatrix} Y_1 \\ Y_2 \end{bmatrix} = \mathcal{X}_k \mathcal{Q} \begin{bmatrix} \widetilde{X} \\ 0 \end{bmatrix} = \frac{1}{\sqrt{2}} \begin{bmatrix} -\mathrm{i}\mathcal{Q}_1^{(22)} X^{(1)} + \mathcal{Q}_2^{(11)} X^{(2)} \\ \mathrm{i}\mathcal{Q}_1^{(12)} X^{(1)} + \mathcal{Q}_2^{(21)} X^{(2)} \\ \mathrm{i}\mathcal{Q}_1^{(22)} X^{(1)} + \mathcal{Q}_2^{(11)} X^{(2)} \\ -\mathrm{i}\mathcal{Q}_1^{(12)} X^{(1)} + \mathcal{Q}_2^{(21)} X^{(2)} \end{bmatrix} \in \mathbb{C}^{4k\times q}.$$

In the equation above, only $Y_1 \in \mathbb{C}^{2k \times q}$ contains the desired eigenvectors of the matrix pencil $s\widetilde{\mathcal{S}} - \widetilde{\mathcal{H}}$ and $Y$ does not have to be computed explicitly. More specifically, we can express $Y_1$ as

$$Y_1 := \frac{1}{\sqrt{2}} \begin{bmatrix} -\mathrm{i}\mathcal{Q}_1^{(22)} X^{(1)} + \mathcal{Q}_2^{(11)} X^{(2)} \\ \mathrm{i}\mathcal{Q}_1^{(12)} X^{(1)} + \mathcal{Q}_2^{(21)} X^{(2)} \end{bmatrix}.$$

**Implementation and Numerical Experiments** To compute the eigenvector matrix $Y_1$ we have implemented a FORTRAN 77 subroutine `MB04BV` that is subject to be included into SLICOT. This routine takes the output of the SLICOT routine `MB04BD` (that performs the generalized symplectic URV decomposition as in (4.25)) as inputs, i. e., the matrices $S_{11}$, $T_{11}$, $H_{11}$, $H_{22}$, $\mathcal{Q}_1$, and $\mathcal{Q}_2$.

In this paragraph we also compare our structure-exploiting approach with the standard one for general eigenvalue problems, namely the QZ algorithm (with eigenvalue reordering) [GL96]. In order to have a fair comparison of the performance of both methods, we have also implemented a FORTRAN 77 subroutine which combines the QZ algorithm with reordering the purely imaginary eigenvalues to the top by using the LAPACK subroutines `DGGES` and `DGGEV`. For testing purposes, MEX-files have been written to call both subroutines from MATLAB.

In order to test the performance of `MB04BV`, we feed it with random examples which have purely imaginary eigenvalues. We randomly generate matrices $E, A \in \mathbb{R}^{n \times n}$, $B \in \mathbb{R}^{n \times m}$, $C \in \mathbb{R}^{m \times n}$, and $D \in \mathbb{R}^{m \times m}$, and define the transfer function $G(s) = C(sE - A)^{-1}B + D$. Define

$$s\widetilde{\mathcal{S}} - \widetilde{\mathcal{H}} = s \left[\begin{array}{cc|cc} E & 0 & 0 & 0 \\ 0 & 0 & 0 & 0 \\ \hline 0 & 0 & E^{\mathsf{T}} & 0 \\ 0 & 0 & 0 & 0 \end{array}\right] - \left[\begin{array}{cc|cc} A & B & 0 & 0 \\ C & D & 0 & \gamma I_m \\ \hline 0 & 0 & -A^{\mathsf{T}} & -C^{\mathsf{T}} \\ 0 & -\gamma I_m & -B^{\mathsf{T}} & -D^{\mathsf{T}} \end{array}\right]. \tag{4.27}$$

If $sE - A \in \mathbb{R}[s]^{n \times n}$ is regular, has no eigenvalues on the imaginary axis and $G \in \mathcal{RH}_\infty^{m \times m}$, then the matrix pencil $s\widetilde{\mathcal{S}} - \widetilde{\mathcal{H}}$ is guaranteed to have purely imaginary eigenvalues if $\inf_{\omega \in \mathbb{R}} \sigma_{\max}(G(\mathrm{i}\omega)) < \gamma < \|G\|_{\mathcal{L}_\infty}$ [BSV12a]. Theoretically, when the distance between $\gamma$ and $\|G\|_{\mathcal{L}_\infty}$ decreases, the difficulty of the example will increase, in the sense that the eigenvalues will be increasingly sensitive to perturbations, as the numerical results will later demonstrate. This is due to the fact, that there are two purely imaginary eigenvalues which almost form a non-trivial Jordan block in the Weierstraß canonical form. Then, the transformation matrices will be ill-conditioned, which leads to a higher sensitivity of these eigenvalues.

Now we compare both approaches by constructing random pencils of the form (4.27) with $n = 100$ and $m = 5$. Table 4.2 shows the performance results of both algorithms when computing both desired eigenvalues and eigenvectors for different values of $\gamma = \|G\|_{\mathcal{L}_\infty}(1-\kappa)$, from the more well- to the more ill-conditioned examples. Each row contains the results for a thousand test runs. The time measured is for both eigenvalue and eigenvector computation. The accuracy of the results is measured by computing the average of the relative residuals given by $\|(\lambda_i\widetilde{\mathcal{S}} - \widetilde{\mathcal{H}})v_i\|_2 / \|v_i\|_2$, where $(\lambda_i, v_i)$ are pairs of computed purely imaginary eigenvalues and their respective eigenvectors. The runtime is given in seconds, using the `tic` and `toc` commands in MATLAB. Furthermore, for the QZ algorithm we have an additional column that indicates the percentage of examples that could not be solved. This is due to the fact that eigenvalues might be perturbed off the imaginary axis and will not be considered as purely imaginary when the distance to the imaginary axis exceeds a certain threshold. For our tests this value is set to 1e–10. First of all, Table 4.2 shows that

Table 4.2: Comparison of the two methods for eigenvector computation; here $\gamma = \|G\|_{\mathcal{L}_\infty}(1-\kappa)$

| | new algorithm | | QZ algorithm | | |
|---|---|---|---|---|---|
| $\kappa$ | runtime | avg. rel. residual | runtime | avg. rel. residual | failure rate |
| $10^{-2}$ | 99.48 | 1.1936e–13 | 154.62 | 1.0388e–13 | 0.0% |
| $10^{-4}$ | 99.53 | 1.5555e–13 | 153.96 | 3.2024e–13 | 0.1% |
| $10^{-6}$ | 99.45 | 1.3882e–13 | 153.74 | 3.2727e–12 | 0.8% |
| $10^{-8}$ | 99.92 | 1.1820e–13 | 153.69 | 1.9054e–11 | 4.6% |
| $10^{-10}$ | 99.42 | 1.3450e–13 | 151.76 | 6.6909e–11 | 28.9% |
| $10^{-12}$ | 99.51 | 1.3827e–13 | 147.13 | 6.5136e–11 | 78.7% |

the QZ algorithm needs about 50% more time to execute than the new algorithm. However, the most important aspect of the new algorithm is the improved reliability. We can see that the failure rate of the QZ algorithm is dramatically increasing when the examples become more ill-conditioned. By failure we mean that the algorithm extracts a different number of eigenvectors, compared to the actual number of purely imaginary eigenvalues. Moreover, even in the cases where the QZ algorithm successfully extracts the eigenvectors, the average relative residual becomes significantly larger when the condition gets worse. On the other hand, the new algorithm performs more reliably and more accurately. Its runtime and accuracy remain at the same level from the "easier" examples to the "harder" ones.

We now briefly describe the nature of QZ algorithm's failure by examining a small example. Consider a randomly generated skew-Hamiltonian/Hamiltonian matrix pencil

$$s\widetilde{\mathcal{S}} - \widetilde{\mathcal{H}} = s\begin{bmatrix} 0.7060 & 0.2769 & 0 & 0 & 0 & 0 \\ 0.0318 & 0.0462 & 0 & 0 & 0 & 0 \\ 0 & 0 & 0 & 0 & 0 & 0 \\ 0 & 0 & 0 & 0.7060 & 0.0318 & 0 \\ 0 & 0 & 0 & 0.2769 & 0.0462 & 0 \\ 0 & 0 & 0 & 0 & 0 & 0 \end{bmatrix}$$

$$- \begin{bmatrix} 0.7431 & 0.6555 & 0.0971 & 0 & 0 & 0 \\ 0.3922 & 0.1712 & 0.8235 & 0 & 0 & 0 \\ 0.6948 & 0.3171 & 0.9502 & 0 & 0 & 0.9502 \\ 0 & 0 & 0 & -0.7431 & -0.3922 & -0.6948 \\ 0 & 0 & 0 & -0.6555 & -0.1712 & -0.3171 \\ 0 & 0 & -0.9502 & -0.0971 & -0.8235 & -0.9502 \end{bmatrix} \in \mathbb{R}[s]^{6\times 6},$$

of the form (4.27) with $\kappa = 10^{-6}$. The spectrum is given by

$$\Lambda\big(\widetilde{\mathcal{S}}, \widetilde{\mathcal{H}}\big) = \{927.5\mathrm{i}, -927.5\mathrm{i}, 1.161, -1.161\}$$

with two eigenvalues at infinity. In the spectrum, only the purely imaginary eigenvalue 927.5i is interesting to us. The new algorithm successfully extracts one eigenvector with relative residual 1.8594e–15.

However, the QZ algorithm fails to extract the eigenvector. Since it does not respect the structure of the pencil, all purely imaginary eigenvalues will be perturbed off the imaginary axis. When we are selecting the purely imaginary eigenvalues, we cannot expect that the real part of an eigenvalue is exactly zero, as it is for the structure-preserving algorithm. Alternatively, what we do is to ask if the real part is smaller than some tolerance. This tolerance is empirically set to 1e–10. This is a rather tight bound, but it illustrates the behavior of the QZ algorithm quite well. However, the real part of this eigenvalue after perturbation is 1.4079e–10. Because the real part is slightly larger than the tolerance, this eigenvalue is not selected and thus no eigenvector is extracted.

Finally, we can conclude that the new method to compute the eigenvectors

a) is more robust, especially for ill-conditioned examples;

b) is comparably accurate for well-conditioned examples, and significantly more accurate for ill-conditioned examples;

c) needs only about 2/3 of time to execute.

### 4.4.9 An Illustrative Example

In this section we present some numerical results of the algorithm for enforcing negative imaginariness. As an example we use the constrained damped mass-spring system described the introduction, see Figure 1.1. Since the input and the output are colocated, i. e., at the same position, such a system has a negative imaginary transfer function, see also [PL10]. The system we consider has $n = 11$ state variables and $m = 1$ input and output. To obtain non-negative imaginary test examples we perturb the state matrix $A$ by a matrix $\widehat{A}$ with small norm. Here, we have a look on two such example systems which have been created by a relatively large perturbation of the matrix $A$.

We analyze the results for different values of the tuning parameter $\alpha$ and present the relative error measured in the $\mathcal{HL}_2$-norm and the number of iterations needed to make the systems negative imaginary.

Table 4.3: Numerical results for enforcing negative imaginariness of the first example

| $\alpha$ | 0.5 | 0.4 | 0.3 | 0.2 | 0.1 | 0.05 | 0.02 | 0.01 |
|---|---|---|---|---|---|---|---|---|
| rel. error | 0.28648 | 0.24260 | 0.17241 | 0.11459 | 0.08087 | 0.07706 | 0.07634 | 0.07480 |
| # iter. | 1 | 5 | 3 | 1 | 10 | 26 | 61 | 132 |

Table 4.4: Numerical results for enforcing negative imaginariness of the second example

| $\alpha$ | 0.5 | 0.4 | 0.3 | 0.2 | 0.1 | 0.05 | 0.02 | 0.01 |
|---|---|---|---|---|---|---|---|---|
| rel. error | — | — | — | — | 0.58480 | 0.47038 | 0.43616 | 0.42904 |
| # iter. | — | — | — | — | 9 | 121 | 276 | 514 |

For the first example, the perturbation of $A$ is still small enough that the algorithm gives reasonable results for all values of $\alpha$. The results are listed in Table 4.3. Note, that for larger $\alpha$ additional violations of negative imaginariness are introduced since the perturbations of the eigenvalues are too large to be captured by the first order perturbation theory. For the second example we increased the size of the perturbation and then the violation of negative imaginariness is so large that the enforcement algorithm fails, if $\alpha$ is too big. We only get reasonable results if we further decrease

$\alpha$ and make the perturbation of the system in each step sufficiently small. See also Table 4.4 in which all results are presented. We observe that when we run the algorithm, it often happens that there occur additional negative imaginariness violations for larger frequencies. This is due to the fact that for these frequencies, the eigenvalues of $\Phi(\mathrm{i}\omega)$ are already very close to zero and thus can be easily perturbed to negative values. Therefore, the algorithm has to enforce negative imaginariness (repeatedly) for these frequencies which drastically increases the iteration numbers for smaller $\alpha$.

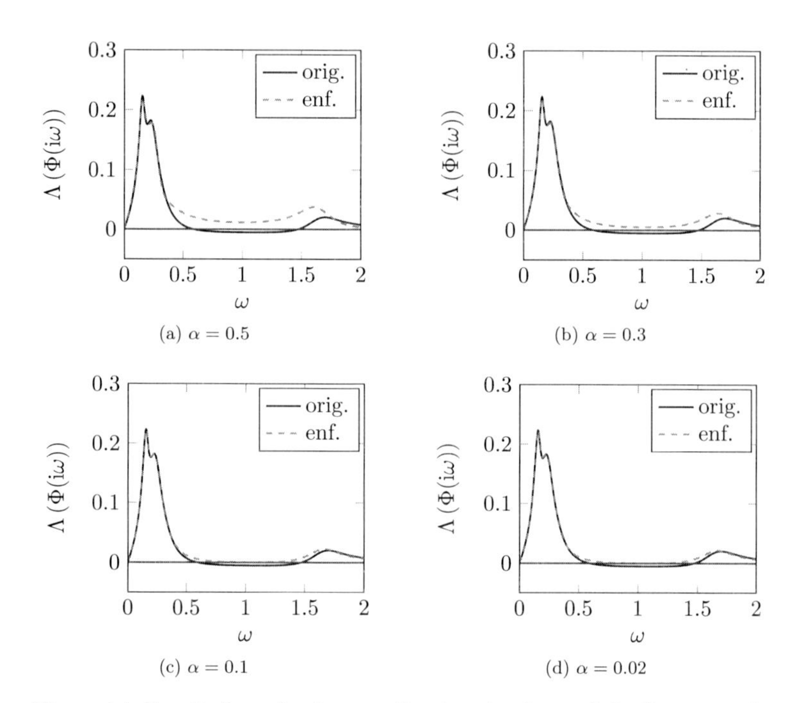


Figure 4.4: Results for enforcing negative imaginariness of the first example

The numerical results are also depicted in Figures 4.4 and 4.5. For the first example we see that for larger values of $\alpha$, we perform a slightly too large perturbation as the eigenvalue curves have some distance from the zero level. However, for smaller values of $\alpha$ this distance gets smaller and the approximation gets better. For the second example one can see that for $\alpha = 0.1$ we have a large error around $\omega = 1.1$ as there is a very high peak for the negative imaginary system. But again, as $\alpha$

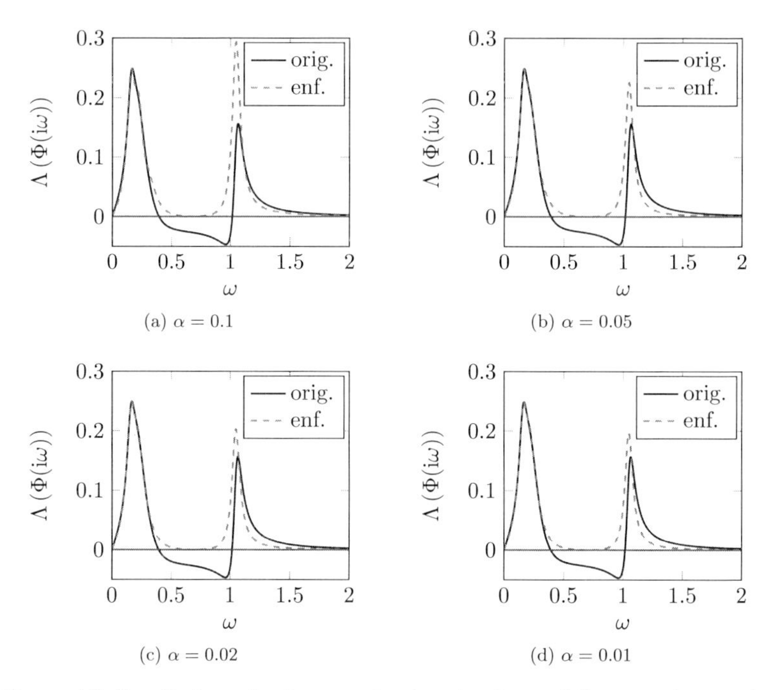


Figure 4.5: Results for enforcing negative imaginariness of the second example

decreases, also the size of the peak gets closer to the one of the original system and the approximation gets better.

## 4.5 Conclusions and Outlook

In this chapter we have introduced the negative imaginary property for transfer functions related to descriptor systems. We have shown equivalent conditions for negative imaginariness in terms of the spectrum of a certain even matrix pencil. We have analyzed the EKCF of this pencil and showed that it has to fulfill a certain block structure. In the second part of the chapter, we have introduced a numerical method for restoring negative imaginariness in the case that it has been lost when applying a system approximation algorithm such as done in model order reduction. This numerical method relies on the structure-preserving computation of the purely imaginary eigenvalues and associated eigenvectors of related skew-Hamiltonian/Hamiltonian matrix pencils. Finally, we have presented some numerical results and we have discussed the behavior of the enforcement algorithm. A future research topic might be the analysis of an negative imaginariness enforcement procedure which also allows the perturbation of other matrices than $B_1$ to obtain more accurate results. For instance it would be interesting to analyze whether the enforcement procedure in [BS13] could be adapted to our problem.

# 5 Computation of the Complex Cyclo-Dissipativity Radius

## 5.1 Introduction

In this chapter we turn back to the concept of cyclo-dissipative systems. When modeling cyclo-dissipative processes, it is important to reflect this property in the structure of the model. Otherwise, simulations could produce physically meaningless results. However, due to modeling errors, introduced, e. g., by model order reduction, linearizations, or uncertainties in the parameters of the system, it could easily happen that cyclo-dissipativity in the model structure is lost. In other words, a physically cyclo-dissipative process is modeled by a non-cyclo-dissipative mathematical model. Then it is necessary to restore this structure by a post-processing procedure, typically known as dissipativity enforcement [BS13] (or passivity enforcement in certain special cases, e. g., [GT04, GTU06], or see Chapter 4 for system with ccw I/O dynamics). On the other hand, even if the model is cyclo-dissipative, it might be close to a non-cyclo-dissipative model. Then it is desirable to assess robustness of cyclo-dissipativity with respect to perturbations of the model. A more precise formulation of this question is:

*What is the distance of a given cyclo-dissipative model*
*to the set of non-cyclo-dissipative models?*

The aim of this chapter is to define the distance to non-cyclo-dissipativity, discuss its properties and an algorithm to compute it.

A first attempt to calculate the cyclo-dissipativity radius is given in [OD05]. There the authors consider the problem of computing the passivity radius for a stable linear time-invariant system $(I_n, A, B, C, D) \in \Sigma_{n,m,m}$. Then they define the perturbed system

$$\begin{aligned} \dot{x}(t) &= (A + \Delta_A)\, x(t) + (B + \Delta_B)\, u(t), \\ y(t) &= (C + \Delta_C)\, x(t) + (D + \Delta_D)\, u(t), \end{aligned} \tag{5.1}$$

with $\Delta_A \in \mathbb{C}^{n\times n}$, $\Delta_B \in \mathbb{C}^{n\times m}$, $\Delta_C \in \mathbb{C}^{m\times n}$, and $\Delta_D \in \mathbb{C}^{m\times m}$ and ask the question what is

$$r_{\text{pas}}(A, B, C, D) := \inf \left\{ \left\| \begin{bmatrix} \Delta_A & \Delta_B \\ \Delta_C & \Delta_D \end{bmatrix} \right\|_2 : \text{system (5.1) is not passive} \right\}. \tag{5.2}$$

This number is also called the *complex passivity radius* of $(I_n, A, B, C, D)$. Then they compute the solution by solving a sequence of structured matrix perturbation problems for Hermitian matrices. However, this approach has some disadvantages:

a) The definition of the passivity radius in (5.2) does not turn over to the corresponding transfer function, defined by $G(s) := C(sI_n - A)^{-1}B + D \in \mathbb{R}(s)^{m\times m}$. Instead of asking for the passivity radius one could ask for the positive realness radius. Under system theoretic aspects they should be the same. To see that they are not, we consider a state-space transformation of the system of the form

$$\begin{bmatrix} \widetilde{A} & \widetilde{B} \\ \widetilde{C} & \widetilde{D} \end{bmatrix} := \begin{bmatrix} W^{-1} & 0 \\ 0 & I_m \end{bmatrix} \begin{bmatrix} A & B \\ C & D \end{bmatrix} \begin{bmatrix} W & 0 \\ 0 & I_m \end{bmatrix}$$

with $W \in \mathrm{Gl}_n(\mathbb{R})$. The associated transfer function remains invariant under such a transformation, i. e., $G(s) = \widetilde{C}\big(sI_n - \widetilde{A}\big)^{-1}\widetilde{B} + \widetilde{D}$. On the other hand, for the associated transformation of the perturbation matrices we generally obtain

$$\begin{aligned} \left\| \begin{bmatrix} \Delta_{\widetilde{A}} & \Delta_{\widetilde{B}} \\ \Delta_{\widetilde{C}} & \Delta_{\widetilde{D}} \end{bmatrix} \right\|_2 &:= \left\| \begin{bmatrix} W^{-1} & 0 \\ 0 & I_m \end{bmatrix} \begin{bmatrix} \Delta_A & \Delta_B \\ \Delta_C & \Delta_D \end{bmatrix} \begin{bmatrix} W & 0 \\ 0 & I_m \end{bmatrix} \right\|_2 \\ &\neq \left\| \begin{bmatrix} \Delta_A & \Delta_B \\ \Delta_C & \Delta_D \end{bmatrix} \right\|_2. \end{aligned}$$

For instance, we have

$$2.5 = \left\| \begin{bmatrix} 1 & 0.5 \\ 2 & 1 \end{bmatrix} \right\|_2 = \left\| \begin{bmatrix} 0.5 & 0 \\ 0 & 1 \end{bmatrix} \begin{bmatrix} 1 & 1 \\ 1 & 1 \end{bmatrix} \begin{bmatrix} 2 & 0 \\ 0 & 1 \end{bmatrix} \right\|_2 \neq \left\| \begin{bmatrix} 1 & 1 \\ 1 & 1 \end{bmatrix} \right\|_2 = 2.$$

This means, that the definition of the complex passivity radius is a property of the dynamical system and not of the transfer function $G(s)$.

b) The definition only takes the special case of passivity into account and does not consider the case of general supply rates.

These observations are the motivation for this chapter. Our contribution is a definition of the passivity radius that is invariant under certain system transformations. Moreover, we consider differential-algebraic systems with arbitrary quadratic supply rates.

This chapter is structured as follows. In Section 5.2 we discuss a spectral characterization of cyclo-dissipativity via an even matrix pencil and give a definition of the complex cyclo-dissipativity radius. Since the problem of computing the cyclo-dissipativity radius is related to structured perturbations of the even pencils, in Section 5.3 we discuss perturbations of the various substructures, in particular, the singular and higher-index structures of the even pencil. In Sections 5.4 we derive a computational procedure to compute the complex cyclo-dissipativity radius. Finally, in Section 5.5 we present some numerical results of the algorithm and summarize the chapter in Section 5.6, where we also point to some open problems.

## 5.2 The Cyclo-Dissipativity Radius

Assume that $\Sigma := (E, A, B, C, D) \in \Sigma_{n,m,p}$ and let $Q_s = Q_s^\mathsf{T} \in \mathbb{R}^{p\times p}$, $S_s \in \mathbb{R}^{p\times m}$, and $R_s = R_s^\mathsf{T} \in \mathbb{R}^{m\times m}$ be given. Let the supply rate $s(\cdot,\cdot)$ be defined as in (3.101). Our definition of the cyclo-dissipativity radius is based on structured perturbations of the form

$$\begin{bmatrix} A & B \\ C & D \end{bmatrix} \rightsquigarrow \begin{bmatrix} A & B \\ C & D \end{bmatrix} + \begin{bmatrix} \mathcal{L}_1 \\ \mathcal{L}_2 \end{bmatrix} \Delta \begin{bmatrix} \mathcal{R}_1 & \mathcal{R}_2 \end{bmatrix}$$

with

- a perturbation matrix $\Delta \in \mathbb{C}^{m_0\times p_0}$ and
- matrices $\mathcal{L}_1 \in \mathbb{R}^{n\times m_0}$, $\mathcal{L}_2 \in \mathbb{R}^{p\times m_0}$, $\mathcal{R}_1 \in \mathbb{R}^{p_0\times n}$, and $\mathcal{R}_2 \in \mathbb{R}^{p_0\times m}$ defining the perturbation structure.

The perturbation structures in $\mathcal{L}$ and $\mathcal{R}$ can be chosen in such a way to analyze the influence of perturbations on individual entries in $\begin{bmatrix} A & B \\ C & D \end{bmatrix}$ which could reflect perturbations of a certain set of parameters of the system.

**Definition 5.2.1.** Define the matrices

$$\mathcal{L} := \begin{bmatrix} \mathcal{L}_1 \\ \mathcal{L}_2 \end{bmatrix} \in \mathbb{R}^{n+p\times m_0} \quad \text{and} \quad \mathcal{R} := \begin{bmatrix} \mathcal{R}_1 & \mathcal{R}_2 \end{bmatrix} \in \mathbb{R}^{p_0\times n+m}.$$

Then the number

$$\begin{aligned} r_{s,\mathbb{C}}(\Sigma, \mathcal{L}, \mathcal{R}) := \inf \big\{ \|\Delta\|_2 \ : \ & \text{the perturbed system is not cyclo-dissipative} \\ & \text{with respect to the supply rate } s(\cdot,\cdot) \text{ and with } \Delta \in \mathbb{C}^{m_0\times p_0} \big\} \end{aligned}$$

is called the *complex cyclo-dissipativity radius* of the system $\Sigma$ with respect to the supply rate $s(\cdot,\cdot)$.

*Remark* 5.2.2 (Real cyclo-dissipativity radii). Note, that even if the matrices defining the system and the supply rate are all real, we use complex perturbations. In general, the computation of the real cyclo-dissipativity radius is much more involved. However, since the set of the real matrices is contained the set complex matrices, we always have the inequality

$$r_{s,\mathbb{C}}(\Sigma, \mathcal{L}, \mathcal{R}) \le r_{s,\mathbb{R}}(\Sigma, \mathcal{L}, \mathcal{R}),$$

where $r_{s,\mathbb{R}}(\Sigma, \mathcal{L}, \mathcal{R})$ denotes the corresponding real cyclo-dissipativity radius.

To characterize the loss of cyclo-dissipativity, we make use of the conditions in Corollary 3.9.5. Therefore, define

$$\mathcal{N}(s) := \begin{bmatrix} 0 & 0 & 0 & -sE+A & B \\ 0 & 0 & -I_p & C & D \\ 0 & -I_p & Q_s & 0 & S_s \\ sE^\mathsf{T}+A^\mathsf{T} & C^\mathsf{T} & 0 & 0 & 0 \\ B^\mathsf{T} & D^\mathsf{T} & S_s^\mathsf{T} & 0 & R_s \end{bmatrix} \in \mathbb{R}[s]^{\ell\times\ell} \tag{5.3}$$

with $\ell := 2(n+p)+m$. Note that from the numerical point of view, it is preferable to use the pencil from Corollary 3.9.5 d) rather than the one from c), because it is constructed by only using the original data without forming explicit matrix-times-its-transpose-like products as for the pencil in c). It is mentioned in [BBMX99] that forming such products might be numerically unstable and should be avoided whenever possible. From Corollary 3.9.5 d) it is clear that cyclo-dissipativity is lost if and only if there exists an $\omega_0 \in \mathbb{R}$ with $\mathrm{i}\omega_0 \notin \Lambda(E,A)$ such that

$$\eta(\mathcal{N}(\mathrm{i}\omega_0)) \neq m,$$

where $\eta(\cdot)$ denotes the sign-sum function as in (3.107). Define

$$K := \begin{bmatrix} K_1 \\ K_2 \end{bmatrix} = \begin{bmatrix} \mathcal{L}^\mathsf{T} & 0_{m_0\times p} & 0_{m_0\times n+m} \\ 0_{p_0\times n+p} & 0_{p_0\times p} & \mathcal{R} \end{bmatrix} \tag{5.4}$$

and consider the perturbed even pencil

$$\mathcal{N}_\Delta(s) := \mathcal{N}(s) + K^\mathsf{T} \begin{bmatrix} 0 & \Delta \\ \Delta^\mathsf{H} & 0 \end{bmatrix} K. \tag{5.5}$$

Then we obtain

$$r_{s,\mathbb{C}}(\Sigma,\mathcal{L},\mathcal{R}) = \inf\big\{ \|\Delta\|_2 : \eta\left(\mathcal{N}_\Delta(\mathrm{i}\omega_0)\right) \neq m \\ \text{for some } \omega_0 \in \mathbb{R} \text{ with } \mathrm{i}\omega_0 \notin \Lambda(E, A+\mathcal{L}_1\Delta\mathcal{R}_1) \text{ and } \Delta \in \mathbb{C}^{m_0\times p_0}\big\}.$$

*Remark* 5.2.3.

a) It is easily verified that $r_{s,\mathbb{C}}(\Sigma,\mathcal{L},\mathcal{R})$ is invariant under generalized state-space transformations, i. e., for $\mathcal{W} := \begin{bmatrix} W & 0 \\ 0 & I_p \end{bmatrix} \in \mathrm{Gl}_{n+p}(\mathbb{R})$, $\mathcal{T} := \begin{bmatrix} T & 0 \\ 0 & I_m \end{bmatrix} \in \mathrm{Gl}_{n+m}(\mathbb{R})$, and

$$\begin{bmatrix} -s\widetilde{E}+\widetilde{A} & \widetilde{B} \\ \widetilde{C} & \widetilde{D} \end{bmatrix} := \mathcal{W} \begin{bmatrix} -sE+A & B \\ C & D \end{bmatrix} \mathcal{T}, \quad \widetilde{\mathcal{L}} := \mathcal{W}\mathcal{L}, \quad \widetilde{\mathcal{R}} := \mathcal{R}\mathcal{T},$$
$$\widetilde{\Sigma} := \left(\widetilde{E},\widetilde{A},\widetilde{B},\widetilde{C},\widetilde{D}\right) \in \Sigma_{n,m,p},$$

it holds that

$$r_{s,\mathbb{C}}(\Sigma, \mathcal{L}, \mathcal{R}) = r_{s,\mathbb{C}}(\widetilde{\Sigma}, \widetilde{\mathcal{L}}, \widetilde{\mathcal{R}}).$$

This follows from the fact that $U^{\mathsf{T}}\mathcal{N}_\Delta(s)U$ has the same Kronecker structure as $\mathcal{N}_\Delta(s)$ for any nonsingular $U \in \mathbb{R}^{\ell\times\ell}$. In our case we have $U := \operatorname{diag}(\mathcal{W}, I_p, \mathcal{T})$. This also leads to a proper definition of the cyclo-dissipativity radius in the frequency domain, since the associated Popov functions are invariant under such state-space transformations.

b) The structure of the perturbations is contained in the matrix $K$. In this chapter, the structure of $K$ does not play any particular role for computational purposes. Therefore, one could also allow different classes of perturbations as long as they are conforming to the perturbation structure of the even pencil, given by (5.5) (with possibly different $K$). However, in this case, statement a) does not necessarily hold true anymore.

c) Under certain conditions, it is also possible to compute the *complex dissipativity radius*. Therefore, assume that $(E, A, B, C, D) \in \Sigma_{n,m,p}$ with the system space $\mathcal{V}_{\text{sys}}$ as in (3.5) is dissipative with respect to the supply rate $s(\cdot,\cdot)$ given by

$$\begin{bmatrix} Q_s & S_s \\ S_s^{\mathsf{T}} & R_s \end{bmatrix} =_{\widetilde{\mathcal{V}}} \begin{bmatrix} \widehat{K}_1^{\mathsf{T}}\widehat{K}_1 & \widehat{K}_1^{\mathsf{T}}\widehat{L}_1 \\ \widehat{L}_1^{\mathsf{T}}\widehat{K}_1 & \widehat{L}_1^{\mathsf{T}}\widehat{L}_1 \end{bmatrix} - \begin{bmatrix} \widehat{K}_2^{\mathsf{T}}\widehat{K}_2 & \widehat{K}_2^{\mathsf{T}}\widehat{L}_2 \\ \widehat{L}_2^{\mathsf{T}}\widehat{K}_2 & \widehat{L}_2^{\mathsf{T}}\widehat{L}_2 \end{bmatrix}$$

with $\widehat{K}_1 \in \mathbb{R}^{m\times p}$, $\widehat{K}_2 \in \mathbb{R}^{p_2\times p}$, $\widehat{L}_1 \in \mathbb{R}^{m\times m}$, $\widehat{L}_2 \in \mathbb{R}^{p_2\times m}$, and

$$\widetilde{\mathcal{V}}_{\text{sys}} := \left\{ \begin{pmatrix} y \\ u \end{pmatrix} \in \mathbb{R}^{p+m} : y = Cx + Du \text{ with } \begin{pmatrix} x \\ u \end{pmatrix} \in \mathcal{V}_{\text{sys}} \right\}$$

such that $\widehat{K}_1 G(s) + \widehat{L}_1 \in \mathrm{Gl}_m(\mathbb{R}(s))$. From the proof of Theorem 3.7.4 and Corollary 3.9.4 b), dissipativity is equivalent to the bounded realness of the rational function $G_{\mathrm{e}}(s) \in \mathbb{R}(s)^{p\times m}$ that is realized by

$$\begin{aligned} \Sigma_{\mathrm{e}} :&= (E_{\mathrm{e}}, A_{\mathrm{e}}, B_{\mathrm{e}}, C_{\mathrm{e}}) \\ &= \left( \begin{bmatrix} E & 0 \\ 0 & 0 \end{bmatrix}, \begin{bmatrix} A & B \\ \widehat{K}_1 C & \widehat{K}_1 D + \widehat{L}_1 \end{bmatrix}, \begin{bmatrix} 0 \\ -I_m \end{bmatrix}, \begin{bmatrix} \widehat{K}_2 C & \widehat{K}_2 D + \widehat{L}_2 \end{bmatrix} \right) \in \Sigma_{n+m,m,p}. \end{aligned}$$

Define the perturbed transfer function

$$\begin{aligned} G_{\mathrm{e},\Delta}(s) := &\begin{bmatrix} \widehat{K}_2(C + \mathcal{L}_2\Delta\mathcal{R}_1) & \widehat{K}_2(D + \mathcal{L}_2\Delta\mathcal{R}_2) + \widehat{L}_2 \end{bmatrix} \\ &\cdot \begin{bmatrix} sE - (A + \mathcal{L}_1\Delta\mathcal{R}_1) & -(B + \mathcal{L}_1\Delta\mathcal{R}_2) \\ -\widehat{K}_1(C + \mathcal{L}_2\Delta\mathcal{R}_1) & -\widehat{K}_1(D + \mathcal{L}_2\Delta\mathcal{R}_2) + \widehat{L}_1 \end{bmatrix}^{-1} \begin{bmatrix} 0 \\ -I_m \end{bmatrix}. \end{aligned}$$

Then the complex dissipativity radius $\widehat{r}_{s,\mathbb{C}}(\Sigma, \mathcal{L}, \mathcal{R})$ is given by

$$\widehat{r}_{s,\mathbb{C}}(\Sigma, \mathcal{L}, \mathcal{R}) = \inf\left\{\|\Delta\|_2 : G_{\mathrm{e},\Delta}(s) \text{ is not bounded real with } \Delta \in \mathbb{C}^{m_0 \times p_0}\right\}.$$

Assume that one of the following statements holds true:

a) $\Lambda(E_{\mathrm{e}}, A_{\mathrm{e}}) \subset \mathbb{C}^-$ and the index of $sE_{\mathrm{e}} - A_{\mathrm{e}}$ is at most one;

b) $\Sigma_{\mathrm{e}}$ is given in a weakly minimal realization, i. e., it is completely controllable and completely observable.

With the assumptions above, it is not possible that an uncontrollable or unobservable mode of the system $\Sigma_{\mathrm{e}}$ in the closed right half-plane is perturbed to a pole of $G_{\mathrm{e},\Delta}(s)$ for an arbitrarily small $\Delta$. Since the pole perturbations are continuous and due to the special structure of bounded real rational functions, the condition that $I_{n+m} - G_{\mathrm{e},\Delta}^{\sim}(\mathrm{i}\omega)G_{\mathrm{e},\Delta}(\mathrm{i}\omega) \geq 0$ for all $\omega \in \mathbb{R}$ is violated before one of the poles of $G_{\mathrm{e},\Delta}(s)$ is moved to the imaginary axis. Therefore, in this case we have

$$\widehat{r}_{s,\mathbb{C}}(\Sigma, \mathcal{L}, \mathcal{R}) = r_{s,\mathbb{C}}(\Sigma, \mathcal{L}, \mathcal{R}).$$

## 5.3 Perturbations of the Singular Part and the Defective Infinite Eigenvalues

Now we study the effect of structured perturbations of the form (5.5) on the even pencil $\mathcal{N}(s) \in \mathbb{R}[s]^{\ell \times \ell}$. There are fundamental differences between perturbations of the part of the pencil which correspond to different block types of the EKCF. Therefore, we distinguish the "exceptional" cases of perturbations of the singular part or defective infinite eigenvalues (i. e., those which correspond to blocks of type E3 of size larger than $1 \times 1$) from perturbations of the remaining part.

### 5.3.1 Perturbation of the Singular Part

In this subsection we consider perturbations of the singular part of the pencil $\mathcal{N}(s)$. For this analysis we assume that $\mathcal{N}(s) \in \mathbb{R}[s]^{\ell \times \ell}$ has no purely imaginary eigenvalues. With regard to Proposition 3.2.2 and Lemma 3.4.1 this equivalent to the fact that $(E, A, B, C, D) \in \Sigma_{n,m,p}$ has no uncontrollable modes on the imaginary axis and $\Phi(s) \in \mathbb{R}(s)^{m \times m}$ has no purely imaginary zeros.

First, we analyze whether and how the zero eigenvalues of a given Hermitian matrix can move by a structured perturbation. The main result is summarized in the following theorem.

**Theorem 5.3.1.** *Let $H \in \mathbb{C}^{\ell \times \ell}$ be a given Hermitian matrix and consider the spectral decomposition*

$$H = \begin{bmatrix} V_1 & V_2 \end{bmatrix} \begin{bmatrix} \Theta & 0 \\ 0 & 0 \end{bmatrix} \begin{bmatrix} V_1 & V_2 \end{bmatrix}^{\mathsf{H}}$$

*with $V_1 \in \mathbb{C}^{\ell \times \ell_1}$, $V_2 \in \mathbb{C}^{\ell \times \ell_2}$ and $\begin{bmatrix} V_1 & V_2 \end{bmatrix}$ having unitary columns, and $\Theta \in \mathbb{R}^{\ell_1 \times \ell_1}$ being a diagonal matrix containing the nonzero eigenvalues of $H$ on its diagonal. Consider a perturbed matrix*

$$\widetilde{H} = H + \begin{bmatrix} K_1 \\ K_2 \end{bmatrix}^{\mathsf{T}} \begin{bmatrix} 0 & \Delta \\ \Delta^{\mathsf{H}} & 0 \end{bmatrix} \begin{bmatrix} K_1 \\ K_2 \end{bmatrix}$$

*with nonzero $K_1 \in \mathbb{R}^{m_0 \times \ell}$, $K_2 \in \mathbb{R}^{p_0 \times \ell}$, and $\Delta \in \mathbb{C}^{m_0 \times p_0}$. Let $H^+$ denote the Moore-Penrose inverse of $H$. Then the following statements are satisfied:*

*a) If $K_1 V_2 = 0$ and $K_2 V_2 = 0$, then $H$ and $\widetilde{H}$ have the same zero eigenvalues for all sufficiently small $\Delta \neq 0$.*

*b) If $K_1 V_2 \neq 0$, and $K_2 V_2 \neq 0$, then there exists an arbitrarily small $\Delta \neq 0$ such that some zero eigenvalues of $H$ can be perturbed to positive and some to negative eigenvalues in $\widetilde{H}$.*

*c) If $K_1 V_2 \neq 0$ and $K_2 V_2 = 0$, then there exists an arbitrarily small $\Delta \neq 0$ such that some zero eigenvalues of $H$ are perturbed to positive (negative) eigenvalues in $\widetilde{H}$ if and only if $K_2 H^+ K_2^{\mathsf{T}} \ngeq 0$ ($K_2 H^+ K_2^{\mathsf{T}} \nleq 0$). If $K_2 H^+ K_2^{\mathsf{T}} \geq 0$ ($K_2 H^+ K_2^{\mathsf{T}} \leq 0$), then the smallest perturbation $\Delta_0$ that perturbs a zero eigenvalue of $H$ to a positive (negative) eigenvalue of $\widetilde{H}$ fulfills*

$$\|\Delta_0\|_2 > \inf_{\Delta \in \mathbb{C}^{m_0 \times p_0}} \left\{ \|\Delta\|_2 : V_1^{\mathsf{H}} \left( H + \begin{bmatrix} K_1 \\ K_2 \end{bmatrix}^{\mathsf{T}} \begin{bmatrix} 0 & \Delta \\ \Delta^{\mathsf{H}} & 0 \end{bmatrix} \begin{bmatrix} K_1 \\ K_2 \end{bmatrix} \right) V_1 \text{ is singular} \right\}. \tag{5.6}$$

*d) If $K_1 V_2 = 0$ and $K_2 V_2 \neq 0$, then there exists an arbitrarily small $\Delta \neq 0$ such that some zero eigenvalues of $H$ are perturbed to positive (negative) eigenvalues in $\widetilde{H}$ if and only if $K_1 H^+ K_1^{\mathsf{T}} \ngeq 0$ ($K_1 H^+ K_1^{\mathsf{T}} \nleq 0$). If $K_1 H^+ K_1^{\mathsf{T}} \geq 0$ ($K_1 H^+ K_1^{\mathsf{T}} \leq 0$), then the smallest perturbation $\Delta_0$ that perturbs a zero eigenvalue of $H$ to a positive (negative) eigenvalue of $\widetilde{H}$ fulfills* (5.6).

*Proof.* Assume that $\Delta = \varepsilon \widetilde{\Delta}$ with $\|\widetilde{\Delta}\|_2 = 1$ and $\varepsilon > 0$ are given and define the matrices

$$\begin{aligned} \Theta_{11} &:= V_1^{\mathsf{H}} K_1^{\mathsf{T}} \widetilde{\Delta} K_2 V_1 + V_1^{\mathsf{H}} K_2^{\mathsf{T}} \widetilde{\Delta}^{\mathsf{H}} K_1 V_1, \\ \Theta_{12} &:= V_1^{\mathsf{H}} K_1^{\mathsf{T}} \widetilde{\Delta} K_2 V_2 + V_1^{\mathsf{H}} K_2^{\mathsf{T}} \widetilde{\Delta}^{\mathsf{H}} K_1 V_2, \\ \Theta_{22} &:= V_2^{\mathsf{H}} K_1^{\mathsf{T}} \widetilde{\Delta} K_2 V_2 + V_2^{\mathsf{H}} K_2^{\mathsf{T}} \widetilde{\Delta}^{\mathsf{H}} K_1 V_2. \end{aligned}$$

This yields

$$\begin{bmatrix} V_1 & V_2 \end{bmatrix}^{\mathsf{H}} \widetilde{H} \begin{bmatrix} V_1 & V_2 \end{bmatrix} = \begin{bmatrix} \Theta + \varepsilon \Theta_{11} & \varepsilon \Theta_{12} \\ \varepsilon \Theta_{12}^{\mathsf{H}} & \varepsilon \Theta_{22} \end{bmatrix}.$$

Since $\Theta$ has no zero eigenvalues, there exists an $\varepsilon_0$ such that $\Theta + \varepsilon\Theta_{11} \in \mathrm{Gl}_{\ell_1}(\mathbb{C})$ for all $0 < \varepsilon < \varepsilon_0$. The Haynsworth inertia additivity formula [Hay68] implies that for all $0 < \varepsilon < \varepsilon_0$ it holds that

$$\mathrm{In}\left(\widetilde{H}\right) = \mathrm{In}\left(\Theta + \varepsilon\Theta_{11}\right) + \mathrm{In}\left(F(\varepsilon)\right) = \mathrm{In}\left(\Theta\right) + \mathrm{In}\left(F(\varepsilon)\right),$$

where $F(\varepsilon) := \varepsilon\Theta_{22} - \varepsilon^2\Theta_{12}^{\mathsf{H}}\left(\Theta + \varepsilon\Theta_{11}\right)^{-1}\Theta_{12}$. Since $F(\varepsilon)$ is a rational function, it can be expanded into a Taylor series at expansion point $\varepsilon_* = 0$, i. e.,

$$F(\varepsilon) = \varepsilon\Theta_{22} - \varepsilon^2\sum_{k=0}^{\infty} M_k\varepsilon^k \tag{5.7}$$

with $M_k = \Theta_{12}^{\mathsf{H}}\left(-\Theta^{-1}\Theta_{11}\right)^k\Theta^{-1}\Theta_{12}$, see [FBK13, p. 73].

Now we prove statement a): We have

$$\begin{bmatrix} V_1 & V_2 \end{bmatrix}^{\mathsf{H}} \widetilde{H} \begin{bmatrix} V_1 & V_2 \end{bmatrix} = \begin{bmatrix} \Theta + \varepsilon\Theta_{11} & 0 \\ 0 & 0 \end{bmatrix}$$

and no perturbation of the zero eigenvalues of $H$ is possible.

Next we show b): By the assumptions we can choose $\widetilde{\Delta}$ such that $\Theta_{22} \neq 0$. From (5.7) we have $F(\varepsilon) = \varepsilon\Theta_{22} + \mathcal{O}(\varepsilon^2)$ and thus we obtain

$$\mathrm{In}(F(\varepsilon)) = \mathrm{In}(\Theta_{22})$$

for all sufficiently small $\varepsilon$. Now the result follows since we can choose $\widetilde{\Delta}$ such that $\Theta_{22}$ has at least one positive eigenvalue. This can be seen from the fact that by choosing $-\widetilde{\Delta}$ as perturbation, the signatures of all eigenvalues of $\Theta_{22}$ change as well.

Now we prove statement c): From the assumptions it follows that $\Theta_{22} = 0$ and $\Theta_{12} = V_1^{\mathsf{H}}K_2^{\mathsf{T}}\widetilde{\Delta}^{\mathsf{H}}K_1V_2$. We obtain $F(\varepsilon) = -\varepsilon^2\Theta_{12}^{\mathsf{H}}\Theta^{-1}\Theta_{12} + \mathcal{O}(\varepsilon^3)$. Moreover, it holds that

$$\begin{aligned}\Theta_0 :&= \Theta_{12}^{\mathsf{H}}\Theta^{-1}\Theta_{12} \\ &= V_2^{\mathsf{H}}K_1^{\mathsf{T}}\widetilde{\Delta}K_2V_1\Theta^{-1}V_1^{\mathsf{H}}K_2^{\mathsf{T}}\widetilde{\Delta}^{\mathsf{H}}K_1V_2 \\ &= V_2^{\mathsf{H}}K_1^{\mathsf{T}}\widetilde{\Delta}K_2H^{+}K_2^{\mathsf{T}}\widetilde{\Delta}^{\mathsf{H}}K_1V_2.\end{aligned}$$

Let $K_2H^{+}K_2^{\mathsf{T}} = P^{\mathsf{H}}\Xi P$ with a diagonal matrix $\Xi \in \mathbb{R}^{p_0\times p_0}$ and a unitary matrix $P \in \mathbb{C}^{p_0\times p_0}$ be a spectral decomposition. With $\widehat{\Delta} = \widetilde{\Delta}P^{\mathsf{H}}$ this yields

$$\Theta_0 = V_2^{\mathsf{H}}K_1^{\mathsf{T}}\widehat{\Delta}\Xi\widehat{\Delta}^{\mathsf{H}}K_1V_2.$$

Assume that $K_1V_2$ has a nonzero entry at the $(p, q)$-th position. Then, by choosing a nonzero value for $\widehat{\Delta}^{\mathsf{H}}$ at the position $(r, p)$ (and zero at the other entries), $\Theta_0$ is a matrix with exactly one nonzero eigenvalue that has the signature of the entry at the $(r,r)$-th position of $\Xi$. Therefore, if $K_2 H^+ K_2^T \not\geq 0$, one of the zero eigenvalues of $H$ can be perturbed to a positive eigenvalue of $\widetilde{H}$ by an arbitrarily small perturbation.

Now assume that $K_2 H^+ K_2^T \geq 0$. Then obviously no eigenvalue of $\Theta_0$ can be negative for any $\widehat{\Delta}$ and no perturbation of the eigenvalues of $H$ in positive direction is possible for sufficiently small $\varepsilon$. Now consider the function

$$\widetilde{F}(\varepsilon) := \Theta_{12}^{\mathsf{H}}(\Theta + \varepsilon\Theta_{11})^{-1}\Theta_{12},$$

which is continuous as long as $\varepsilon < \varepsilon_0 := \inf_{\varepsilon > 0}\{\varepsilon : \Theta + \varepsilon\Theta_{11} \text{ is singular}\}$. Moreover, we have $\operatorname{rank}\left(\widetilde{F}(\varepsilon)\right) = \operatorname{rank}(\Theta_{12})$ for all $\varepsilon < \varepsilon_0$. However, due to the continuity of $\widetilde{F}(\varepsilon)$ and the constancy of the rank, also $\operatorname{In}\left(\widetilde{F}(\varepsilon)\right)$ is constant for all $\varepsilon < \varepsilon_0$, otherwise there would exist an $\widehat{\varepsilon} \in (0, \varepsilon_0)$ such that $\operatorname{rank}\left(\widetilde{F}(\widehat{\varepsilon})\right) < \operatorname{rank}(\Theta_{12})$ which is a contradiction and proves the claim.

The proof of statement d) is completely analogous to the one of c). □

The above theorem directly leads to the following corollary, describing the loss of cyclo-dissipativity due to perturbations of the singular part of $\mathcal{N}(s)$.

**Corollary 5.3.2.** *Assume that* $\Sigma := (E, A, B, C, D) \in \Sigma_{n,m,p}$ *and let* $Q_s = Q_s^\mathsf{T} \in \mathbb{R}^{p\times p}$, $S_s \in \mathbb{R}^{p\times m}$, *and* $R_s = R_s^\mathsf{T} \in \mathbb{R}^{m\times m}$ *be given. Let the system be cyclo-dissipative with respect to* $s(\cdot,\cdot)$ *as defined in* (3.101) *and let* $\mathcal{N}(s) \in \mathbb{R}[s]^{\ell\times\ell}$ *and* $K$ *be defined as in* (5.3) *and* (5.4), *respectively. Assume that* $\mathcal{N}(s)$ *has no purely imaginary eigenvalues. For all* $\omega \in \mathbb{R}$ *consider the spectral decomposition*

$$\mathcal{N}(\mathrm{i}\omega) = \begin{bmatrix} V_1(\omega) & V_2(\omega)\end{bmatrix} \begin{bmatrix} \Theta(\omega) & 0 \\ 0 & 0 \end{bmatrix} \begin{bmatrix} V_1(\omega) & V_2(\omega)\end{bmatrix}^\mathsf{H}$$

*with analytic functions* $V_1 : \mathbb{R} \to \mathbb{C}^{\ell\times\ell_1}$, $V_2 : \mathbb{R} \to \mathbb{C}^{\ell\times\ell_2}$ *and* $\begin{bmatrix} V_1(\cdot) & V_2(\cdot)\end{bmatrix}$ *having pointwise unitary columns, and* $\Theta : \mathbb{R} \to \mathbb{R}^{\ell_1\times\ell_1}$ *being a analytic diagonal matrix-valued function containing the nonzero eigenvalues of* $\mathcal{N}(\mathrm{i}\cdot)$ *on its diagonal. Then the following statements hold true:*

*a) If there exists an* $\omega_0 \in \mathbb{R}$ *such that*

*i)* $K_1 V_2(\omega_0) \neq 0$ *and* $K_2 V_2(\omega_0) \neq 0$*; or*

*ii)* $K_1 V_2(\omega_0) \neq 0$ *and* $K_2 V_2(\omega_0) = 0$ *and* $K_2 \mathcal{N}(\mathrm{i}\omega_0)^+ K_2^\mathsf{T} \not\leq 0$*; or*

*iii)* $K_1 V_2(\omega_0) = 0$ *and* $K_2 V_2(\omega_0) \neq 0$ *and* $K_1 \mathcal{N}(\mathrm{i}\omega_0)^+ K_1^\mathsf{T} \not\leq 0$,

*then the cyclo-dissipativity radius is zero.*

*b) On the other hand, if none of the above conditions is satisfied for all* $\omega_0 \in \mathbb{R}$, *then the zero eigenvalues of* $\mathcal{N}(\mathrm{i}\omega_0)$ *remain zero in* $\mathcal{N}_\Delta(\mathrm{i}\omega_0)$ *for all* $\omega_0 \in \mathbb{R}$ *and any sufficiently small perturbation* $\Delta \in \mathbb{C}^{m_0\times p_0}$, *where* $\mathcal{N}_\Delta(s) \in \mathbb{C}[s]^{\ell\times\ell}$ *is as in* (5.5).

*Proof.* The sign-sum function can be changed by an arbitrarily small perturbation if for one $\omega_0 \in \mathbb{R}$, some of the zero eigenvalues of $\mathcal{N}(\mathrm{i}\omega_0)$ can be perturbed to negative values. Due to Theorem 5.3.1 applied to $\mathcal{N}(\mathrm{i}\omega_0)$, this is the case if and only if one of the conditions a) i)–iii) is fulfilled.

Statement b) follows directly from Theorem 5.3.1 a), c), and d). □

*Remark* 5.3.3. If $\mathcal{N}(s) \in \mathbb{R}[s]^{\ell\times\ell}$ has a purely imaginary eigenvalue $\mathrm{i}\omega_0$, then in principal, we would have to check the conditions of Theorem 5.3.1 for $\mathcal{N}(\mathrm{i}\omega_0)$ as well. However, in this case, the situation is more subtle. For instance, consider the even matrix pencil

$$\mathcal{N}(s) = \begin{bmatrix} 0 & 0 & 0 & 1 & 0 \\ 0 & 0 & s & 0 & 0 \\ 0 & -s & 0 & 0 & 0 \\ 1 & 0 & 0 & 0 & 0 \\ 0 & 0 & 0 & 0 & 0 \end{bmatrix} \in \mathbb{R}[s]^{5\times 5}$$

with $0 \in \mathfrak{Z}(\mathcal{N})$, $\eta(\mathcal{N}(\mathrm{i}\omega)) = 1$ for all $\omega \in \mathbb{R}\setminus\{0\}$, and $\eta(\mathcal{N}(0)) = 3$. This pencil is not directly connected to cyclo-dissipativity, but certain effects can still be observed.

We have

$$\mathcal{N}(\mathrm{i}\omega) = \frac{1}{2}\begin{bmatrix} 1 & 1 & 0 & 0 & 0 \\ 0 & 0 & 1 & \mathrm{i} & 0 \\ 0 & 0 & -\mathrm{i} & -1 & 0 \\ 1 & -1 & 0 & 0 & 0 \\ 0 & 0 & 0 & 0 & \sqrt{2} \end{bmatrix} \begin{bmatrix} 1 & 0 & 0 & 0 & 0 \\ 0 & -1 & 0 & 0 & 0 \\ 0 & 0 & \omega & 0 & 0 \\ 0 & 0 & 0 & -\omega & 0 \\ 0 & 0 & 0 & 0 & 0 \end{bmatrix} \begin{bmatrix} 1 & 0 & 0 & 1 & 0 \\ 1 & 0 & 0 & -1 & 0 \\ 0 & 1 & \mathrm{i} & 0 & 0 \\ 0 & -\mathrm{i} & -1 & 0 & 0 \\ 0 & 0 & 0 & 0 & \sqrt{2} \end{bmatrix},$$

i. e., using the notation of Corollary 5.3.2, for $\omega \neq 0$ we have

$$\Theta(\omega) = \begin{bmatrix} 1 & 0 & 0 & 0 \\ 0 & -1 & 0 & 0 \\ 0 & 0 & \omega & 0 \\ 0 & 0 & 0 & -\omega \end{bmatrix}, \quad V_1(\omega) = \frac{1}{\sqrt{2}}\begin{bmatrix} 1 & 1 & 0 & 0 \\ 0 & 0 & 1 & \mathrm{i} \\ 0 & 0 & -\mathrm{i} & -1 \\ 1 & -1 & 0 & 0 \\ 0 & 0 & 0 & 0 \end{bmatrix}, \quad V_2(\omega) = \begin{bmatrix} 0 \\ 0 \\ 0 \\ 0 \\ 1 \end{bmatrix}.$$

Now assume that

$$K = \begin{bmatrix} K_1 \\ K_2 \end{bmatrix} = \begin{bmatrix} 0 & 1 & 0 & 0 & 0 \\ 0 & 0 & 0 & 0 & 1 \end{bmatrix}.$$

Then we have $K_1 V_2(\cdot) \equiv 0$, but it holds that $K_2 V_2(\cdot) \equiv 1$. Moreover, for all $\omega \in \mathbb{R}\setminus\{0\}$ we have $K_1 \mathcal{N}(\mathrm{i}\omega)^+ K_1^\mathsf{T} = 0$, i. e., according to Theorem 5.3.1 d), the sign-sum function remains constant for all $\omega \neq 0$. Indeed, it holds that

$$\mathcal{N}_\Delta(s) = \begin{bmatrix} 0 & 0 & 0 & 1 & 0 \\ 0 & 0 & s & 0 & \Delta \\ 0 & -s & 0 & 0 & 0 \\ 1 & 0 & 0 & 0 & 0 \\ 0 & \Delta^{\mathsf{H}} & 0 & 0 & 0 \end{bmatrix}.$$

The EKCF of the subpencil

$$\widetilde{\mathcal{N}}_\Delta(s) = \begin{bmatrix} 0 & s & \Delta \\ -s & 0 & 0 \\ \Delta^{\mathsf{H}} & 0 & 0 \end{bmatrix}$$

contains only a block of type E4 of size $3 \times 3$ for all $\Delta \in \mathbb{C}\setminus\{0\}$. Thus, by Lemma 2.1.15 f) it holds that $\operatorname{In}\left(\widetilde{\mathcal{N}}_\Delta(\mathrm{i}\omega)\right) = (1,1,1)$ for all $\omega \in \mathbb{R}$. Hence, we can conclude that $\eta(\mathcal{N}_\Delta(\mathrm{i}\omega)) = \eta(\mathcal{N}(\mathrm{i}\omega)) = 1$ for all $\omega \in \mathbb{R} \setminus \{0\}$.

On the other hand, for $\omega = 0$ we obtain $\eta(\mathcal{N}(0)) = 3$. However, for any $\Delta \in \mathbb{C}\setminus\{0\}$, we have $\eta(\mathcal{N}_\Delta(0)) = 1$ and $0 \notin \mathfrak{Z}(\mathcal{N}_\Delta)$. In particular, in the context of a cyclo-dissipative system this means that, if the original system is cyclo-dissipative, then the perturbed system remains cyclo-dissipative for any nonzero perturbation $\Delta \in \mathbb{C}$, even though there exists a point $\omega_0 \in \mathbb{R}$ with $\eta(\mathcal{N}(\mathrm{i}\omega_0)) \neq \eta(\mathcal{N}_\Delta(\mathrm{i}\omega_0))$.

To check whether there exists an $\omega_0$ that fulfills one of the conditions i)–iii) in Corollary 5.3.2 a), we make the following considerations. First we check whether $K_1V_2(\cdot) \equiv 0$.

If the EKCF of $\mathcal{N}(s)$ has no blocks of type E4 of size larger than $1 \times 1$, then $V_2(\cdot)$ is constant and if $K_1V_2(\omega_0) = 0$ holds for one point $\omega_0$, then it holds that for all of them. However, if there are larger blocks of type E4, $V_2(\omega_0)$ depends on $\omega_0$ and the situation is more complicated. Let $\mathcal{D}(s) \in \mathbb{C}[s]^{2j-1 \times 2j-1}$ be a block of type E4. Then $\mathcal{D}(\mathrm{i}\omega_0)$ has exactly one zero eigenvalue with eigenvector

$$v_j(\omega_0) = \begin{pmatrix} 0 & \dots & 0 & (\mathrm{i}\omega_0)^j & (\mathrm{i}\omega_0)^{j-1} & \dots & (\mathrm{i}\omega_0)^0 \end{pmatrix}^{\mathsf{T}} \in \mathbb{C}^{2j-1}.$$

This means that we have the representation

$$K_1V_2(\omega) = \sum_{i=0}^{w} (\mathrm{i}\omega)^i \widetilde{Q}_i,$$

where $\widetilde{Q}_i \in \mathbb{R}^{m_0 \times \ell_2}$, $i = 1, \dots, w$, and the largest block of type E4 in the EKCF of $\mathcal{N}(s)$ is of size $(2w-1) \times (2w-1)$. Indeed, $w$ is the same as in the even staircase form (2.1). If $K_1V_2(\omega_0) \neq 0$ for an $\omega_0 \in \mathbb{R}$, then

$$Q(s) := \sum_{i=0}^{w} s^i \widetilde{Q}_i \in \mathbb{R}[s]^{m_0 \times \ell_2}$$

is not the zero polynomial and hence $Q(\mathrm{i}\omega_0) = 0$ can only hold in a finite number of points (namely in at most $w$ many). Therefore, the condition $K_1 V_2(\cdot) \equiv 0$ can be checked with at most $w$ function evaluations. The same technique can also be used to check the condition $K_2 V_2(\cdot) \equiv 0$.

Now assume that $K_1 V_2(\cdot) \equiv 0$ and $K_2 V_2(\cdot) \not\equiv 0$. It remains to check whether $\mathcal{M}_1(\omega) := K_1 \mathcal{N}(\mathrm{i}\omega)^+ K_1^\mathsf{T} \not\leq 0$ holds true for an $\omega \in \mathbb{R}$. It holds that

$$\mathcal{M}_1(\omega) = K_1 V_1(\omega) \Theta(\omega)^{-1} V_1(\omega)^\mathsf{H} K_1^\mathsf{T},$$

i. e., $\mathcal{M}_1(\cdot)$ has a zero at $\omega_0 \in \mathbb{R}$ if and only if there is a rank drop in $K_1 V_1(\cdot)$ at $\omega_0$. On the other hand, this condition is equivalent to

$$\begin{aligned}
\operatorname{rank} \begin{bmatrix} \Theta(\omega_0) & V_1(\omega_0)^\mathsf{H} K_1^\mathsf{T} \\ K_1 V_1(\omega_0) & 0 \end{bmatrix} &= \operatorname{rank} \begin{bmatrix} \Theta(\omega_0) & 0 & V_1(\omega_0)^\mathsf{H} K_1^\mathsf{T} \\ 0 & 0 & V_2(\omega_0)^\mathsf{H} K_1^\mathsf{T} \\ K_1 V_1(\omega_0) & K_1 V_2(\omega_0) & 0 \end{bmatrix} \\
&= \operatorname{rank} \begin{bmatrix} \mathcal{N}(\mathrm{i}\omega_0) & K_1^\mathsf{T} \\ K_1 & 0 \end{bmatrix} \\
&< \operatorname{rank}_{\mathbb{R}(s)} \begin{bmatrix} \mathcal{N}(s) & K_1^\mathsf{T} \\ K_1 & 0 \end{bmatrix},
\end{aligned}$$

i. e., $\mathrm{i}\omega_0$ is an eigenvalue of the (possibly singular) even pencil $\mathcal{H}_1(s) := \begin{bmatrix} \mathcal{N}(s) & K_1^\mathsf{T} \\ K_1 & 0 \end{bmatrix} \in \mathbb{R}[s]^{\ell+m_0 \times \ell+m_0}$. Now let $\mathrm{i}\alpha_1 < \mathrm{i}\alpha_2 < \ldots < \mathrm{i}\alpha_k$ be the purely imaginary eigenvalues of $\mathcal{H}_1(s)$ with positive imaginary part and set $\alpha_0 = 0$ and $\alpha_{k+1} = \infty$. Due to the piecewise continuity of the spectrum of $\mathcal{M}_1(\cdot)$ it is only necessary to verify whether it holds that $\mathcal{M}_1(\widehat{\alpha}_j) \not\leq 0$ for some $\widehat{\alpha}_j \in (\alpha_j, \alpha_{j+1})$, $j = 0, \ldots, k$. The same strategy can also be applied to check the same condition for $\mathcal{M}_2(\omega) := K_2 \mathcal{N}(\mathrm{i}\omega)^+ K_2^\mathsf{T} \not\leq 0$.

### 5.3.2 Perturbation of the Defective Infinite Eigenvalues

In this subsection we consider perturbations of the defective infinite eigenvalues of $\mathcal{N}(s) \in \mathbb{R}[s]^{\ell \times \ell}$, i. e., those that correspond to blocks of type E3 of size larger than $1 \times 1$ in the EKCF. Therefore, we need the following theorem which is similar to Theorem 5.3.1 in some aspects. Note that in the following we only consider $\mathcal{N}(\mathrm{i}\omega)$ for $\omega \to \infty$, since for even pencils with real coefficients it holds that $\Lambda(\mathcal{N}(\mathrm{i}\omega)) = \Lambda(\mathcal{N}(-\mathrm{i}\omega))$.

**Theorem 5.3.4.** *Let $H : (\omega_0, \infty) \to \mathbb{C}^{\ell \times \ell}$ be a given analytic Hermitian matrix-valued function. Consider the pointwise spectral decomposition*

$$H(\omega) = \begin{bmatrix} V_1(\omega) & V_2(\omega) \end{bmatrix} \begin{bmatrix} \Theta(\omega) & 0 \\ 0 & \Omega(\omega) \end{bmatrix} \begin{bmatrix} V_1(\omega) & V_2(\omega) \end{bmatrix}^\mathsf{H}$$

*with analytic functions* $V_1 : (\omega_0, \infty) \to \mathbb{C}^{\ell \times \ell_1}$, $V_2 : (\omega_0, \infty) \to \mathbb{C}^{\ell \times \ell_2}$ *and* $\begin{bmatrix} V_1(\cdot) & V_2(\cdot) \end{bmatrix}$ *having pointwise unitary columns. Furthermore,* $\Theta : (\omega_0, \infty) \to \mathbb{R}^{\ell_1 \times \ell_1}$ *and* $\Omega : (\omega_0, \infty) \to \mathbb{R}^{\ell_2 \times \ell_2}$ *are analytic diagonal matrix-valued functions containing the eigenvalues of* $H(\cdot)$ *on their diagonal with the additional properties that* $\left\| \Theta(\omega)^{-1} \right\|_2 < h < \infty$ *for all* $\omega \in (\omega_0, \infty)$ *and* $\lim_{\omega \to \infty} \Omega(\omega) = 0$.

*Consider a perturbed analytic matrix-valued function*

$$\widetilde{H}(\omega) = H(\omega) + \begin{bmatrix} K_1 \\ K_2 \end{bmatrix}^\mathsf{T} \begin{bmatrix} 0 & \Delta \\ \Delta^\mathsf{H} & 0 \end{bmatrix} \begin{bmatrix} K_1 \\ K_2 \end{bmatrix}$$

*with nonzero* $K_1 \in \mathbb{R}^{m_0 \times \ell}$, $K_2 \in \mathbb{R}^{p_0 \times \ell}$, *and* $\Delta \in \mathbb{C}^{m_0 \times p_0}$.

*Furthermore, define* $V_{1,\infty} := \lim_{\omega \to \infty} V_1(\omega)$, $V_{2,\infty} := \lim_{\omega \to \infty} V_2(\omega)$, *and* $H_\infty^+ := \lim_{\omega \to \infty} H(\omega)^+$. *Then the following statements are satisfied:*

*a) If* $K_1 V_{2,\infty} = 0$ *and* $K_2 V_{2,\infty} = 0$, *then* $H(\omega)$ *and* $\widetilde{H}(\omega)$ *have the same zero eigenvalues for* $\omega \to \infty$ *for all sufficiently small* $\Delta \neq 0$.

*b) If* $K_1 V_{2,\infty} \neq 0$ *and* $K_2 V_{2,\infty} \neq 0$, *then there exists an arbitrarily small* $\Delta \neq 0$ *such that for* $\omega \to \infty$, *some zero eigenvalues of* $H(\omega)$ *can be perturbed to positive and some negative eigenvalues in* $\widetilde{H}(\omega)$.

*c) If* $K_1 V_{2,\infty} \neq 0$ *and* $K_2 V_{2,\infty} = 0$, *then there exists an arbitrarily small* $\Delta \neq 0$ *such that for* $\omega \to \infty$, *some zero eigenvalues of* $H(\omega)$ *are perturbed to positive (negative) eigenvalues in* $\widetilde{H}(\omega)$ *if and only if* $K_2 H_\infty^+ K_2^\mathsf{T} \ngeq 0$ $(K_2 H_\infty^+ K_2^\mathsf{T} \nleq 0)$. *If* $K_2 H_\infty^+ K_2^\mathsf{T} \geq 0$ $(K_2 H_\infty^+ K_2^\mathsf{T} \leq 0)$, *then the smallest perturbation* $\Delta_0$ *that perturbs a zero eigenvalue of* $H(\omega)$ *to a positive (negative) eigenvalue of* $\widetilde{H}(\omega)$ *for* $\omega \to \infty$ *fulfills*

$$\|\Delta_0\|_2 > \inf_{\Delta \in \mathbb{C}^{m_0 \times p_0}} \left\{ \|\Delta\|_2 : V_{1,\infty}^\mathsf{H} \left( H(\omega) + \begin{bmatrix} K_1 \\ K_2 \end{bmatrix}^\mathsf{T} \begin{bmatrix} 0 & \Delta \\ \Delta^\mathsf{H} & 0 \end{bmatrix} \begin{bmatrix} K_1 \\ K_2 \end{bmatrix} \right) V_{1,\infty} \right.$$
$$\left. \text{has a zero eigenvalue for } \omega \to \infty \right\}. \quad (5.8)$$

*d) If* $K_1 V_{2,\infty} = 0$ *and* $K_2 V_{2,\infty} \neq 0$, *then there exists an arbitrarily small* $\Delta \neq 0$ *such that for* $\omega \to \infty$, *some zero eigenvalues of* $H(\omega)$ *are perturbed to positive (negative) eigenvalues in* $\widetilde{H}(\omega)$ *if and only if* $K_1 H_\infty^+ K_1^\mathsf{T} \ngeq 0$ $(K_1 H_\infty^+ K_1^\mathsf{T} \nleq 0)$. *If* $K_1 H_\infty^+ K_1^\mathsf{T} \geq 0$ $(K_1 H_\infty^+ K_1^\mathsf{T} \leq 0)$, *then the smallest perturbation* $\Delta_0$ *that perturbs a zero eigenvalue of* $H(\omega)$ *to a positive (negative) eigenvalue of* $\widetilde{H}(\omega)$ *for* $\omega \to \infty$ *fulfills* (5.8).

*Proof.* Assume that $\Delta = \varepsilon\widetilde{\Delta}$ with $\|\widetilde{\Delta}\|_2 = 1$ and $\varepsilon > 0$ are given and define the following analytic matrix-valued functions for $\omega \in (\omega_0, \infty)$:

$$\begin{aligned}
\Theta_{11}(\omega) &:= V_1^{\mathsf{H}}(\omega)K_1^{\mathsf{T}}\widetilde{\Delta}K_2V_1(\omega) + V_1(\omega)^{\mathsf{H}}K_2^{\mathsf{T}}\widetilde{\Delta}^{\mathsf{H}}K_1V_1(\omega),\\
\Theta_{12}(\omega) &:= V_1^{\mathsf{H}}(\omega)K_1^{\mathsf{T}}\widetilde{\Delta}K_2V_2(\omega) + V_1(\omega)^{\mathsf{H}}K_2^{\mathsf{T}}\widetilde{\Delta}^{\mathsf{H}}K_1V_2(\omega),\\
\Theta_{22}(\omega) &:= V_2^{\mathsf{H}}(\omega)K_1^{\mathsf{T}}\widetilde{\Delta}K_2V_2(\omega) + V_2(\omega)^{\mathsf{H}}K_2^{\mathsf{T}}\widetilde{\Delta}^{\mathsf{H}}K_1V_2(\omega).
\end{aligned} \tag{5.9}$$

This yields

$$\begin{bmatrix}V_1(\omega) & V_2(\omega)\end{bmatrix}^{\mathsf{H}} \widetilde{H}(\omega) \begin{bmatrix}V_1(\omega) & V_2(\omega)\end{bmatrix} = \begin{bmatrix}\Theta(\omega) + \varepsilon\Theta_{11}(\omega) & \varepsilon\Theta_{12}(\omega)\\ \varepsilon\Theta_{12}^{\mathsf{H}}(\omega) & \Omega(\omega) + \varepsilon\Theta_{22}(\omega)\end{bmatrix}$$

for all $\omega \in (\omega_0, \infty)$. Since all eigenvalues of $\Theta(\omega)$ are uniformly bounded away from zero for all $\omega > \omega_1 > \omega_0$, there exists an $\varepsilon_0$ such that $\Theta(\omega) + \varepsilon\Theta_{11}(\omega)$ is nonsingular for all $(\varepsilon, \omega) \in (0, \varepsilon_0) \times (\omega_1, \infty)$. The Haynsworth inertia additivity formula [Hay68] yields that for all such $(\varepsilon, \omega)$ it holds that

$$\operatorname{In}\left(\widetilde{H}(\omega)\right) = \operatorname{In}\left(\Theta(\omega) + \varepsilon\Theta_{11}(\omega)\right) + \operatorname{In}\left(F(\varepsilon,\omega)\right) = \operatorname{In}\left(\Theta(\omega)\right) + \operatorname{In}\left(F(\varepsilon,\omega)\right)$$

with $F(\varepsilon,\omega) := \Omega(\omega) + \varepsilon\Theta_{22}(\omega) - \varepsilon^2\Theta_{12}^{\mathsf{H}}(\omega)\left(\Theta(\omega) + \varepsilon\Theta_{11}(\omega)\right)^{-1}\Theta_{12}(\omega)$.

Now we prove statements a)–d).

First we show a): Since the functions $\Omega(\cdot)$, $K_1V_2(\cdot)$, and $K_2V_2(\cdot)$ are analytic and tend to zero for $\omega \to \infty$ and $\Theta_{11}(\cdot)$ is analytic and bounded, each of them can be expanded into a Taylor series at expansion point $\omega_* = \infty$. This yields

$$\begin{aligned}
\Omega(\omega) &= \sum_{k=-\infty}^{-1} L_k\omega^k, \quad & \Theta_{11}(\omega) &= \sum_{k=-\infty}^{0} M_k\omega^k,\\
\Theta_{12}(\omega) &= \sum_{k=-\infty}^{-1} N_k\omega^k, \quad & \Theta_{22}(\omega) &= \sum_{k=-\infty}^{-2} P_k\omega^k.
\end{aligned} \tag{5.10}$$

Note that the upper summation indices for $\Theta_{11}(\cdot)$, $\Theta_{12}(\cdot)$, and $\Theta_{22}(\cdot)$ follow directly from the respective definitions in (5.9) and the assumptions of a). Furthermore, since $\|\Theta(\cdot)\|_2^{-1}$ is uniformly bounded from above, we have

$$\Theta(\omega)^{-1} = \sum_{k=-\infty}^{0} Q_k\omega^k.$$

Therefore, by plugging in the expansions for $\Theta(\cdot)^{-1}$ and $\Theta_{11}(\cdot)$ it holds that

$$\begin{aligned}
\left(\Theta(\omega) + \varepsilon\Theta_{11}(\omega)\right)^{-1} &= \sum_{k=0}^{\infty}\left(\Theta(\omega)^{-1}\Theta_{11}(\omega)\right)^k\Theta(\omega)^{-1}\varepsilon^k\\
&= \sum_{k,j=0}^{\infty} R_{k,j}\varepsilon^k\omega^{-j}
\end{aligned}$$

for $\varepsilon \in (0, \varepsilon_0)$. Altogether this yields

$$\begin{aligned} F(\varepsilon, \omega) &= \sum_{k=-\infty}^{-1} L_k \omega^k + \varepsilon \sum_{k=-\infty}^{-2} P_k \omega^k \\ &\quad - \varepsilon^2 \left( \sum_{k=-\infty}^{-1} N_k^{\mathsf{H}} \omega^k \right) \left( \sum_{k,j=0}^{\infty} R_{k,j} \varepsilon^k \omega^{-j} \right) \left( \sum_{k=-\infty}^{-1} N_k \omega^k \right) \\ &= \sum_{k=-\infty}^{-1} L_k \omega^k + \varepsilon \mathcal{O}\left(\omega^{-2}\right), \end{aligned} \tag{5.11}$$

and therefore, $F(\varepsilon, \omega)$ tends to zero for $\omega \to \infty$ and all $\varepsilon \in (0, \varepsilon_0)$.

Now we show b): We can use the same line of argumentation as in a). However, we can choose a perturbation $\widetilde{\Delta}$ such that instead of (5.10) it holds that

$$\begin{aligned} \Theta_{12}(\omega) &= \sum_{k=-\infty}^{0} N_k \omega^k, \\ \Theta_{22}(\omega) &= \sum_{k=-\infty}^{0} P_k \omega^k, \quad P_0 \neq 0. \end{aligned}$$

Hence we get

$$F(\varepsilon, \omega) = \sum_{k=-\infty}^{-1} L_k \omega^k + \varepsilon \sum_{k=-\infty}^{0} P_k \omega^k + \mathcal{O}\left(\varepsilon^2\right),$$

i. e., as in the proof of Theorem 5.3.1 we can achieve perturbations of the zero eigenvalues of $H(\omega)$ for $\omega \to \infty$ into positive as well as negative direction for an arbitrarily small perturbation $\Delta$.

Next we show statement c): We approach the problem as in a). However, we can choose a perturbation $\widetilde{\Delta}$ such that instead of (5.10) it holds that

$$\begin{aligned} \Theta_{12}(\omega) &= \underbrace{V_1^{\mathsf{H}}(\omega) K_1^{\mathsf{T}} \widetilde{\Delta} K_2 V_2(\omega)}_{\to 0 \text{ for } \omega \to \infty} + V_1^{\mathsf{H}}(\omega) K_2^{\mathsf{T}} \widetilde{\Delta}^{\mathsf{H}} K_1 V_2(\omega) \\ &= \sum_{k=-\infty}^{0} N_k \omega^k, \quad N_0 = \lim_{\omega \to \infty} V_1^{\mathsf{H}}(\omega) K_2^{\mathsf{T}} \widetilde{\Delta}^{\mathsf{H}} K_1 V_2(\omega), \\ \Theta_{22}(\omega) &= \sum_{k=-\infty}^{-1} P_k \omega^k. \end{aligned}$$

Therefore, by comparing with (5.11) we obtain the expansion

$$F(\varepsilon, \omega) = \sum_{k=-\infty}^{-1} L_k \omega^k + \varepsilon \sum_{k=-\infty}^{-1} P_k \omega^k - \varepsilon^2 \sum_{k,j=0}^{\infty} \widetilde{R}_{k,j} \varepsilon^k \omega^{-j}$$

with

$$\begin{aligned}\widetilde{R}_{0,0} &= \lim_{\omega\to\infty}\left(V_2^{\mathsf{H}}(\omega)K_1^{\mathsf{H}}\widetilde{\Delta}K_2V_1(\omega)\Theta(\omega)^{-1}V_1^{\mathsf{H}}(\omega)K_2^{\mathsf{T}}\widetilde{\Delta}^{\mathsf{H}}K_1V_2(\omega)\right)\\ &= V_{2,\infty}^{\mathsf{H}}K_1^{\mathsf{H}}\widetilde{\Delta}K_2H_\infty^+K_2^{\mathsf{T}}\widetilde{\Delta}^{\mathsf{H}}K_1V_{2,\infty}.\end{aligned}$$

Now we can use the same argumentation as in Theorem 5.3.1 to obtain the result.

Finally, d) is proved completely analogously to the previous statement. □

The following corollary now makes a statement about the loss of cyclo-dissipativity due to perturbations of the defective infinite eigenvalues of $\mathcal{N}(s) \in \mathbb{R}[s]^{\ell\times\ell}$.

**Corollary 5.3.5.** *Assume that* $\Sigma := (E, A, B, C, D) \in \Sigma_{n,m,p}$ *and let* $Q_s = Q_s^{\mathsf{T}} \in \mathbb{R}^{p\times p}$, $S_s \in \mathbb{R}^{p\times m}$, *and* $R_s = R_s^{\mathsf{T}} \in \mathbb{R}^{m\times m}$ *be given. Let the system be cyclo-dissipative with respect to* $s(\cdot,\cdot)$ *as defined in* (3.101) *and let* $\mathcal{N}(s) \in \mathbb{R}[s]^{\ell\times\ell}$ *and* $K$ *be defined as in* (5.3) *and* (5.4), *respectively. For all* $\omega \in \mathbb{R}$ *consider the spectral decomposition*

$$\mathcal{N}(\mathrm{i}\omega) = \begin{bmatrix}V_1(\omega) & V_2(\omega)\end{bmatrix}\begin{bmatrix}\Theta(\omega) & 0\\ 0 & \Omega(\omega)\end{bmatrix}\begin{bmatrix}V_1(\omega) & V_2(\omega)\end{bmatrix}^{\mathsf{H}}$$

*with analytic functions* $V_1 : \mathbb{R} \to \mathbb{C}^{\ell\times\ell_1}$, $V_2 : \mathbb{R} \to \mathbb{C}^{\ell\times\ell_2}$ *and* $\begin{bmatrix}V_1(\cdot) & V_2(\cdot)\end{bmatrix}$ *having pointwise unitary columns, and* $\Theta : \mathbb{R} \to \mathbb{R}^{\ell_1\times\ell_1}$, $\Omega : \mathbb{R} \to \mathbb{R}^{\ell_2\times\ell_2}$ *being analytic diagonal matrix-valued functions with the additional properties that* $\left\|\Theta(\omega)^{-1}\right\|_2 < h < \infty$ *for all sufficiently large* $\omega \in \mathbb{R}$ *and* $\lim_{\omega\to\infty}\Omega(\omega) = 0$. *Let*

$$\begin{aligned}\mathcal{I}_j &:= \{i \in \mathbb{N} : \text{the } i\text{-th column of } K_jV_{2,\infty} \text{ is nonzero}\}\\ &= \left\{i_{j,1}, i_{j,2}, \ldots, i_{j,\widehat{\ell}_j}\right\},\end{aligned}$$

*and define the matrices*

$$\mathcal{Q}_1 := \left[e_{i_{1,1}}^{(m_0)}, \ldots, e_{i_{1,\widehat{\ell}_1}}^{(m_0)}\right], \quad \mathcal{Q}_2 := \left[e_{i_{2,1}}^{(p_0)}, \ldots, e_{i_{2,\widehat{\ell}_2}}^{(p_0)}\right],$$

*where* $e_i^{(k)}$ *denotes the* $i$*-th unit vector of length* $k$. *Moreover, define the matrices* $\mathcal{N}_\infty^+ := \lim_{\omega\to\infty}\mathcal{N}(\mathrm{i}\omega)^+$ *and* $V_{2,\infty} := \lim_{\omega\to\infty}V_2(\omega)$. *For an* $\omega_0 \in \mathbb{R}$ *define the inertias*

$$\left(\pi_+^{\mathcal{I}_j}, \pi_0^{\mathcal{I}_j}, \pi_-^{\mathcal{I}_j}\right) := \mathrm{In}(\mathcal{Q}_j\Omega(\omega_0)\mathcal{Q}_j^{\mathsf{T}}).$$

*Then the following statements hold true:*

*a) If it holds that*

*i)* $K_1V_{2,\infty} \neq 0$ *and* $K_2V_{2,\infty} \neq 0$; *or*

*ii)* $K_1V_{2,\infty} \neq 0$ *and* $K_2V_{2,\infty} = 0$ *and*

*1) if* $\pi_+^{\mathcal{I}_1} + \pi_0^{\mathcal{I}_1} \neq 0$ *then* $K_2 \mathcal{N}_\infty^+ K_2^\mathsf{T} \nleq 0$*, or*

*2) if* $\pi_-^{\mathcal{I}_1} \neq 0$ *then* $K_2 \mathcal{N}_\infty^+ K_2^\mathsf{T} \ngeq 0$*; or*

*iii)* $K_1 V_{2,\infty} = 0$ *and* $K_2 V_{2,\infty} \neq 0$ *and*

*1) if* $\pi_+^{\mathcal{I}_2} + \pi_0^{\mathcal{I}_2} \neq 0$ *then* $K_1 \mathcal{N}_\infty^+ K_1^\mathsf{T} \nleq 0$*, or*

*2) if* $\pi_-^{\mathcal{I}_2} \neq 0$ *then* $K_1 \mathcal{N}_\infty^+ K_1^\mathsf{T} \ngeq 0$*,*

*then the cyclo-dissipativity radius is zero.*

*b) If on the other hand, none of the above conditions is satisfied, then with* $\mathcal{N}_\Delta(s) \in \mathbb{C}[s]^{\ell \times \ell}$ *as in (5.5) it holds that* $\eta(\mathcal{N}(\mathrm{i}\omega)) = \eta(\mathcal{N}_\Delta(\mathrm{i}\omega))$ *for all sufficiently large* $\omega \in \mathbb{R}$ *and all sufficiently small perturbations* $\Delta \in \mathbb{C}^{m_0 \times p_0}$*.*

*Proof.* First we show a). The sign-sum function $\eta(\mathcal{N}(\mathrm{i}\cdot))$ can be changed if for $\omega \to \infty$

- one of the constant zero eigenvalues of $\Omega(\omega)$ can be perturbed to a negative value;
- one of the eigenvalues of $\Omega(\omega)$ approaching zero from above can be perturbed to a negative value;
- one of the eigenvalues of $\Omega(\omega)$ approaching zero from below can be perturbed to a positive value.

But this is the case if and only if the conditions in a) are fulfilled. We show how to construct the corresponding perturbation matrix for case a) ii) 1):

Similarly as in the proof of Theorem 5.3.1 consider the matrix

$$\begin{aligned}\Theta_0 :&= V_{2,\infty}^\mathsf{H} K_1^\mathsf{T} \widetilde{\Delta} K_2 \mathcal{N}_\infty^+ K_2^\mathsf{T} \widetilde{\Delta}^\mathsf{H} K_1 V_{2,\infty} \\ &= V_{2,\infty}^\mathsf{H} K_1^\mathsf{T} \widehat{\Delta} \Xi \widehat{\Delta}^\mathsf{H} K_1 V_{2,\infty}\end{aligned}$$

with a spectral decomposition $K_2 \mathcal{N}_\infty^+ K_2^\mathsf{T} = P^\mathsf{H} \Xi P$ and a unitary matrix $P \in \mathbb{C}^{p_0 \times p_0}$, a diagonal matrix $\Xi \in \mathbb{R}^{p_0 \times p_0}$, and $\widehat{\Delta} := \widetilde{\Delta} P^\mathsf{H}$.

Assume that $\mathcal{Q}_1 \Omega(\omega) \mathcal{Q}_1^\mathsf{T}$ for sufficiently large $\omega$ has a nonnegative eigenvalue at position $(q,q)$. Since by construction $K_1 V_{2,\infty} \mathcal{Q}_1^\mathsf{T}$ has only nonzero columns, it has a nonzero $(p,q)$-th entry for some $p$. Finally, $\Xi$ has at least one positive eigenvalue, assume at position $(r,r)$. Now we can construct a desired perturbation as in the proof of Theorem 5.3.1 by setting the $(r,p)$-th element of $\widehat{\Delta}^\mathsf{H}$ to a nonzero value and the others to zero.

Statement b) is a direct consequence of Theorem 5.3.4 a), c), and d). □

To check the above conditions, we evaluate $\Theta(\omega_\infty)$, $\Omega(\omega_\infty)$, and $V_2(\omega_\infty)$ for sufficiently large $\omega_\infty$. Since $V_2(\omega)$ converges for $\omega \to \infty$, we can get a good estimate of $V_{2,\infty}$ in this way. From $\Theta(\omega_\infty)$, $\Omega(\omega_\infty)$, and $V_2(\omega_\infty)$ we can determine all information that is needed to apply Corollary 5.3.5.

We will illustrate the above result with the following example.

**Example 5.3.6.** Consider the even matrix pencil

$$\mathcal{N}(s) = \begin{bmatrix} s\mathrm{i} & 1 \\ 1 & 0 \end{bmatrix} \in \mathbb{C}[s]^{2\times 2}.$$

Note that we take a pencil with complex coefficients, since a pencil with real coefficients and similar properties would be at least of size $4 \times 4$ and it would be difficult to determine the eigenvalues of $\mathcal{N}(\mathrm{i}\omega)$ analytically. However, the main features of our theory also apply to complex pencils. For this example it holds that

$$\mathcal{N}(\mathrm{i}\omega) = \begin{bmatrix} \gamma(\omega) & \delta(\omega) \\ \gamma(\omega)(\alpha(\omega)-\beta(\omega)) & \delta(\omega)(\alpha(\omega)+\beta(\omega)) \end{bmatrix}$$
$$\cdot \begin{bmatrix} -\alpha(\omega)-\beta(\omega) & 0 \\ 0 & -\alpha(\omega)+\beta(\omega) \end{bmatrix} \begin{bmatrix} \gamma(\omega) & \delta(\omega) \\ \gamma(\omega)(\alpha(\omega)-\beta(\omega)) & \delta(\omega)(\alpha(\omega)+\beta(\omega)) \end{bmatrix}^{\mathsf{H}}$$

with

$$\alpha(\omega) = \frac{\omega}{2}, \quad \beta(\omega) = \sqrt{\frac{\omega^2}{4}+1},$$

and the normalization functions

$$\gamma(\omega) = \left(1 + (\alpha(\omega)-\beta(\omega))^2\right)^{-1/2}, \quad \delta(\omega) = \left(1 + (\alpha(\omega)+\beta(\omega))^2\right)^{-1/2}.$$

Using the notation of Corollary 5.3.5, we have

$$\Theta(\omega) = -\alpha(\omega)-\beta(\omega), \quad \Omega(\omega) = -\alpha(\omega)+\beta(\omega),$$
$$V_1(\omega) = \begin{bmatrix} \gamma(\omega) \\ \gamma(\omega)(\alpha(\omega)-\beta(\omega)) \end{bmatrix}, \quad V_2(\omega) = \begin{bmatrix} \delta(\omega) \\ \delta(\omega)(\alpha(\omega)+\beta(\omega)) \end{bmatrix},$$
$$V_{2,\infty} = \lim_{\omega\to\infty} \begin{bmatrix} \delta(\omega) \\ \delta(\omega)(\alpha(\omega)+\beta(\omega)) \end{bmatrix} = \begin{bmatrix} 0 \\ 1 \end{bmatrix}.$$

a) Assume that

$$K = \begin{bmatrix} K_1 \\ K_2 \end{bmatrix} = \begin{bmatrix} 1 & 0 \\ 0 & 1 \end{bmatrix}.$$

Then we obtain $K_1 V_{2,\infty} = 0$ and $K_2 V_{2,\infty} = 1$. Moreover, we have $\left(\pi_+^{\mathcal{I}_2}, \pi_0^{\mathcal{I}_2}, \pi_-^{\mathcal{I}_2}\right) = (1,0,0)$, and

$$K_1 \mathcal{N}_\infty^+ K_1^\mathsf{T} = \begin{bmatrix} 1 & 0 \end{bmatrix} \begin{bmatrix} 0 & 0 \\ 0 & 0 \end{bmatrix} \begin{bmatrix} 1 \\ 0 \end{bmatrix} = 0.$$

In other words, the only possible condition in Corollary 5.3.5 a) iii) 1) is not satisfied and hence, no perturbation of the defective infinite eigenvalues is possible. Indeed, we have

$$\mathcal{N}_\Delta(s) = \begin{bmatrix} s\mathrm{i} & 1+\Delta^\mathsf{H} \\ 1+\Delta & 0 \end{bmatrix},$$

whose EKCF is the same as for $\mathcal{N}(s)$ for all $\Delta \in \mathbb{C}$ with $\|\Delta\|_2 < 1$.

b) Now assume that

$$K = \begin{bmatrix} K_1 \\ K_2 \end{bmatrix} = \begin{bmatrix} 0 & 1 \\ 0 & 1 \end{bmatrix}.$$

Then we obtain $K_1 V_{2,\infty} = 1$ and $K_2 V_{2,\infty} = 1$, i. e., Corollary 5.3.5 a) i) holds true. This is confirmed by considering

$$\mathcal{N}_\Delta(s) = \begin{bmatrix} s\mathrm{i} & 1 \\ 1 & \Delta + \Delta^\mathsf{H} \end{bmatrix}.$$

Assume that $\rho := \Delta + \Delta^\mathsf{H}$. Then, the eigenvalues of $\mathcal{N}_\Delta(\mathrm{i}\omega)$ are

$$\lambda_1(\omega) = \frac{-\omega + \rho}{2} + \sqrt{\frac{(\omega+\rho)^2}{4} + 1},$$
$$\lambda_2(\omega) = \frac{-\omega + \rho}{2} - \sqrt{\frac{(\omega+\rho)^2}{4} + 1}.$$

Note that the signature of $\lambda_1(\omega)$ changes for $\omega = -1/\rho$, so for any $\rho \neq 0$, the defective infinite eigenvalues are perturbed.

## 5.4 Computation of the Complex Cyclo-Dissipativity Radius

Now we turn to perturbation of the "nonzero" part of the pencil $\mathcal{N}(s) \in \mathbb{R}[s]^{\ell \times \ell}$, i. e., we consider perturbations of the nonzero eigenvalues of $\mathcal{N}(\mathrm{i}\cdot)$. From now on we assume $\mathcal{N}(s)$ has no purely imaginary eigenvalues and that there does not exist an arbitrarily small perturbation of the singular part or the defective infinite eigenvalues as discussed above. From the perturbation bounds (5.6) and (5.8) it remains to calculate the smallest perturbation that moves one of the nonzero eigenvalues of $\mathcal{N}(\mathrm{i}\omega)$ to zero for an $\omega \in \mathbb{R}$.

Since the sign-sum function can only change at purely imaginary eigenvalues, we obtain

$$\begin{aligned} r_{s,\mathbb{C}}(\Sigma, \mathcal{L}, \mathcal{R}) = \inf \big\{ \|\Delta\|_2 : &\operatorname{rank}(\mathcal{N}_\Delta(\mathrm{i}\omega_0)) < \operatorname{rank}_{\mathbb{C}(s)}(\mathcal{N}_\Delta(s)) \\ &\text{with } \Delta \in \mathbb{C}^{m_0 \times p_0} \text{ for some } \mathrm{i}\omega_0 \notin \Lambda\left(E, A + \mathcal{L}_1 \Delta \mathcal{R}_1\right) \big\}. \end{aligned} \tag{5.12}$$

From (5.6) and (5.8), the condition (5.12) is equivalent to the existence of an $\mathrm{i}\omega_0 \notin \Lambda\left(E, A + \mathcal{L}_1 \Delta \mathcal{R}_1\right)$ such that

$$\det\left( \Theta(\omega_0) - V_1^\mathsf{H}(\omega_0) K^\mathsf{T} \begin{bmatrix} 0 & \Delta \\ \Delta^\mathsf{H} & 0 \end{bmatrix} K V_1(\omega_0) \right) = 0. \tag{5.13}$$

Similarly as in [OD05], the determinant condition in (5.13) is equivalent to

$$\det\left( I_{m_0+p_0} - \begin{bmatrix} 0 & \Delta \\ \Delta^\mathsf{H} & 0 \end{bmatrix} \mathcal{M}(\omega_0) \right) = 0,$$

where $\mathcal{M}(\omega) := KV_1(\omega)\Theta(\omega)^{-1}V_1^{\mathsf{H}}(\omega)K^{\mathsf{T}} = K\mathcal{N}(\mathrm{i}\omega)^{+}K^{\mathsf{T}}$. This multiplicative perturbation structure will be used to derive an algorithm for the computation of the complex cyclo-dissipativity radius.

First, we focus on multiplicative perturbations of a *fixed* Hermitian matrix $\mathcal{M} = \mathcal{M}(\omega_0)$, i.e., we want to compute

$$q_{\mathbb{C}}(\mathcal{M}) := \inf\left\{\|\Delta\|_2 : \det\left(I_{m_0+p_0} - \begin{bmatrix} 0 & \Delta \\ \Delta^{\mathsf{H}} & 0 \end{bmatrix}\mathcal{M}\right) = 0 \text{ with } \Delta \in \mathbb{C}^{m_0 \times p_0}\right\}. \tag{5.14}$$

The solution of this subproblem will then be subsequently used to find the solution of the general problem by

$$r_{s,\mathbb{C}}(\Sigma, \mathcal{L}, \mathcal{R}) = \inf_{\omega\in\mathbb{R}} q_{\mathbb{C}}(\mathcal{M}(\omega)).$$

Consider now the subproblem of computing (5.14). A solution of this problem has already been given [HQ98]. There, it is shown that when partitioning $\mathcal{M} = \begin{bmatrix} \mathcal{M}_{11} & \mathcal{M}_{12} \\ \mathcal{M}_{12}^{\mathsf{H}} & \mathcal{M}_{22} \end{bmatrix}$ with $\mathcal{M}_{11} \in \mathbb{C}^{m_0\times m_0}$, $\mathcal{M}_{12} \in \mathbb{C}^{m_0\times p_0}$, and $\mathcal{M}_{22} \in \mathbb{C}^{p_0\times p_0}$ it holds that

$$\begin{aligned} &\det\left(I_{m_0+p_0} - \begin{bmatrix} 0 & \Delta \\ \Delta^{\mathsf{H}} & 0 \end{bmatrix}\begin{bmatrix} \mathcal{M}_{11} & \mathcal{M}_{12} \\ \mathcal{M}_{12}^{\mathsf{H}} & \mathcal{M}_{22} \end{bmatrix}\right) \\ &= \det\left(I_{m_0+p_0} - \begin{bmatrix} \sqrt{\gamma} I_{m_0} & 0 \\ 0 & I_{p_0}/\sqrt{\gamma} \end{bmatrix}^{-1}\begin{bmatrix} 0 & \Delta \\ \Delta^{\mathsf{H}} & 0 \end{bmatrix}\begin{bmatrix} \mathcal{M}_{11} & \mathcal{M}_{12} \\ \mathcal{M}_{12}^{\mathsf{H}} & \mathcal{M}_{22} \end{bmatrix}\begin{bmatrix} \sqrt{\gamma} I_{m_0} & 0 \\ 0 & I_{p_0}/\sqrt{\gamma} \end{bmatrix}\right) \\ &= \det\left(I_{m_0+p_0} - \begin{bmatrix} 0 & \Delta \\ \Delta^{\mathsf{H}} & 0 \end{bmatrix}\begin{bmatrix} \gamma\mathcal{M}_{11} & \mathcal{M}_{12} \\ \mathcal{M}_{12}^{\mathsf{H}} & \mathcal{M}_{22}/\gamma \end{bmatrix}\right) \end{aligned}$$

for all $\gamma > 0$. Then the following theorem can be proven.

**Theorem 5.4.1.** *[HQ98, Thm. 1] Let*

$$\mathcal{F}(\gamma) := \begin{bmatrix} \gamma\mathcal{M}_{11} & \mathcal{M}_{12} \\ \mathcal{M}_{12}^{\mathsf{H}} & \mathcal{M}_{22}/\gamma \end{bmatrix}.$$

*Furthermore, assume that* $\mathrm{In}(\mathcal{M}) = (\pi_+, \pi_0, \pi_-)$. *Define the numbers*

$$r_+ = \begin{cases} \left(\inf\limits_{\gamma>0}\{\lambda_1(\mathcal{F}(\gamma))\}\right)^{-1} & \text{if } \pi_+ > 0, \\ \infty & \text{if } \pi_+ = 0, \end{cases}$$

$$r_- = \begin{cases} \left(\inf\limits_{\gamma>0}\{-\lambda_{m_0+p_0}(\mathcal{F}(\gamma))\}\right)^{-1} & \text{if } \pi_- > 0, \\ \infty & \text{if } \pi_- = 0, \end{cases}$$

*where* $\lambda_j(\mathcal{F}(\gamma))$, $j = 1, \ldots, m_0 + p_0$, *denotes the* $j$*-th largest eigenvalue of* $\mathcal{F}(\gamma)$. *Then it holds that*

$$q_{\mathbb{C}}(\mathcal{M}) = \min\{r_+, r_-\}. \tag{5.15}$$

In the following lemma we show that for our problem the minimum in (5.15) is indeed attained for some $\gamma_0 \in (0, \infty)$.

**Lemma 5.4.2.** *Let* $\mathcal{M} = K\mathcal{N}^+K^\mathsf{T} = KV\Theta^{-1}V^\mathsf{H}K^\mathsf{T} \in \mathbb{C}^{m_0+p_0\times m_0+p_0}$ *for some Hermitian* $\mathcal{N} \in \mathbb{C}^{\ell\times\ell}$, *and some* $\Theta \in \mathrm{Gl}_{\ell_1}(\mathbb{R})$, $V \in \mathbb{C}^{m_0+p_0\times \ell_1}$, *and* $K \in \mathbb{R}^{\ell\times m_0+p_0}$ *be given. Then the minimum in* (5.15) *is attained for some* $\gamma_0 \in (0, \infty)$.

*Proof.* Let $K^\mathsf{T} = \begin{bmatrix} K_1^\mathsf{T} & K_2^\mathsf{T} \end{bmatrix}$ be a partitioning with $K_1 \in \mathbb{R}^{m_0\times\ell}$, $K_2 \in \mathbb{C}^{p_0\times\ell}$. We now consider row compressions of $K_1V$ and $K_2V$, i.e., let $U_1 \in \mathbb{C}^{m_0\times m_0}$, $U_2 \in \mathbb{C}^{p_0\times p_0}$ be unitary matrices such that

$$K_1V = U_1\begin{bmatrix} \widetilde{K}_1 \\ 0 \end{bmatrix}, \quad K_2V = U_2\begin{bmatrix} \widetilde{K}_2 \\ 0 \end{bmatrix},$$

where $\widetilde{K}_1$ and $\widetilde{K}_2$ have full row-rank. Therefore, we get

$$\begin{aligned}
\mathcal{M} &= \begin{bmatrix} U_1 & 0 \\ 0 & U_2 \end{bmatrix} \begin{bmatrix} \widetilde{K}_1 \\ 0 \\ \widetilde{K}_2 \\ 0 \end{bmatrix} \Theta^{-1} \begin{bmatrix} \widetilde{K}_1^\mathsf{T} & 0 & \widetilde{K}_2^\mathsf{T} & 0 \end{bmatrix} \begin{bmatrix} U_1^\mathsf{T} & 0 \\ 0 & U_2^\mathsf{T} \end{bmatrix} \\
&= \begin{bmatrix} U_1 & 0 \\ 0 & U_2 \end{bmatrix} \begin{bmatrix} \widetilde{K}_1\Theta^{-1}\widetilde{K}_1^\mathsf{T} & 0 & \widetilde{K}_1\Theta^{-1}\widetilde{K}_2^\mathsf{T} & 0 \\ 0 & 0 & 0 & 0 \\ \widetilde{K}_2\Theta^{-1}\widetilde{K}_1^\mathsf{T} & 0 & \widetilde{K}_2\Theta^{-1}\widetilde{K}_2^\mathsf{T} & 0 \\ 0 & 0 & 0 & 0 \end{bmatrix} \begin{bmatrix} U_1^\mathsf{T} & 0 \\ 0 & U_2^\mathsf{T} \end{bmatrix}.
\end{aligned}$$

Now, $q_{\mathbb{C}}(\mathcal{M})$ and $q_{\mathbb{C}}(\widetilde{\mathcal{M}})$ with

$$\widetilde{\mathcal{M}} = \begin{bmatrix} \widetilde{K}_1\Theta^{-1}\widetilde{K}_1^\mathsf{T} & \widetilde{K}_1\Theta^{-1}\widetilde{K}_2^\mathsf{T} \\ \widetilde{K}_2\Theta^{-1}\widetilde{K}_1^\mathsf{T} & \widetilde{K}_2\Theta^{-1}\widetilde{K}_2^\mathsf{T} \end{bmatrix}$$

are attained for the same $\gamma_0$, since the zero rows and columns do not give any additional information. However, $\widetilde{K}_1\Theta^{-1}\widetilde{K}_1^\mathsf{T}$ and $\widetilde{K}_2\Theta^{-1}\widetilde{K}_2^\mathsf{T}$ are both nonsingular. Using [HQ98, Lem. 5] and the argumentation of the proof of [HQ98, Claim 1(b)], it follows that there are eigenvalues of $\mathcal{F}(\gamma)$ tending to $\pm\infty$ for $\gamma \to 0$ and $\gamma \to \infty$. This is a contradiction and shows that $\gamma_0 \in (0, \infty)$. □

The following statement is about the construction of an optimal perturbation matrix $\Delta \in \mathbb{C}^{m_0 \times p_0}$ that achieves (5.14). It summarizes arguments from the proof of [HQ98, Claim 1(a)].

**Theorem 5.4.3.** *Let the minimum in* (5.15) *be attained at* $\gamma_0 \in (0, \infty)$ *with the optimal eigenvalue* $\lambda_0 = \max\{\lambda_1(\mathcal{F}(\gamma_0)), -\lambda_{m_0+p_0}(\mathcal{F}(\gamma_0))\}$. *Then there exists an eigenvector* $v_0 \in \mathbb{C}^{m_0+p_0}$ *of* $\mathcal{F}(\gamma_0)$ *corresponding to the eigenvalue* $\lambda_0$ *such that*

$$\mathcal{F}(\gamma_0)v_0 = \lambda_0 v_0, \quad \text{and} \quad v_0^{\mathsf{H}} \frac{\mathrm{d}\mathcal{F}(\gamma_0)}{\mathrm{d}\gamma} v_0 = 0.$$

*Partition* $v_0 = \begin{pmatrix} v_{01}^{\mathsf{H}} & v_{02}^{\mathsf{H}} \end{pmatrix}^{\mathsf{H}}$ *according to the block structure of* $\mathcal{M}$. *Then an optimal perturbation is given by* $\Delta = \lambda_0^{-1} v_{01} v_{02}^{\mathsf{H}} / v_{01}^{\mathsf{H}} v_{01}$.

Finally, the following theorem shows that at least one of the functions $\lambda_1(\mathcal{F}(\gamma))$ or $-\lambda_{m_0+p_0}(\mathcal{F}(\gamma))$ (whichever has a larger supremum) is unimodal. This means that any local supremum is simultaneously a global supremum. Therefore, the solution of the optimization problem can be numerically easily tracked.

**Theorem 5.4.4.** *[HQ98, Prop. 1] Let* $\mathcal{F}$, $r_+$, *and* $r_-$ *be given as in Theorem 5.4.1. Then the following statements hold true:*

*(a) If* $r_+ \leq r_-$, *then* $\lambda_1(\mathcal{F}(\gamma))$ *is unimodal and any local infimum of* $-\lambda_{m_0+p_0}(\mathcal{F}(\gamma))$ *is equal or smaller than* $(r_+)^{-1}$.

*(b) If* $r_- \leq r_+$, *then* $-\lambda_{m_0+p_0}(\mathcal{F}(\gamma))$ *is unimodal and any local infimum of* $\lambda_1(\mathcal{F}(\gamma))$ *is equal or smaller than* $(r_-)^{-1}$.

In the next part we will discuss how to actually compute the complex cyclo-dissipativity radius. We will do this by using a level-set method that has already been applied in [OD05], or in [SDT96] to compute the real structured stability radius which is a generalization of the real stability radius of [QBR+95]. We follow the line of argumentation of [OD05].

Let

$$\begin{aligned} \mathcal{F}(\gamma, \omega) &= \begin{bmatrix} \sqrt{\gamma} I_{m_0} & 0 \\ 0 & I_{p_0}/\sqrt{\gamma} \end{bmatrix} \mathcal{M}(\omega) \begin{bmatrix} \sqrt{\gamma} I_{m_0} & 0 \\ 0 & I_{p_0}/\sqrt{\gamma} \end{bmatrix} \\ &= \begin{bmatrix} \gamma \mathcal{M}_{11}(\omega) & \mathcal{M}_{12}(\omega) \\ \mathcal{M}_{12}^{\mathsf{H}}(\omega) & \mathcal{M}_{22}(\omega)/\gamma \end{bmatrix} \end{aligned}$$

be given with inertia $\mathrm{In}(\mathcal{F}(\gamma,\omega)) = (\pi_+(\omega), \pi_0(\omega), \pi_-(\omega))$ (which is independent from $\gamma$). Furthermore, define

$$\varphi_+(\gamma,\omega) := \begin{cases} \lambda_1(\mathcal{F}(\gamma,\omega)) & \text{if } \pi_+(\omega) > 0, \\ 0 & \text{if } \pi_+(\omega) = 0, \end{cases}$$

$$\varphi_-(\gamma,\omega) := \begin{cases} -\lambda_{m_0+p_0}(\mathcal{F}(\gamma,\omega)) & \text{if } \pi_-(\omega) > 0, \\ 0 & \text{if } \pi_-(\omega) = 0, \end{cases}$$

and

$$\widehat{\varphi}_+(\omega) := \left(\inf_{\gamma>0} \varphi_+(\gamma,\omega)\right)^{-1}, \quad \widehat{\varphi}_-(\omega) := \left(\inf_{\gamma>0} \varphi_-(\gamma,\omega)\right)^{-1},$$

with the convention that $0^{-1} := \infty$. By definition of the complex cyclo-dissipativity radius we have

$$r_{s,\mathbb{C}}(\Sigma,\mathcal{L},\mathcal{R}) = \inf_{\omega\in\mathbb{R}} \min\left\{\widehat{\varphi}_+(\omega), \widehat{\varphi}_-(\omega)\right\}.$$

Initially, we evaluate $\xi_0 := \min\left\{\widehat{\varphi}_+(\omega_0), \widehat{\varphi}_-(\omega_0)\right\}$ for some $\omega_0 \in \mathbb{R}$. Let $\gamma_0$ be a minimizing value of $\gamma$. This value is determined by computing the points $\gamma_+$, $\gamma_- > 0$ that fulfill

$$\left.\frac{\partial\lambda_1(\mathcal{F}(\gamma,\omega_0))}{\partial\gamma}\right|_{\gamma=\gamma_+} = 0 \quad \text{and} \quad \left.\frac{\partial\lambda_{m+p}(\mathcal{F}(\gamma,\omega_0))}{\partial\gamma}\right|_{\gamma=\gamma_-} = 0, \tag{5.16}$$

respectively. The two partial derivatives can be explicitly determined and then $\gamma_+$, $\gamma_-$ can be computed using a bisection scheme or Newton's method, see also [OD05] for details. Finally, we set

$$\gamma_0 = \begin{cases} \gamma_+ & \text{if } \xi_0 = \widehat{\varphi}_+(\omega_0), \\ \gamma_- & \text{if } \xi_0 = \widehat{\varphi}_-(\omega_0). \end{cases} \tag{5.17}$$

Furthermore, by definition we have

$$\varphi_+(\gamma_0,\omega)^{-1} \leq \widehat{\varphi}_+(\omega), \quad \varphi_-(\gamma_0,\omega)^{-1} \leq \widehat{\varphi}_-(\omega),$$

and therefore, we obtain

$$\min\left\{\varphi_+(\gamma_0,\omega)^{-1}, \varphi_-(\gamma_0,\omega)^{-1}\right\} \leq \min\left\{\widehat{\varphi}_+(\omega), \widehat{\varphi}_-(\omega)\right\} \tag{5.18}$$

for all $\omega \in \mathbb{R}$.

We now compute the intervals for $\omega$ in which it holds that

$$\xi_0 > \min\left\{\varphi_+(\gamma_0,\omega)^{-1}, \varphi_-(\gamma_0,\omega)^{-1}\right\}. \tag{5.19}$$

This condition is fulfilled if one of the conditions

a) $\xi_0 > \varphi_+(\gamma_0, \omega)^{-1}$; or

b) $\xi_0 > \varphi_-(\gamma_0, \omega)^{-1}$

is satisfied.

The intervals that fulfill condition a) can be found by computing the values of $\omega \in \mathbb{R}$ that fulfill

$$\det\left(\xi_0^{-1} I_{m_0+p_0} - \Gamma_0 K \mathcal{N}(\mathrm{i}\omega)^+ K^\mathsf{T} \Gamma_0\right) = 0 \quad \text{with} \quad \Gamma_0 = \begin{bmatrix} \sqrt{\gamma_0} I_{m_0} & 0 \\ 0 & I_{p_0}/\sqrt{\gamma_0} \end{bmatrix}.$$

This is equivalent to the condition

$$\det\left(\Theta(\omega) - V_1^\mathsf{H}(\omega) K^\mathsf{T} \Gamma_0 \xi_0 \Gamma_0 K V_1(\omega)\right) = 0,$$

or in other terms

$$\operatorname{rank}\left(\mathcal{N}(\mathrm{i}\omega) - K^\mathsf{T} \Gamma_0 \xi_0 \Gamma_0 K\right) < \operatorname{rank}_{\mathbb{R}(s)}\left(\mathcal{N}(s) - K^\mathsf{T} \Gamma_0 \xi_0 \Gamma_0 K\right). \tag{5.20}$$

This is an even eigenvalue problem in two parameters. By exploiting the Schur complement structure of $\mathcal{N}(s) - K^\mathsf{T}\Gamma_0\xi_0\Gamma_0 K$, we obtain a larger even eigenvalue problem for the extended even pencil

$$\mathcal{H}_+(s, \xi_0, \gamma_0) := \begin{bmatrix} \mathcal{N}(s) & K^\mathsf{T}\Gamma_0 \\ \Gamma_0 K & \xi_0^{-1} I_{m_0+p_0} \end{bmatrix} \in \mathbb{R}[s]^{\ell+m_0+p_0 \times \ell+m_0+p_0},$$

whose *purely imaginary eigenvalues* we must compute. This pencil has the same finite eigenvalues as $\mathcal{N}(s) - K^\mathsf{T}\Gamma_0\xi_0\Gamma_0 K$, but more infinite eigenvalues. The advantage of this extended formulation is that it is not necessary to explicitly form matrix-times-its-transpose like products as in the reduced formulation (5.20). As mentioned in [BBMX99], forming such products might be numerically unstable and should be avoided whenever possible.

Similarly, the intervals that fulfill condition b) are determined by the values of $\omega \in \mathbb{R}$ that satisfy

$$\det\left(-\xi_0^{-1} I_{m_0+p_0} - \Gamma_0 K \mathcal{N}(\mathrm{i}\omega)^+ K^\mathsf{T} \Gamma_0\right) = 0.$$

This leads to another extended even eigenvalue problem, namely for the even pencil

$$\mathcal{H}_-(s, \xi_0, \gamma_0) := \begin{bmatrix} \mathcal{N}(s) & K^\mathsf{T}\Gamma_0 \\ \Gamma_0 K & -\xi_0^{-1} I_{m_0+p_0} \end{bmatrix} \in \mathbb{R}[s]^{\ell+m_0+p_0 \times \ell+m_0+p_0},$$

whose *purely imaginary eigenvalues* have to be computed.

By taking the union of the intervals obtained for subproblems a) and b), we obtain the so-called level-set satisfying (5.19). Due to the relation (5.18), this level-set must contain the values $\omega_* \in \mathbb{R}$ for which the complex cyclo-dissipativity radius is attained, i. e.,

$$r_{s,\mathbb{C}}(\Sigma, \mathcal{L}, \mathcal{R}) = \min\left\{\widehat{\varphi}_+(\omega_*), \widehat{\varphi}_-(\omega_*)\right\}.$$

In the next step we choose a new frequency $\omega_1$ such that

$$\xi_1 := \min\left\{\widehat{\varphi}_+(\omega_1), \widehat{\varphi}_-(\omega_1)\right\} < \xi_0$$

and continue as before until the level-set is empty. In this case we have found the solution. It can be shown that this algorithm converges *globally*. Similarly as for other related level-set methods [BB90b, BS90, SDT96, GDV98, LTD00, OD05, BSV12a, BSV12b], the convergence is at least quadratic, but depending on the strategy for updating the test frequencies, it can also be faster. Note also that we can ensure a maximum relative error of $\varepsilon$, assuming that we have exact arithmetics.

In Algorithm 5.1 we summarize the complete procedure.

## 5.5 Numerical Results

### 5.5.1 Illustrative Examples

To demonstrate the effectiveness of Algorithm 5.1, we compute the complex cyclo-dissipativity radius for the example from Chapter 1 given by (1.4) with $g = 5$,

$$\begin{aligned} k_1 = \ldots = k_4 = \kappa_2 = \ldots = \kappa_4 = 2, \quad \kappa_1 = \kappa_5 = 4, \\ d_1 = \ldots = d_4 = \delta_2 = \ldots = \delta_4 = 5, \quad \delta_1 = \delta_5 = 10, \end{aligned}$$

and

$$Q_s = -1, \quad S_s = 0, \quad R_s = 1, \quad \mathcal{L} = \begin{bmatrix} B \\ 0 \end{bmatrix}, \quad \mathcal{R} = \begin{bmatrix} C & 0 \end{bmatrix}.$$

With this set of matrices $Q$, $S$, $R$ we measure the distance of the system to the closest non-contractive system, see Definition 3.9.16. Moreover, a system is contractive if its transfer function $G(s) = C(sE - A)^{-1}B + D \in \mathbb{R}(s)^{p\times m}$ is bounded real (see Remark 3.9.22) which is equivalent to a $\mathcal{H}_\infty$-norm smaller or equal than one, i. e.,

$$\|G\|_{\mathcal{H}_\infty} \leq 1.$$

Since $\Lambda(E, A) \subset \mathbb{C}^-$ and by Remark 5.2.3 c), the complex cyclo-dissipativity radius coincides with the complex dissipativity radius. Our test system satisfies

$$\|G\|_{\mathcal{H}_\infty} = 0.1589,$$

so we expect the dissipativity radius to be rather large.

**Algorithm 5.1** Computation of the complex cyclo-dissipativity radius

**Input:** System $\Sigma := (E, A, B, C, D) \in \Sigma_{n,m,p}$ which is cyclo-dissipative with respect to the supply rate (3.101), matrices $\mathcal{L}$, $\mathcal{R}$ defining the perturbation structure, desired relative accuracy $\varepsilon$.

**Output:** Complex cyclo-dissipativity radius $\xi_* := r_{s,\mathbb{C}}(\Sigma, \mathcal{L}, \mathcal{R})$, optimal frequency $\omega_*$, optimal scaling parameter $\gamma_*$, optimal perturbation $\Delta_*$.

1: Compute all eigenvalues $\lambda_1, \ldots, \lambda_k$ of $\mathcal{N}(s) \in \mathbb{R}[s]^{\ell \times \ell}$ as in (5.3).
2: Choose an eigenvalue $\lambda_0$ among $\lambda_1, \ldots, \lambda_k$ with smallest absolute real part and nonnegative imaginary part and set $\omega_0 := \operatorname{Im}(\lambda_0)$.
3: Compute $\xi_0 := \min\{\widehat{\varphi}_+(0), \widehat{\varphi}_-(0)\}$ with corresponding optimal scaling parameter $\gamma_0$ using (5.16) and (5.17).
4: Compute $\xi_{\omega_0} := \min\{\widehat{\varphi}_+(\omega_0), \widehat{\varphi}_-(\omega_0)\}$ with corresponding optimal scaling parameter $\gamma_{\omega_0}$ using (5.16) and (5.17).
5: Compute $\xi_\infty := \min\{\widehat{\varphi}_+(\omega_\infty), \widehat{\varphi}_-(\omega_\infty)\}$ for sufficiently large $\omega_\infty$ with corresponding optimal scaling parameter $\gamma_\infty$ using using (5.16) and (5.17) to estimate $\lim_{\omega \to \infty} \min\{\widehat{\varphi}_+(\omega), \widehat{\varphi}_-(\omega)\}$.
6: Set $\xi := \min\{\xi_0, \xi_{\omega_0}, \xi_\infty\}$ and select $\omega \in \{0, \omega_0, \omega_\infty\}$, and $\gamma \in \{\gamma_0, \gamma_{\omega_0}, \gamma_\infty\}$ accordingly.
7: **repeat**
8: Set $\widehat{\xi} := \xi(1 - 2\varepsilon)$.
9: Compute all eigenvalues of $\mathcal{H}_+(s, \widehat{\xi}, \gamma), \mathcal{H}_-(s, \widehat{\xi}, \gamma) \in \mathbb{R}[s]^{\ell+m_0+p_0 \times \ell+m_0+p_0}$.
10: **if** no finite, purely imaginary eigenvalues **then**
11: Set $\xi_* := \xi$, $\gamma_* := \gamma$, $\omega_* := \omega$.
12: Construct $\Delta_*$ as in Theorem 5.4.3.
13: Break.
14: **else**
15: Set $\{\mathrm{i}\omega_1, \ldots, \mathrm{i}\omega_k\}$ = purely imaginary eigenvalues of $\mathcal{H}_+(s, \widehat{\xi}, \gamma)$ and $\mathcal{H}_-(s, \widehat{\xi}, \gamma)$ satisfying $0 \leq \omega_1 < \ldots < \omega_k < \infty$.
16: Compute new test frequencies as $m_j := \sqrt{\omega_j \omega_{j+1}}$ for $j = 1, \ldots, k-1$.
17: Compute $\xi_j := \min\{\widehat{\varphi}_+(m_j), \widehat{\varphi}_-(m_j)\}$ with corresponding optimal scaling parameter $\gamma_j$ using (5.16) and (5.17) for $j = 1, \ldots, k-1$.
18: Set $\xi := \min\{\xi_1, \ldots, \xi_{\ell-1}\}$ and select $\omega \in \{m_1, \ldots, m_{k-1}\}$ and $\gamma \in \{\gamma_1, \ldots, \gamma_{k-1}\}$ accordingly.
19: **end if**
20: **until** Break

Table 5.1: Intermediate results of the dissipativity radius computation for the first example

| iteration # | $\xi$ | $\gamma$ | level-set, $\omega_* \in$ |
|---|---|---|---|
| 1 | 5.290577 | 1.000000 | $(0.137948, 0.156058)$ |
| 2 | 5.289442 | 1.000000 | $(0.146724, 0.148263)$ |

A reduction of the corresponding even pencil $\mathcal{N}(s) \in \mathbb{R}[s]^{25\times 25}$ to even staircase form reveals the following structures. Using the notation of Theorems 2.1.16 and 2.1.17, we get the values

$$\begin{aligned} w = 2, \quad s_1 = 2, \quad s_2 = 0, \quad q_1 = 2, \quad q_2 = 0,\\ \pi_{+,1} = 1, \quad \pi_{-,1} = 2, \quad r_1 = 3,\\ \pi_{+,2} = 2, \quad \pi_{-,2} = 3, \quad r_2 = 5,\\ \pi_{+,3} = 2, \quad \pi_{-,3} = 3, \quad r_3 = 5. \end{aligned} \tag{5.21}$$

The application of Theorem 2.1.17 yields that $\mathcal{N}(s)$ has no blocks of type E4, but 5 blocks of type E3 (three $1 \times 1$ blocks among which 2 have positive and 1 has negative sign-characteristic and two $3 \times 3$ blocks among which 1 has positive and 1 has negative sign-characteristic). Therefore, $V_{2,\infty}$ has two columns. For this example, $KV_{2,\infty} = 0$, which means that no arbitrarily small perturbation of the defective infinite eigenvalues, that destroys dissipativity, is possible.

We get the following test frequencies and initial values:

- at $\omega = 0.000000$: $\xi = 9.259962$, $\gamma = 1.000000$;
- at $\omega = 0.156058$: $\xi = 5.440666$, $\gamma = 1.000000$;
- at $\omega = 4.50360\text{e}{+}12$: $\xi = 4.05648\text{e}{+}27$, $\gamma = 1.000000$.

The smallest value for $\xi$ is attained at $\omega = 0.156058$. Therefore, Algorithm 5.1 gives the results presented in Table 5.1. As final result we obtain

$$r_{s,\mathbb{C}}(\Sigma, \mathcal{L}, \mathcal{R}) = 5.289442, \quad \text{at} \quad \omega_* = 0.147491.$$

As a second example we use the same matrices as above but replace $B$ by $6.2893 \cdot B$. When doing this we obtain a system with

$$\|G\|_{\mathcal{H}_\infty} = 0.9996,$$

so we expect a small value for the dissipativity radius. Indeed, we get the following results. First, the structure of the even staircase form is the same as in (5.21).

Table 5.2: Intermediate results of the dissipativity radius computation for the second example

| iteration # | $\xi$ | $\gamma$ | level-set, $\omega_* \in$ |
|---|---|---|---|
| 1 | 2.0304218649893e–04 | 1.000000 | $(0.137948337, 0.156058240)$ |
| 2 | 2.2582134423309e–05 | 1.000000 | $(0.146724145, 0.148262833)$ |
| 3 | 2.2572391984637e–05 | 1.000000 | $(0.147491482, 0.147502786)$ |
| 4 | 2.2572391984415e–05 | 1.000000 | $(0.147497100, 0.147497169)$ |
| 5 | 2.2572391984474e–05 | 1.000000 | $(0.147497126, 0.147497144)$ |

Moreover, we also get $KV_{2,\infty} = 0$. Thus, no perturbation of the defective infinite eigenvalues, that destroys dissipativity, is possible. As test frequencies and initial values we get

- at $\omega = 0.000000$: $\xi = 0.631336$, $\gamma = 1.000000$;
- at $\omega = 0.156058$: $\xi = 0.024067$, $\gamma = 1.000000$;
- at $\omega = 4.50360\text{e}{+}12$: $\xi = 6.44981\text{e}{+}26$, $\gamma = 1.000000$.

The smallest value for $\xi$ is attained at $\omega = 0.156058$. Then Algorithm 5.1 gives the results presented in Table 5.2. As final result we obtain

$$r_{s,\mathbb{C}}(\Sigma, \mathcal{L}, \mathcal{R}) = 2.25724\text{e}{-}05, \quad \text{at} \quad \omega_* = 0.147497,$$

which means that the system is indeed close to a non-dissipative one.

## 5.5.2 Limitations of the Method

The algorithm presented is this paper has various limitations and shortcomings that we want to comment on.

In the most general setting, the algorithm has to deal with possibly singular pencils $\mathcal{N}(s)$, $\mathcal{H}_+(s, \xi, \gamma)$, and $\mathcal{H}_-(s, \xi, \gamma)$. In order to safely compute the eigenvalues, it is necessary to reduce these pencils to even staircase form and compute the eigenvalues of the regular index-one part. As already mentioned in Subsection 2.1.2, this reduction heavily relies on the numerical determination of ranks of several intermediate matrices. If there is no clear gap in the small singular values, then the rank decision strongly depends on the truncation tolerance that has been chosen. Therefore, the even staircase form and all their invariants might also strongly depend on the drop tolerance.

Another problem occurs when the system under consideration is very close to a non-cyclo-dissipative system. To illustrate the problems in this case we take the

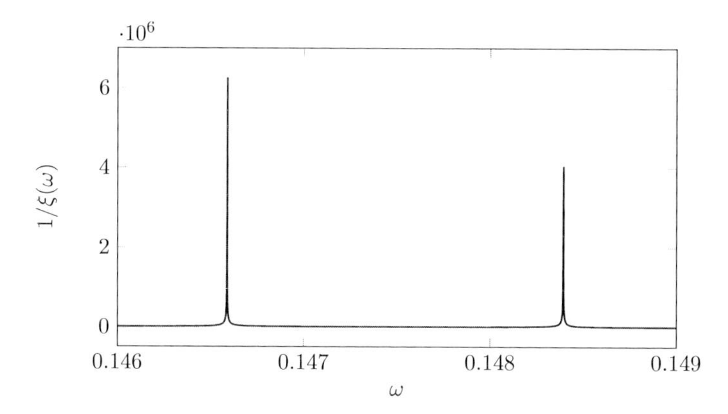


Figure 5.1: Plot of $1/\xi(\omega)$ for an almost non-cyclo-dissipative system

Table 5.3: Computed level-sets for all intermediate $\xi$ values for an almost non-cyclo-dissipative system

| iteration # | $\xi$ | level-set, $\omega_* \in$ |
|---|---|---|
| 1 | 2.379047e–02 | $(0.13794834, 0.15605824)$ |
| 2 | 6.723999e–05 | $(0.14647426, 0.14672415) \cup (0.14826283, 0.14850728)$ |
| 3 | 4.566576e–06 | $(0.14658205, 0.14659945) \cup (0.14838501, 0.14840202)$ |
| 4 | 2.212123e–08 | $(0.14659066, 0.14659075) \cup (0.14839352, 0.14839359)$ |
| 5 | 2.640649e–10 | $(0.14659049, 0.14659071) \cup (0.14839355, 0.14839377)$ |

initial example from above and replace $B$ by $6.2910 \cdot B$. Then we plot the values $1/\xi(\omega)$ (that are all attained at $\gamma(\omega) = 1$), see Figure 5.1. Assume that we have chosen a sufficiently small level $\xi_0 > r_{s,\mathbb{C}}(\Sigma, \mathcal{L}, \mathcal{R})$. Then the pencils $\mathcal{H}_+(s, \xi_0, 1)$ and $\mathcal{H}_-(s, \xi_0, 1)$ have two pairs of eigenvalues in which the eigenvalues are extremely close together. In theory, the level set for $\xi_1 < \xi_0$ is contained in the level set for $\xi_0$. However, due to round-off errors in the eigenvalue computation this is not the case in practice. Therefore, it is not possible to resolve the intervals that contain the optimal frequency $\omega_*$ with sufficient accuracy. For illustration we list the computed level sets in Table 5.3. Even if we cannot give the cyclo-dissipativity radius very accurately in this case, we still obtain rather small intervals, that are at least very close to the optimal frequency. In this case we can, e. g., do a sampling in a small neighborhood of the level-set and still obtain a result, however without guaranteed accuracy. See also [LV13] for more details on this strategy for a related problem.

## 5.6 Summary and Open Problems

In this chapter we have presented an algorithm to compute the distance of a cyclo-dissipative system to the closest non-cyclo-dissipative system under structured perturbations. Since cyclo-dissipativity can be characterized via a spectral condition of even matrix pencils we have studied the perturbation theory of such pencils. We have discussed different substructures of such pencils, especially the singular and higher-index parts have been in focus. We have given computable conditions to check whether there are arbitrarily small perturbations that destroy cyclo-dissipativity of the system. Finally, to compute the cyclo-dissipativity radius we have solved a sequence of eigenvalue optimization problems. We have demonstrated the behavior of the algorithm for some examples and have discussed numerical problems arising when the system is very close to a non-cyclo-dissipative one.

There are various open problems and possible future research directions. For instance, it would be desirable to generalize this algorithm in order to compute the *real* cyclo-dissipativity radius. We have intensively tried this, but there occur various problems, similarly as in [HQ97, HQ98]. The main problem is that instead of the largest and smallest eigenvalues of $\mathcal{F}(\gamma)$ we have to consider the second-largest and second-smallest eigenvalues of a matrix $\mathcal{G}(\alpha, \beta)$ in two parameters. In this case it is not obvious how to show whether one of these eigenvalue functions is unimodal in both arguments. Moreover, there exist difficulties in constructing the optimal perturbation for some situations, for instance if the optimum is attained for one of the parameters $\alpha$, $\beta$ going to zero or infinity.

Therefore, an alternative approach is to study the effect of perturbations of the even pencils by using structured pseudospectra and optimizing over them. This has already been done in various papers of different authors, for instance [GO11, GL13, GKL14b]. The main challenge in this context is to take the special block structure of the perturbation matrix into account.

We further want to remark that in this context one could also consider the "dual" problem, namely, finding the smallest perturbation that makes a non-cyclo-dissipative system cyclo-dissipative. This is the problem of cyclo-dissipativity enforcement, see, for instance [BS13] or Chapter 4 for a related problem.

# 6 Computation of the $\mathcal{H}_\infty$-Norm for Large-Scale Descriptor Systems

## 6.1 Introduction

Consider a dynamical system $(E, A, B, C, D) \in \Sigma_{n,m,p}$ with transfer function $G \in \mathcal{RH}_\infty^{p\times m}$. The $\mathcal{H}_\infty$-norm of a transfer function (see Definition 2.2.20) is a popular tool to measure the distance of transfer functions which is of importance in several applications.

Assume for instance that $(\widetilde{E}, \widetilde{A}, \widetilde{B}, \widetilde{C}, \widetilde{D}) \in \Sigma_{r,m,p}$ with transfer function $\widetilde{G} \in \mathcal{RH}_\infty^{p\times m}$ is a reduced order model of $(E, A, B, C, D)$. Then the transfer function of the error system is given by

$$G_{\mathrm{err}}(s) = G(s) - \widetilde{G}(s) = \begin{bmatrix} C & -\widetilde{C} \end{bmatrix} \left( s \begin{bmatrix} E & 0 \\ 0 & \widetilde{E} \end{bmatrix} - \begin{bmatrix} A & 0 \\ 0 & \widetilde{A} \end{bmatrix} \right)^{-1} \begin{bmatrix} B \\ \widetilde{B} \end{bmatrix} + D - \widetilde{D} \in \mathcal{RH}_\infty^{p\times m}.$$

The value of $\|G_{\mathrm{err}}\|_{\mathcal{H}_\infty}$ can be now interpreted as the worst-case error for $G(s)$ evaluated on the imaginary axis.

Another field of application can be found in robust control where the $\mathcal{H}_\infty$-norm attains the role of a robustness measure. Consider a static output feedback controller $K \in \mathbb{R}^{m\times p}$ that stabilizes the system $(E, A, B, C) \in \Sigma_{n,m,p}$. This leads to the closed-loop dynamics

$$E\dot{x}(t) = A_K x(t) := (A + BKC)\, x(t).$$

In robust control we are interested in the robustness of the closed-loop system with respect to perturbations in the controller $K$. In other words, we want to know how much we can maximally perturb $K$ such that the perturbed closed-loop system

$$E\dot{x}(t) = A_{K+\Delta} x(t) = (A_K + B\Delta C)\, x(t)$$

is guaranteed to remain stable. To quantify robustness of a dynamical system, unstructured (i. e., for $B = C = I_n$) and structured stability radii for matrices were introduced, first by Hinrichsen and Pritchard in [HP86a, HP86b, HP90]. The generalization to matrix pencils is a nontrivial issue due to the fact that the influence of the infinite eigenvalues as well as perturbations that make the pencil singular have to be studied. For structured perturbations this has already been considered in [Du08]. In particular, it provides a relationship between the $\mathcal{H}_\infty$-norm and the structured complex stability radius of the pencil $sE - A$. We have found an easier and more intuitive proof of this relation, outlined in Lemma 6.2.2 and Proposition 6.2.3. A further generalization of the structured complex stability radius has been analyzed in [DTL11], allowing simultaneous structured perturbations of $A$ and $E$. However, we do not consider perturbations of $E$ since we would not have a relation to the $\mathcal{H}_\infty$-norm anymore. Another recent survey paper on robust stability of descriptor systems and stability radii of matrix pencils is [DLM13].

Numerical methods for computing the $\mathcal{H}_\infty$-norm are well-established. Most of them are based on relations between the $\mathcal{H}_\infty$-norm and the spectrum of certain Hamiltonian matrices or pencils. For an overview, we refer to [Bye88, BBK89, BS90, BB90b, BSV12a, BSV12b]. We briefly summarize the most general result presented in [BSV12a]. In [BSV12a] it is shown that if $sE - A \in \mathbb{R}[s]^{n\times n}$ is regular, has no purely imaginary eigenvalues and $\inf_{\omega\in\mathbb{R}} \sigma_{\max}(G(\mathrm{i}\omega)) < \gamma$, it holds that $\|G\|_{\mathcal{H}_\infty} < \gamma$ if and only if the even pencil

$$s\mathcal{E} - \mathcal{A}(\gamma) := \left[\begin{array}{cc|cc} 0 & -sE^{\mathsf{T}} - A^{\mathsf{T}} & -C^{\mathsf{T}} & 0 \\ sE - A & 0 & 0 & -B \\ \hline -C & 0 & \gamma I_p & -D \\ 0 & -B^{\mathsf{T}} & -D^{\mathsf{T}} & \gamma I_m \end{array}\right] \in \mathbb{R}[s]^{2n+m+p\times 2n+m+p} \quad (6.1)$$

has no purely imaginary eigenvalues. Based on this fact, the algorithm chooses an initial guess $\gamma < \|G\|_{\mathcal{H}_\infty}$ and iterates over $\gamma$ in a suitable way until $s\mathcal{E} - \mathcal{A}(\gamma)$ has no purely imaginary eigenvalues. This iteration can be implemented in a globally quadratically converging way. The drawback of the algorithm is the decision in each step whether there are purely imaginary eigenvalues. It is important to find *all* of them since otherwise the algorithm could fail. In [BSV12a] this issue is addressed by using a structure-preserving method for a skew-Hamiltonian/Hamiltonian pencil related to $s\mathcal{E} - \mathcal{A}(\gamma)$, which prevents the purely imaginary eigenvalues from moving off the imaginary axis as long as their pairwise distance is sufficiently large (see also Subsection 2.1.3). However, this method computes a full structured factorization of the pencil in each step. Due to its cubic complexity it is infeasible for large-scale problems.

Therefore, there is a need to develop computational methods for determining the $\mathcal{H}_\infty$-norm for large-scale problems. Up to now, the only method that took large-scale systems into account is presented in [CGD04, CGVD07] and uses the bounded real lemma to estimate the $\mathcal{H}_\infty$-norm of a discrete-time state-space system which is required to be given in a minimal realization. This algorithm checks a sequence of LMIs for feasibility. This is done by deciding if a so-called Chandrasekhar iteration converges. However, this test lacks of reliability, in particular if the iterates are approaching the $\mathcal{H}_\infty$-norm. Hence, only an estimation of the norm value can be given by this algorithm.

In this chapter we will present and compare two methods that are based on completely different ideas. The first method [BV14] exploits the relationship between the $\mathcal{H}_\infty$-norm and the complex $\mathcal{H}_\infty$-radius of the transfer function $G(s)$. This is the spectral norm of the smallest complex perturbation $\Delta$ such that the perturbed transfer function

$$G_\Delta(s) := C(sE - (A + B\Delta C))^{-1}B \notin \mathcal{RH}_{\infty,\mathbb{C}}^{p\times m}, \tag{6.2}$$

where $\mathcal{RH}_{\infty,\mathbb{C}}^{p\times m}$ denotes the normed space

$$\mathcal{RH}_{\infty,\mathbb{C}}^{p\times m} := \left\{ G(s) \in \mathbb{C}(s)^{p\times m} : \mathfrak{P}(G) \subset \mathbb{C}^- \text{ and } \sup_{\lambda\in\mathbb{C}^+} \|G(\lambda)\|_2 < \infty \right\}.$$

The algorithm is based on *$\varepsilon$-pseudopole sets* for $G(s)$. This means that we consider all perturbations $\Delta$ with $\|\Delta\|_2 < \varepsilon$ and analyze how the poles of (6.2) might move. To compute the complex $\mathcal{H}_\infty$-radius we have to find the $\varepsilon$-pseudopole set that touches the imaginary axis. This calculation is carried out by a nested iteration. The inner iteration is adapted from [GO11] and computes the $\varepsilon$-pseudopole set abscissa for a *fixed value of* $\varepsilon$, that is the real part of the rightmost point in the $\varepsilon$-pseudopole set. This iteration relies on the fact that the entire pseudopole set can be realized by rank-1 perturbations, so an optimal perturbation can be efficiently computed. In the outer iteration, the value of $\varepsilon$ is updated by Newton's method in order to drive the $\varepsilon$-pseudopole set abscissa to zero, similarly as in [GGO13]. Since the inner iteration might only converge to a local maximizer, we discuss a way to obtain good initial values that indeed allows convergence to a global maximizer in most cases. These initial values are obtained by computing some *dominant poles* [RM06b, RM06a], i. e., those poles of $G(s)$ that generate the largest local maxima of $\|G(\mathrm{i}\omega)\|_2$ for $\omega \in \mathbb{R}$.

The second method [LV13] that we present goes back to the original algorithm using the even pencils in (6.1). Actually, we would have to compute *all* purely imaginary eigenvalues of these pencils to obtain *all* intervals for $\omega$ in which $\|G(\mathrm{i}\omega)\|_2 > \gamma$. However, we can relax this requirement when we make again use of the dominant poles of $G(s)$. These are used to calculate shifts for a structure-preserving method for even eigenvalue problem [MSS12] which computes eigenvalues close to these shifts. In this way we do not necessarily compute all frequency intervals with $\|G(\mathrm{i}\omega)\|_2 > \gamma$, but we can still obtain the frequency interval that contains the optimal frequency $\omega_*$ at which the $\mathcal{H}_\infty$-norm is attained.

This chapter is divided into three sections. In Section 6.2 we discuss the first method which is based on optimization over pseudopole sets. Results on the relationship between the $\mathcal{H}_\infty$-norm and the complex $\mathcal{H}_\infty$-radius for descriptor systems are discussed in Subsection 6.2.1. In Subsection 6.2.2 we describe how to compute the pseudopole set abscissa for a transfer function which is the key ingredient of the algorithm. A large part of this section will also be devoted to fixed point analysis. In Subsection 6.2.3 we describe how to use Newton's method to compute the complex $\mathcal{H}_\infty$-radius. Subsection 6.2.4 is devoted to the analysis of the differences between the method presented here and the reference [GGO13] since both are based on similar ideas. In particular, we show that both methods are *not* equivalent in the context of standard state-space systems. In Subsection 6.2.5 we present a study of numerical examples. In particular, we compare our method with existing algorithms. Furthermore, we analyze drawbacks and limitations.

In Section 6.3 we present a modification of the standard algorithm via optimization over $\gamma$ in (6.1). We give details on the computation of the eigenvalues of the even pencil and compare the method with the pseudopole set approach.

Finally, in Section 6.4 we give conclusions and point towards possible future research directions.

## 6.2 The Pseudopole Set Approach

In this section we drive the first method which implements an optimization procedure over pseudopole sets. Let $(E, A, B, C, D) \in \Sigma_{n,m,p}$ with $G \in \mathcal{RH}_\infty^{p \times m}$ be given. Throughout the whole section we assume w. l. o. g. that $D = 0$. Otherwise we could use the realization

$$\left( \begin{bmatrix} E & 0 \\ 0 & 0 \end{bmatrix}, \begin{bmatrix} A & 0 \\ 0 & -I_p \end{bmatrix}, \begin{bmatrix} B \\ D \end{bmatrix}, \begin{bmatrix} C & I_p \end{bmatrix} \right) \in \Sigma_{m+p,m,p} \tag{6.3}$$

to achieve this form. Moreover, we assume that $m, p \ll n$ and that *all* matrices $E$, $A$, $B$, $C$ are sparse.

### 6.2.1 Complex $\mathcal{H}_\infty$-Radius and $\mathcal{H}_\infty$-Norm

First we establish a connection between the $\mathcal{H}_\infty$-radius and the $\mathcal{H}_\infty$-norm of a transfer function $G(s)$. To do so, we have consider the controllability and observability concepts in Subsection 2.2.2 in more detail.

Recall that the system $(E, A, B, C) \in \Sigma_{n,m,p}$ is

a) *completely controllable*, if $\operatorname{rank} \begin{bmatrix} \lambda E - A & B \end{bmatrix} = n$ for all $\lambda \in \mathbb{C}$ and furthermore $\operatorname{rank} \begin{bmatrix} E & B \end{bmatrix} = n$;

b) *completely observable*, if $\operatorname{rank} \begin{bmatrix} \lambda E - A \\ C \end{bmatrix} = n$ for all $\lambda \in \mathbb{C}$ and furthermore $\operatorname{rank} \begin{bmatrix} E \\ C \end{bmatrix} = n$.

We can also define these concepts for single eigenvalues of the $sE - A \in \mathbb{R}[s]^{n\times n}$ as follows. A descriptor system $(E, A, B, C) \in \Sigma_{n,m,p}$ is called

a) *controllable at* $\lambda \in \mathbb{C}$ if $\operatorname{rank} \begin{bmatrix} \lambda E - A & B \end{bmatrix} = n$;

b) *controllable at infinity* if $\operatorname{rank} \begin{bmatrix} E & B \end{bmatrix} = n$;

c) *observable* at $\lambda \in \mathbb{C}$ if $\operatorname{rank} \begin{bmatrix} \lambda E - A \\ C \end{bmatrix} = n$;

d) *observable at infinity* if $\operatorname{rank} \begin{bmatrix} E \\ C \end{bmatrix} = n$;

otherwise it is called uncontrollable or unobservable at $\lambda$, respectively. Note that in the above definitions one can also consider each individual block of the WCF of $sE - A$ separately in case of multiple eigenvalues. This is possible by considering the corresponding eigenvectors. Let $x$, $y \in \mathbb{C}^n$ be right and left eigenvectors corresponding to an individual block of type K1 or K2 in the WCF of $sE - A$. For an eigenvalue $\lambda \in \mathbb{C}$ it holds that

$$y^{\mathsf{H}} \begin{bmatrix} \lambda E - A & B \end{bmatrix} = \begin{pmatrix} 0 & y^{\mathsf{H}} B \end{pmatrix}, \quad \begin{bmatrix} \lambda E - A \\ C \end{bmatrix} x = \begin{pmatrix} 0 \\ Cx \end{pmatrix}.$$

Therefore, we say that such a block is controllable if $B^{\mathsf{T}} y \neq 0$, and observable if $Cx \neq 0$, otherwise we call it uncontrollable or unobservable. Similarly we can treat eigenvalues at infinity. We use this definition when talking about controllability and observability of eigenvalues.

For an asymptotically stable system $\Sigma := (E, A, B, C) \in \Sigma_{n,m,p}$ define the numbers

$$\begin{aligned} q_{\mathbb{C}}^{\mathrm{f}}(\Sigma) &:= \inf \left\{ \|\Delta\|_2 : \Lambda(E, A + B\Delta C) \cap \mathrm{i}\mathbb{R} \neq \emptyset \text{ with } \Delta \in \mathbb{C}^{m\times p} \right\}, \\ q_{\mathbb{C}}^{\infty}(\Sigma) &:= \inf \big\{ \|\Delta\|_2 : sE - (A + B\Delta C) \text{ with } \Delta \in \mathbb{C}^{m\times p} \text{ is a singular pencil or} \\ &\qquad \text{has controllable and observable defective infinite eigenvalues} \big\}. \end{aligned}$$

Then we define the *structured complex stability radius of a matrix pencil* $sE - A$ with respect to $B$ and $C$ by

$$q_{\mathbb{C}}(\Sigma) := \min \left\{ q_{\mathbb{C}}^{\mathrm{f}}(\Sigma), q_{\mathbb{C}}^{\infty}(\Sigma) \right\}.$$

The value of $q_{\mathbb{C}}^{\mathrm{f}}(\Sigma)$ is the size of the smallest structured perturbation that makes the system unstable. The interpretation of $q_{\mathbb{C}}^{\infty}(\Sigma)$ is more involved. *Defective* infinite eigenvalues do not make the system unstable. However, if there are controllable and observable ones, we can construct an arbitrarily small structured perturbation such that the system will be unstable. This means that systems with controllable and observable defective infinite eigenvalues are on the "boundary to instability". If the perturbed matrix pencil becomes singular, then a part of the system dynamics is "free" and thus there exist essentially unbounded solution trajectories for the perturbed system.

In practice, it is desirable to make $q_\mathbb{C}(\Sigma)$ as large as possible in order to guarantee a very high robustness against perturbations. Later in this section we show that for a stable system $\Sigma := (E, A, B, C) \in \Sigma_{n,m,p}$ with transfer function $G \in \mathcal{RH}_\infty^{p\times m}$, it holds that

$$q_\mathbb{C}(\Sigma) = \begin{cases} 1/\|G\|_{\mathcal{H}_\infty} & \text{if } G(s) \not\equiv 0, \\ \infty & \text{if } G(s) \equiv 0, \end{cases}$$

so a large value of $q_\mathbb{C}(\Sigma)$ corresponds to a small $\mathcal{H}_\infty$-norm of the transfer function $G(s)$.

We also introduce the complex $\mathcal{H}_\infty$-radius for a transfer function $G \in \mathcal{RH}_\infty^{p\times m}$ (called complex structured stability radius in [BV14]). For $\Delta \in \mathbb{C}^{m\times p}$ we define the perturbed transfer function

$$G_\Delta(s) := C\left(sE - (A + B\Delta C)\right)^{-1} B \in \mathbb{C}(s)^{p\times m} \tag{6.4}$$

and the numbers

$$\begin{aligned} r_\mathbb{C}^\mathrm{f}(G) &:= \inf\left\{\|\Delta\|_2 : \mathfrak{P}(G_\Delta) \cap \mathrm{i}\mathbb{R} \neq \emptyset \text{ with } \Delta \in \mathbb{C}^{m\times p}\right\}, \\ r_\mathbb{C}^\infty(G) &:= \inf\left\{\|\Delta\|_2 : G_\Delta(s) \text{ as in (6.4) with } \Delta \in \mathbb{C}^{m\times p} \text{ is improper} \right. \\ &\qquad\qquad\qquad\qquad\qquad\qquad\qquad \left. \text{or not well-defined}\right\}. \end{aligned}$$

Then the *complex $\mathcal{H}_\infty$-radius of a transfer function $G(s)$* is defined by

$$r_\mathbb{C}(G) := \min\left\{r_\mathbb{C}^\mathrm{f}(G), r_\mathbb{C}^\infty(G)\right\}.$$

*Remark* 6.2.1. In fact, $r_\mathbb{C}(G)$ is the structured distance of a function $G \in \mathcal{RH}_\infty^{p\times m}$ to the set of functions which are not in $\mathcal{RH}_{\infty,\mathbb{C}}^{p\times m}$, i. e.,

$$r_\mathbb{C}(G) = \inf\left\{\|\Delta\|_2 : G_\Delta \notin \mathcal{RH}_{\infty,\mathbb{C}}^{p\times m} \text{ with } G_\Delta(s) \text{ as in (6.4) and } \Delta \in \mathbb{C}^{m\times p}\right\}.$$

For stable or weakly minimal descriptor systems $\Sigma := (E, A, B, C) \in \Sigma_{n,m,p}$ with transfer function $G \in \mathcal{RH}_\infty^{p\times m}$, we have $q_\mathbb{C}^\mathrm{f}(\Sigma) = r_\mathbb{C}^\mathrm{f}(G)$ and $q_\mathbb{C}(\Sigma) = r_\mathbb{C}(G)$. However, $G(s)$ can also be realized by an unstable descriptor system when all unstable eigenvalues are uncontrollable or unobservable. In this case, the definitions of $q_\mathbb{C}^\mathrm{f}(\Sigma)$ and $q_\mathbb{C}(\Sigma)$ do not make sense whereas those of $r_\mathbb{C}^\mathrm{f}(G)$ and $r_\mathbb{C}(G)$ do. It is very important to well distinguish between these definitions.

Next we prove an important relationship between the $\mathcal{H}_\infty$-norm and the complex $\mathcal{H}_\infty$-radius of a transfer function $G(s)$.

**Lemma 6.2.2.** *[BV14, Lem. 3.1] Let $(E, A, B, C) \in \Sigma_{n,m,p}$ with transfer function $G \in \mathcal{RH}_\infty^{p\times m}$ be given. Then it holds that*

$$r_{\mathbb{C}}^{\infty}(G) = \begin{cases} 1/\lim\limits_{\omega\to\infty} \sigma_{\max}(G(\mathrm{i}\omega)) & \text{if } G(s) \not\equiv 0, \\ \infty & \text{if } G(s) \equiv 0. \end{cases}$$

*Proof.* If $G(s) \equiv 0$, we cannot make the system improper by any structured perturbation, and therefore $r_{\mathbb{C}}^{\infty}(G) = \infty$. Consider the non-trivial case. We can assume w. l. o. g. that we have a weakly minimal realization of a proper $G(s)$ given in WCF, i. e.,

$$\Sigma = \left( \begin{bmatrix} I_r & 0 \\ 0 & E_{22} \end{bmatrix}, \begin{bmatrix} A_{11} & 0 \\ 0 & I_{n-r} \end{bmatrix}, \begin{bmatrix} B_1 \\ B_2 \end{bmatrix}, \begin{bmatrix} C_1 & C_2 \end{bmatrix} \right) \in \Sigma_{n,m,p}$$

with $A_{11} \in \mathbb{C}^{r\times r}$ and a nilpotent $E_{22} \in \mathbb{R}^{n-r\times n-r}$. Note that for a weakly minimal system with proper transfer function, the nilpotent matrix $E_{22}$ in the Weierstraß canonical form is zero or void. This follows from [Dai89, Thm. 2-6.2 and Lem. 2-6.2]. Using this realization, it holds that

$$\lim_{\omega\to\infty} G(\mathrm{i}\omega) = \begin{cases} -C_2B_2 & \text{if } n \neq r, \\ 0 & \text{if } n = r. \end{cases}$$

If $n = r$, i. e., $E_{22}$ is void, then $r_{\mathbb{C}}^{\infty}(G) = \infty$. If $n \neq r$, we consider structured perturbations of the matrix pencil which lead to

$$\begin{aligned} sE_{\min} - A_{\min}^{\Delta} :&= s\begin{bmatrix} I_r & 0 \\ 0 & 0 \end{bmatrix} - \left( \begin{bmatrix} A_{11} & 0 \\ 0 & I_{n-r} \end{bmatrix} + \begin{bmatrix} B_1 \\ B_2 \end{bmatrix} \Delta \begin{bmatrix} C_1 & C_2 \end{bmatrix} \right) \\ &= s\begin{bmatrix} I_r & 0 \\ 0 & 0 \end{bmatrix} - \begin{bmatrix} A_{11} + B_1\Delta C_1 & B_1\Delta C_2 \\ B_2\Delta C_1 & I_{n-r} + B_2\Delta C_2 \end{bmatrix} \in \mathbb{C}[s]^{n\times n}, \end{aligned}$$

where $\Delta \in \mathbb{C}^{m\times p}$. Now we distinguish whether the pencil $sE_{\min} - A_{\min}^{\Delta}$ is singular or not. If it is regular, then in [BSV12a, Thm. 3] it is shown that the perturbed transfer function is improper if and only if $I_{n-r} + B_2\Delta C_2$ is singular. If $sE_{\min} - A_{\min}^{\Delta}$ is a singular pencil, then $\begin{bmatrix} B_2\Delta C_1 & I_{n-r} + B_2\Delta C_2 \end{bmatrix}$ or $\begin{bmatrix} B_1\Delta C_2 \\ I_{n-r} + B_2\Delta C_2 \end{bmatrix}$ do not have full row or column rank, respectively. Hence, $I_{n-r} + B_2\Delta C_2$ is also singular in this case.

Therefore, we have to determine the value of

$$\begin{aligned} p_{\mathbb{C}} :&= \inf\left\{ \|\Delta\|_2 : I_{n-r} + B_2\Delta C_2 \text{ is singular with } \Delta \in \mathbb{C}^{m\times p} \right\} \\ &= \inf\left\{ \|\Delta\|_2 : -I_{n-r} + B_2\Delta C_2 \text{ is singular with } \Delta \in \mathbb{C}^{m\times p} \right\}. \end{aligned}$$

We consider the complex $\mathcal{H}_\infty$-radius of the transfer function

$$G_\infty(s) := C_2(sI_{n-r} + I_{n-r})^{-1}B_2 \in \mathcal{RH}_\infty^{p\times m}.$$

By employing [HP86b, Prop. 2.1] we obtain

$$\begin{aligned} r_{\mathbb{C}}(G_\infty) &= \frac{1}{\max\limits_{\omega\in\mathbb{R}} \sigma_{\max}(C_2((\mathrm{i}\omega+1)I_{n-r})^{-1}B_2)} \\ &= \frac{1}{\sigma_{\max}(C_2B_2)} \\ &= \frac{1}{\lim\limits_{\omega\to\infty} \sigma_{\max}(G(\mathrm{i}\omega))}. \end{aligned} \tag{6.5}$$

Since the maximum in (6.5) is attained at $\omega = 0$, we have $p_{\mathbb{C}} = r_{\mathbb{C}}(G_\infty)$. This shows the assertion. □

**Proposition 6.2.3.** *[BV14, Prop. 3.2] Let $(E, A, B, C) \in \Sigma_{n,m,p}$ with transfer function $G \in \mathcal{RH}_\infty^{p\times m}$ be given. Then it holds that*

$$r_{\mathbb{C}}(G) = \begin{cases} 1/\left\|G\right\|_{\mathcal{H}_\infty} & \text{if } G(s) \not\equiv 0, \\ \infty & \text{if } G(s) \equiv 0. \end{cases} \tag{6.6}$$

*Proof.* The proof is similar to the corresponding one for state-space systems in [HP86b]. First we analyze the case that the value of the $\mathcal{H}_\infty$-norm is attained at some finite $\omega \in \mathbb{R}$.

Assume that for some $\Delta \in \mathbb{C}^{m\times p}$, $0 \neq x \in \mathbb{C}^m$, and $\omega \in \mathbb{R}$ we have

$$(A + B\Delta C)x = \mathrm{i}\omega Ex,$$

or equivalently

$$x = (\mathrm{i}\omega E - A)^{-1}B\Delta Cx.$$

Since $G \in \mathcal{RH}_\infty^{p\times m}$, we have $v := Cx \neq 0$, i. e., it holds that

$$v = G(\mathrm{i}\omega)\Delta v. \tag{6.7}$$

If $G(s) \equiv 0$ this leads to a contradiction and so $r^{\mathrm{f}}_{\mathbb{C}}(G) = \infty$, otherwise (6.7) implies $\left\|G(\mathrm{i}\omega)\right\|_2 \left\|\Delta\right\|_2 \geq 1$.

Now suppose that $\left\|G\right\|_{\mathcal{H}_\infty}$ is attained at $\omega_0$, i. e., $\left\|G(\mathrm{i}\omega_0)\right\|_2 = \left\|G\right\|_{\mathcal{H}_\infty}$. Let

$$G(\mathrm{i}\omega_0) = \sum_{j=1}^{k} \sigma_j v_j w_j^{\mathsf{H}}$$

be a singular value decomposition of $G(\mathrm{i}\omega_0)$ with $v_j \in \mathbb{C}^p$, $w_j \in \mathbb{C}^m$, $\left\|v_j\right\|_2 = \left\|w_j\right\|_2 = 1$, for $j = 1, \ldots, k := \min\{m, p\}$, and $\left\|G(\mathrm{i}\omega_0)\right\|_2 = \sigma_1 \geq \sigma_2 \geq \ldots \geq \sigma_k \geq 0$. With $\Delta := \sigma_1^{-1} w_1 v_1^{\mathsf{H}}$ it follows that

$$G(\mathrm{i}\omega_0)\Delta v_1 = C(\mathrm{i}\omega_0 E - A)^{-1}B\Delta v_1 = v_1.$$

Defining $x := (\mathrm{i}\omega_0 E - A)^{-1} B\Delta v_1$ leads to $Cx = v_1$ and hence $x \neq 0$. This yields

$$x := (\mathrm{i}\omega_0 E - A)^{-1} B\Delta Cx,$$

and consequently

$$(A + B\Delta C)\, x = \mathrm{i}\omega_0 Ex.$$

From

$$\begin{bmatrix} \mathrm{i}\omega_0 E - (A + B\Delta C) \\ C \end{bmatrix} x = \begin{pmatrix} 0 \\ Cx \end{pmatrix} = \begin{pmatrix} 0 \\ v_1 \end{pmatrix}$$

with $v_1 \neq 0$, we conclude that $\mathrm{i}\omega_0$ is an observable mode of the perturbed system $(E, A + B\Delta C, B, C)$. Similarly we can prove controllability of $\mathrm{i}\omega_0$.

From that we conclude $\|\Delta\|_2 = 1/\,\|G\|_{\mathcal{H}_\infty}$, where $\Delta$ is a perturbation of infimal norm such that $\mathfrak{P}(G_\Delta) \cap \mathrm{i}\mathbb{R} \neq \emptyset$.

This shows that $\|G\|_{\mathcal{H}_\infty} = \|G(\mathrm{i}\omega)\|_2$ for some $\omega \in \mathbb{R}$ if and only if $r_{\mathbb{C}}(G) = r_{\mathbb{C}}^{\mathrm{f}}(G)$. The case that the norm value is attained at infinity is covered by Lemma 6.2.2. □

For the remainder of this section we need the following definitions.

**Definition 6.2.4** (Pseudopole sets)**.** [BV14, Def. 3.3]

a) The *$\varepsilon$-pseudopole set of the transfer function* $G(s)$ is defined by

$$\mathfrak{P}_\varepsilon(G) = \left\{\lambda \in \mathbb{C} : \lambda \in \mathfrak{P}(G_\Delta) \text{ for some } \Delta \in \mathbb{C}^{m\times p} \text{ with } \|\Delta\|_2 < \varepsilon\right\}.$$

Elements of $\mathfrak{P}_\varepsilon(G)$ are called *$\varepsilon$-pseudopoles*.

b) The $\varepsilon$-pseudopole set $\mathfrak{P}_\varepsilon(G)$ is called *regular* if there exists no $\Delta \in \mathbb{C}^{m\times p}$ with $\|\Delta\|_2 \leq \varepsilon$ such that $G_\Delta(s)$ is improper or not well-defined.

c) For a regular $\varepsilon$-pseudopole set, the *$\varepsilon$-pseudopole set abscissa* (called structured $\varepsilon$-pseudospectral abscissa in [BV14]) is given by

$$\alpha_\varepsilon(G) := \sup\left\{\mathrm{Re}(\lambda) : \lambda \in \mathfrak{P}_\varepsilon(G)\right\}.$$

*Remark* 6.2.5 (Pseudopole sets and structured pseudospectra, regularity).

a) The notion of $\varepsilon$-pseudopole sets is strongly related to the concept of *structured $\varepsilon$-pseudospectra.* For $\Sigma = (E, A, B, C) \in \Sigma_{n,m,p}$ these are defined by

$$\Lambda_\varepsilon(\Sigma) := \left\{\lambda \in \mathbb{C} : \lambda \in \Lambda(E, A + B\Delta C) \text{ for some } \Delta \in \mathbb{C}^{m\times p} \text{ with } \|\Delta\|_2 < \varepsilon\right\}.$$

With the set $\Psi(\Sigma)$ of uncontrollable or unobservable modes of the system $\Sigma$, we have the relation

$$\Lambda_\varepsilon(\Sigma) = \mathfrak{P}_\varepsilon(G) \cup \Psi(\Sigma).$$

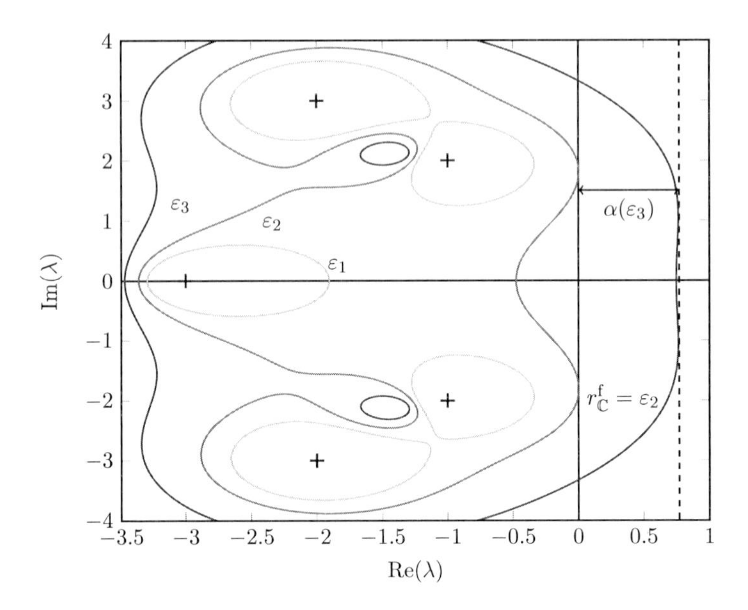


Figure 6.1: Poles (blue crosses) and pseudopole sets of a transfer function for different perturbation levels

b) The definition of regularity of a $\varepsilon$-pseudopole set is strongly related to the so-called *admissibility* of perturbations [BN93, DLM13], i. e., a regular $\varepsilon$-pseudopole set can only be generated by admissible perturbations. From the definition it is clear that regularity is equivalent to $\varepsilon < 1/\lim_{\omega\to\infty}\sigma_{\max}(G(\mathrm{i}\omega))$.

A graphical interpretation of the terms defined in Definition 6.2.4 is given in Figure 6.1.

It is also obvious that $\alpha_{r^{\mathrm{f}}_{\mathbb{C}}(G)}(G) = 0$. So the main idea of our algorithm is to find the (unique) root of the function $\alpha(\varepsilon) := \alpha_\varepsilon(G)$. To get an efficient algorithm, we need to evaluate $\alpha(\varepsilon)$ for different values of $\varepsilon$ in a cheap way. Then we can employ, e. g., Newton's method to compute the actual root.

### 6.2.2 Computation of the $\varepsilon$-Pseudopole Set Abscissa

**Derivation of the Basic Algorithm** In this paragraph we derive a fast algorithm for computing $\alpha(\varepsilon)$. The following fundamental results are generalizations of the corresponding ones in [Rie94].

**Lemma 6.2.6.** *[BV14, Lem. 4.1] Let $(E, A, B, C) \in \Sigma_{n,m,p}$ with transfer function $G(s) \in \mathbb{R}(s)^{p \times m}$ be given. Let $\lambda \in \mathbb{C} \setminus \mathfrak{P}(G)$ be given and $\varepsilon > 0$. Then the following statements are equivalent:*

*a)* $\lambda \in \mathfrak{P}_\varepsilon(G)$.

*b)* $\sigma_{\max}(G(\lambda)) > \varepsilon^{-1}$.

*c) There exist vectors $v \in \mathbb{C}^m$ and $w \in \mathbb{C}^p$ with $\|v\|_2 < 1$ and $\|w\|_2 < 1$ such that $\lambda \in \mathfrak{P}\left(G_{\varepsilon v w^{\mathsf{H}}}\right)$.*

*Proof.* First we show that a) implies b): From $\lambda \in \mathfrak{P}_\varepsilon(G)$ it follows that there exist a matrix $\Delta \in \mathbb{C}^{m \times p}$ with $\|\Delta\|_2 < \varepsilon$ and a vector $x \in \mathbb{C}^n$ such that

$$(\lambda E - (A + B\Delta C))\, x = 0.$$

This is equivalent to

$$(\lambda E - A)\, x = B\Delta C x$$

and therefore

$$Cx = C\,(\lambda E - A)^{-1}\, B\Delta C x.$$

Now we can estimate

$$\|Cx\|_2 \leq \left\|C\,(\lambda E - A)^{-1}\, B\right\|_2 \|\Delta\|_2\, \|Cx\|_2\,,$$

and hence

$$\varepsilon^{-1} < \|\Delta\|_2^{-1} \leq \|G(\lambda)\|_2\,.$$

Next we show that b) implies c): Let $\sigma_{\max}(G(\lambda)) > \varepsilon^{-1}$. Define $\sigma := \sigma_{\max}(G(\lambda))$ with corresponding singular vectors $v \in \mathbb{C}^m$, $w \in \mathbb{C}^p$ satisfying $\|v\|_2 = \|w\|_2 = 1$. Then we have

$$G(\lambda)v = \sigma w, \quad w^{\mathsf{H}} G(\lambda) = \sigma v^{\mathsf{H}}, \quad \sigma > \varepsilon^{-1}. \tag{6.8}$$

Multiplying the first equation of (6.8) by $w^{\mathsf{H}}$ from the left and by $w^{\mathsf{H}}C$ from the right yields

$$w^{\mathsf{H}} C(\lambda E - A)^{-1} B v w^{\mathsf{H}} C = \sigma w^{\mathsf{H}} w w^{\mathsf{H}} C = \sigma w^{\mathsf{H}} C.$$

By setting $y^{\mathsf{H}} := w^{\mathsf{H}} C(\lambda E - A)^{-1}$ we obtain

$$y^{\mathsf{H}} B v w^{\mathsf{H}} C = \sigma y^{\mathsf{H}}(\lambda E - A).$$

It holds that $y^{\mathsf{H}} \neq 0$ since $w^{\mathsf{H}} C \neq 0$, otherwise we would have $\sigma = 0$ which is excluded since $\varepsilon > 0$. Therefore, $\lambda E - \widehat{A} := \lambda E - \left(A + \sigma^{-1} B v w^{\mathsf{H}} C\right)$ is singular. It remains to show that $\lambda$ is indeed a pole of the perturbed transfer function

$$G_{\varepsilon v w^{\mathsf{H}}}(s) = C\left(sE - \widehat{A}\right)^{-1} B,$$

i. e., we have to prove controllability and observability of $\lambda$. Since $y$ is a left eigenvector of $\lambda E - \widehat{A}$ it holds that

$$y^{\mathsf{H}} \begin{bmatrix} \lambda E - \widehat{A} & B \end{bmatrix} = \begin{pmatrix} 0 & y^{\mathsf{H}} B \end{pmatrix} = \begin{pmatrix} 0 & w^{\mathsf{H}} C (\lambda E - A)^{-1} B \end{pmatrix} = \begin{pmatrix} 0 & \sigma v^{\mathsf{H}} \end{pmatrix}.$$

Since $\sigma v^{\mathsf{H}} \neq 0$, $\lambda$ is a controllable mode. Observability can be proven in an analogous manner and is therefore omitted. This yields statement c) by noting that

$$\sigma^{-1} v w^{\mathsf{H}} = \varepsilon \widetilde{v} \widetilde{w}^{\mathsf{H}} \quad \text{with} \quad \widetilde{v} = \frac{1}{\sqrt{\varepsilon \sigma}} v, \ \widetilde{w} = \frac{1}{\sqrt{\varepsilon \sigma}} w,$$

where $\varepsilon \sigma > 1$ by definition.

The implication "c) $\Rightarrow$ a)" is trivial since with $\Delta := \varepsilon v w^{\mathsf{H}}$ we obtain $\lambda \in \mathfrak{P}(G)$. $\square$

By employing the same techniques as in the proof of the previous lemma we can also show the following result.

**Corollary 6.2.7.** *[BV14, Cor. 4.2] Let $(E, A, B, C) \in \Sigma_{n,m,p}$ with transfer function $G(s) \in \mathbb{R}(s)^{p \times m}$ be given. Assume that $\varepsilon > 0$ and $\lambda \in \mathbb{C} \setminus \mathfrak{P}(G)$. Then the following statements are equivalent:*

*a) $G(\lambda)$ has a (not necessarily maximum) singular value $\varepsilon^{-1}$ with right and left singular vectors $v \in \mathbb{C}^m$ and $w \in \mathbb{C}^p$ satisfying $\|v\|_2 = \|w\|_2 = 1$.*

*b) The number $\lambda$ is a controllable and observable mode of the perturbed system $(E, A + \varepsilon B v w^{\mathsf{H}} C), B, C)$ with associated right and left eigenvectors $x$ and $y$ of $sE - (A + \varepsilon B v w^{\mathsf{H}} C) \in \mathbb{C}[s]^{n \times n}$ given by*

$$x = (\lambda E - A)^{-1} B v, \quad \textit{and} \quad y = (\lambda E - A)^{-\mathsf{H}} C^{\mathsf{T}} w. \tag{6.9}$$

From Lemma 6.2.6 we can conclude that

$$\mathfrak{P}_\varepsilon(G) = \mathfrak{P}(G) \cup \left\{\lambda \in \mathbb{C} : \sigma_{\max}\left(G(\lambda)\right) > \varepsilon^{-1}\right\}$$

with boundary

$$\partial \mathfrak{P}_\varepsilon(G) = \left\{\lambda \in \mathbb{C} : \sigma_{\max}\left(G(\lambda)\right) = \varepsilon^{-1}\right\}. \tag{6.10}$$

In other words, also the rightmost structured $\varepsilon$-pseudopole is arbitrarily close to the curve $\partial\mathfrak{P}_\varepsilon(G)$. Thus, our strategy consists of computing a sequence of suitable structured rank-1 perturbed pencils $sE - \left(A + \varepsilon Bvw^{\mathsf{H}}C\right) \in \mathbb{C}[s]^{n\times n}$ such that one of the perturbed eigenvalues converges to the rightmost $\varepsilon$-pseudopole of $G(s)$. A similar technique has already been successfully applied to compute the pseudospectral abscissa of a matrix, see [GO11]. For computational purposes we assume that $\varepsilon$ is chosen such that the corresponding $\varepsilon$-pseudopole set is regular. In this way we guarantee that we only consider admissible perturbations of *finite* eigenvalues and that $\alpha(\varepsilon)$ is *finite*. Inadmissible perturbations are covered by evaluating $\lim_{\omega\to\infty}\sigma_{\max}(G(\mathrm{i}\omega))$ which will be done separately.

With regard to Proposition 4.4.1 we see the following. Let $sE - A \in \mathbb{R}[s]^{n\times n}$ be a given regular matrix pencil and let $x,\, y \in \mathbb{C}^n$ be right and left eigenvectors corresponding to a simple finite eigenvalue $\lambda = \frac{y^{\mathsf{H}}Ax}{y^{\mathsf{H}}Ex}$. Let $sE - \left(A + tBvw^{\mathsf{H}}C\right) \in \mathbb{C}[s]^{n\times n}$ be a perturbed regular matrix pencil with eigenvalue $\widetilde{\lambda}(t)$. Then it holds that

$$\widetilde{\lambda}(t) = \lambda + t\frac{y^{\mathsf{H}}Bvw^{\mathsf{H}}Cx}{y^{\mathsf{H}}Ex} + \mathcal{O}\left(t^2\right).$$

Furthermore, this directly yields

$$\left.\frac{\mathrm{d}\widetilde{\lambda}(t)}{\mathrm{d}t}\right|_{t=0} = \frac{y^{\mathsf{H}}Bvw^{\mathsf{H}}Cx}{y^{\mathsf{H}}Ex}.$$

Now, we describe how such rank-1 perturbations can be constructed in an optimal way. Therefore, let $\lambda$ be a simple eigenvalue of the pencil $sE - A$ with corresponding right and left eigenvectors $x,\, y \in \mathbb{C}^n$ satisfying $y^{\mathsf{H}}Ex > 0$. Let $v \in \mathbb{C}^m$ and $w \in \mathbb{C}^p$ with $\|v\|_2 = \|w\|_2 = 1$ be given vectors. Then it holds that

$$\begin{aligned}\operatorname{Re}\left(\left.\frac{\mathrm{d}\widetilde{\lambda}(t)}{\mathrm{d}t}\right|_{t=0}\right) &= \frac{\operatorname{Re}\left(y^{\mathsf{H}}Bvw^{\mathsf{H}}Cx\right)}{y^{\mathsf{H}}Ex} \\ &\leq \frac{\left\|y^{\mathsf{H}}B\right\|_2\|Cx\|_2}{y^{\mathsf{H}}Ex}. \end{aligned} \tag{6.11}$$

Equality in (6.11) holds for $v = B^{\mathsf{T}}y/\left\|B^{\mathsf{T}}y\right\|_2$, $w = Cx/\left\|Cx\right\|_2$. Hence, local maximal growth in $\operatorname{Re}\left(\widetilde{\lambda}(t)\right)$ as $t$ increases from 0 is achieved for this choice of $v$ and $w$. In this way we generate the initial perturbation. Next we consider subsequent perturbations. Let therefore $sE - \widehat{A} := sE - \left(A + \varepsilon B\widehat{v}\widehat{w}^{\mathsf{H}}C\right) \in \mathbb{C}[s]^{n\times n}$ with a simple eigenvalue $\widehat{\lambda}$ and associated right and left eigenvectors $\widehat{x},\, \widehat{y} \in \mathbb{C}^n$ with $\widehat{y}^{\mathsf{H}}E\widehat{x} > 0$ be the perturbed matrix pencil. In addition, let vectors $v \in \mathbb{C}^m$, $w \in \mathbb{C}^p$ with $\|v\|_2 = \|w\|_2 = 1$ be given. We consider the family of perturbations of the matrix pencil $sE - \widehat{A}$ of the form

$$sE - \left(\widehat{A} + tB\left(vw^{\mathsf{H}} - \widehat{v}\widehat{w}^{\mathsf{H}}\right)C\right),$$

which are structured $\varepsilon$-norm rank-1 perturbations of $sE - A$ for $t = 0$ and $t = \varepsilon$. For the perturbed eigenvalue, for simplicity called again $\widetilde{\lambda}$, we obtain

$$\operatorname{Re}\left( \left.\frac{\mathrm{d}\widetilde{\lambda}(t)}{\mathrm{d}t}\right|_{t=0} \right) = \frac{\operatorname{Re}\left(\widehat{y}^{\mathsf{H}} B \left(vw^{\mathsf{H}} - \widehat{v}\widehat{w}^{\mathsf{H}}\right) C\widehat{x}\right)}{\widehat{y}^{\mathsf{H}} E\widehat{x}}$$
$$\leq \frac{\left\|\widehat{y}^{\mathsf{H}} B\right\|_2 \|C\widehat{x}\|_2 - \operatorname{Re}\left(\widehat{y}^{\mathsf{H}} B\widehat{v}\widehat{w}^{\mathsf{H}} C\widehat{x}\right)}{\widehat{y}^{\mathsf{H}} E\widehat{x}}. \tag{6.12}$$

Similarly to the above considerations, equality in (6.12) holds for $v = B^{\mathsf{T}}\widehat{y}/\left\|B^{\mathsf{T}}\widehat{y}\right\|_2$, $w = C\widehat{x}/\left\|C\widehat{x}\right\|_2$. Therefore, the basic algorithm consists of successively choosing an eigenvalue and constructing the perturbations described above by using the corresponding eigenvectors. However, an important question is how to actually choose these eigenvalues. This will be discussed in the next paragraph.

**Choice of the Eigenvalues** Recall that we want to construct structured $\varepsilon$-norm rank-1 perturbations of the pencil $sE - A$ such that one of the perturbed eigenvalues converges to the rightmost $\varepsilon$-pseudopole of the transfer function $G(s)$. Intuitively, in each step one would choose the rightmost eigenvalue of the perturbed pencil to construct the next perturbation. However, this might not be a good choice. Note that the perturbability of an eigenvalue $\lambda \in \mathbb{C}$ with right and left normalized eigenvectors $x$ and $y$ highly depends on the values of $\left\|B^{\mathsf{T}}y\right\|_2$ and $\left\|Cx\right\|_2$. If these values are small, no large perturbation is possible. We recall that these values are strongly related to the controllability and observability concepts introduced in Subsection 6.2.1. Roughly speaking, the "larger" the values of $\left\|B^{\mathsf{T}}y\right\|_2$, the "larger" is the distance of the system to uncontrollability at $\lambda$. So, large values of $\left\|B^{\mathsf{T}}y\right\|_2$ indicate a good controllability at $\lambda$. Analogous considerations can also be made for observability.

Consequently, for our algorithm we look for eigenvalues that have both sufficiently large real part and a high controllability and observability. An algorithm which unites both concepts is the *(subspace accelerated MIMO) dominant pole algorithm (SAMDP)*, introduced by Rommes and Martins [RM06b, RM06a, RS08, Rom08]. This algorithm can be shown to converge locally superlinearly to the desired eigenvalues. It has actually been designed to find the poles which have the highest influence on the frequency response of the transfer function $G(s)$. Assume that $sE - A \in \mathbb{R}[s]^{n\times n}$ has only simple eigenvalues $\lambda_k$ with left and right eigenvectors $y_k$, $x_k \in \mathbb{C}^n$, normalized such that $y_k^{\mathsf{H}} E x_k = 1$. Then we have the representation

$$G(s) = \sum_{k=1}^{n} \frac{R_k}{s - \lambda_k} + R_\infty \tag{6.13}$$

with the *residues*

$$R_k = Cx_k y_k^{\mathsf{H}} B \quad \text{and} \quad R_\infty = \lim_{\omega\to\infty} G(\mathrm{i}\omega).$$

Then it holds that

$$\|R_k\|_2 = \lambda_{\max}\left(Cx_k y_k^{\mathsf{H}} BB^{\mathsf{T}} y_k x_k^{\mathsf{H}} C^{\mathsf{T}}\right)^{1/2} = \|Cx_k\|_2 \|B^{\mathsf{T}} y_k\|_2$$

is a measure for simultaneous controllability and observability of $\lambda_k$. We observe that if $\lambda_j$ is close to the imaginary axis and $\|R_j\|_2$ is large, then for $\omega \approx \operatorname{Im}(\lambda_j)$ we obtain

$$G(\mathrm{i}\omega) \approx \frac{R_j}{-\operatorname{Re}(\lambda_j)} + \sum_{\substack{k=1\\k\neq j}}^{n} \frac{R_k}{\mathrm{i}\omega - \lambda_k} + R_\infty, \tag{6.14}$$

and therefore, $\|G(\mathrm{i}\omega)\|_2$ is large, too. These considerations give the motivation for the following definition. We call an eigenvalue $\lambda_j \in \Lambda(E, A)$ *dominant pole* of $G(s)$, if

$$\frac{\|R_k\|_2}{|\operatorname{Re}(\lambda_k)|} < \frac{\|R_j\|_2}{|\operatorname{Re}(\lambda_j)|}, \quad k = 1, \ldots, n, \quad k \neq j. \tag{6.15}$$

The most dominant poles can be determined by SAMDP and are essentially what we are looking for. However, we also deal with positive $\varepsilon$-pseudopole set abscissae. By using the definition (6.15), the eigenvalues tend to loose dominance as soon as they have crossed the imaginary axis into the right half-plane. Then in subsequent iterations eigenvalues in the left half-plane tend to be determined as most dominant. This is of course an undesired behavior since this could lead to convergence problems when the rightmost structured $\varepsilon$-pseudopole is "far" in the right half-plane. Therefore, we also propose an alternative dominance measure which does not have this drawback. We call an eigenvalue $\lambda_j \in \Lambda(E, A)$ *exponentially dominant pole* of $G(s)$, if

$$\|R_k\|_2 \exp(\beta \operatorname{Re}(\lambda_k)) < \|R_j\|_2 \exp(\beta \operatorname{Re}(\lambda_j)), \quad k = 1, \ldots, n, \quad k \neq j. \tag{6.16}$$

The parameter $\beta$ is a weight factor which determines the trade-off between the influence of the residues and the real parts of the eigenvalues. In our numerical experiments it turned out that the dominance defined by (6.15) or (6.16) with rather large values of $\beta$ (high weight on the real part) are good choices for many examples. Since SAMDP delivers the poles which have the highest influence on the frequency response of a system and due to the relation (6.6), we can determine good initial estimates for $r_{\mathbb{C}}^{\mathrm{f}}(G)$. We compute some of the dominant poles $\lambda_k$, $k = 1, \ldots, \ell$ and determine an estimate $r_{\mathbb{C}}^{\mathrm{est}}(G)$ as

$$r_{\mathbb{C}}^{\mathrm{est}}(G) = 1 / \max_{1 \le k \le \ell} \sigma_{\max}\left(G\left(\mathrm{i}\omega_k\right)\right) \tag{6.17}$$

with $\omega_k = \operatorname{Im}(\lambda_k)$, $k = 1, \ldots, \ell$.

**Algorithm 6.1** Computation of the pseudopole set abscissa

**Input:** System $\Sigma = (E, A, B, C) \in \Sigma_{n,m,p}$ with transfer function $G(s) \in \mathbb{R}(s)^{p\times m}$, perturbation level $\varepsilon < 1/\lim_{\omega\to\infty} \sigma_{\max}(G(\mathrm{i}\omega))$, tolerance on relative change $\tau$, dominance measure as in (6.15) or (6.16) for all dominant pole computations.

**Output:** $\alpha_\varepsilon(G)$, right and left eigenvectors $x_*$, $y_* \in \mathbb{C}^n$ associated to the optimal pseudopole.

1: Compute a dominant pole $\lambda_0$ of $G(s)$ with associated right and left eigenvectors $x_0$, $y_0 \in \mathbb{C}^n$ of $sE - A$.

2: Compute the perturbation $\widehat{A} = A + \varepsilon \frac{BB^\mathsf{T} y_0 x_0^\mathsf{H} C^\mathsf{T} C}{\left\|B^\mathsf{T} y_0\right\|_2 \left\|C x_0\right\|_2}$.

3: **for** $j = 1, 2, \ldots$ **do**

4: Compute a dominant pole $\lambda_j$ of $\widehat{G}(s) := C\left(sE - \widehat{A}\right)^{-1} B$ with associated right and left eigenvectors $x_j$, $y_j \in \mathbb{C}^n$ of $sE - \widehat{A}$.

5: **if** $\left|\mathrm{Re}\,(\lambda_j) - \mathrm{Re}\,(\lambda_{j-1})\right| < \tau \left|\mathrm{Re}\,(\lambda_j)\right|$ **then**

6: Set $k = j$.

7: Break.

8: **end if**

9: Compute the perturbation $\widehat{A} = A + \varepsilon \frac{BB^\mathsf{T} y_j x_j^\mathsf{H} C^\mathsf{T} C}{\left\|B^\mathsf{T} y_j\right\|_2 \left\|C x_j\right\|_2}$.

10: **end for**

11: Set $\alpha_\varepsilon(G) = \mathrm{Re}\,(\lambda_k)$, $x_* = x_k$, $y_* = y_k$.

**Algorithmic Details** In this paragraph we present some pseudocode of the algorithms that we have derived. Algorithm 6.1 summarizes the procedure for the computation of the $\varepsilon$-pseudopole set abscissa. In our implementation we always initialize Algorithm 6.1 by setting $x_0$ and $y_0$ to the eigenvectors returned by the previous evaluation of $\alpha(\varepsilon)$ (if there is one). This accelerates the computation drastically since the eigenvectors used in the outer iteration converge as well.

We mention the drawback that the algorithm does not necessarily converge to the globally rightmost value on the boundary of the $\varepsilon$-pseudopole set $\partial\mathfrak{P}_\varepsilon(G)$ in (6.10). Mostly it does but in some rare situations the algorithm converges only to a local maximizer. This especially happens in the first iteration of the root-finding algorithm when no good estimates of the optimal eigenvectors are available. Therefore, sometimes one has to try several dominant poles to find the global maximizer in the beginning. Note that we could also follow multiple poles to the boundary of the corresponding pseudopole set in order to increase the chance to find the globally rightmost point. But due to the much higher complexity and a comparably small gain we only follow one pole. In fact, by choosing a dominant pole of the original transfer function as starting pole usually gives the desired result as shown in Subsection 6.2.5.

**Fixed Point Analysis** This section is devoted to the analysis of the fixed points of the iteration given by Algorithm 6.1. The following two lemmas will be needed for our considerations.

**Lemma 6.2.8.** *[GO11] Let $t \in \mathbb{R}$ and consider the $p \times m$ matrix family $C(t) = C_0 + tC_1$. Let $\sigma(t)$ be a singular value of $C(t)$ converging to a simple nonzero singular value $\sigma_0$ of $C_0$ as $t \to 0$. Then, $\sigma(t)$ is analytic near $t = 0$ and*

$$\left.\frac{\mathrm{d}\sigma(t)}{\mathrm{d}t}\right|_{t=0} = w_0^H C_1 v_0,$$

*where $v_0$ and $w_0$ with $\|v_0\|_2 = \|w_0\|_2 = 1$ are, respectively, the right and left singular vectors of $C_0$ corresponding to $\sigma_0$.*

**Lemma 6.2.9.** *[FBK13, p. 73] Let $\lambda_* \in \mathbb{C} \setminus \mathfrak{P}(G)$ be given. Then $G(s)$ can be expanded into a Laurent series at $\lambda_*$ as*

$$\begin{aligned} G(\lambda) &= C(I_n - (\lambda - \lambda_*)(\lambda_* E - A)^{-1}E)^{-1} (\lambda_* E - A)^{-1} B \\ &= M_0 + M_1(\lambda - \lambda_*) + M_2(\lambda - \lambda_*)^2 + \dots \end{aligned}$$

*with the* moments $M_j = C\left(-(\lambda_* E - A)^{-1}E\right)^j (\lambda_* E - A)^{-1}B$.

Besides the above we will make the following assumption [GGO13, Assmp. 2.19] throughout this section.

**Assumption 6.2.10.** [BV14, Assmp. 4.6] Let $(E, A, B, C) \in \Sigma_{n,m,p}$ with transfer function $G(s) \in \mathbb{R}(s)^{p \times m}$ be given. Moreover, let $\varepsilon > 0$ be given such that the associated $\varepsilon$-pseudopole set of $G(s)$ is regular. Let $\lambda_* \in \mathbb{C}$ be a locally rightmost point of $\mathfrak{P}_\varepsilon(G)$. Then we assume that

a) the largest singular value $\varepsilon^{-1}$ of $G(\lambda_*)$ is simple;

b) if $v_* \in \mathbb{C}^m$ and $w_* \in \mathbb{C}^p$ are the corresponding right and left singular vectors with $\|v_*\|_2 = \|w_*\|_2 = 1$, then the pole $\lambda_*$ of the perturbed transfer function $G_{\varepsilon v_* w_*^{\mathsf{H}}}(s)$ is simple. (That $\lambda_*$ is a pole follows from Corollary 6.2.7.)

Note, that using similar arguments as in [BLO03, p. 362], the first part of Assumption 6.2.10 is generically true, i. e., it holds true for almost all systems $(E, A, B, C) \in \Sigma_{n,m,p}$. However, it is not difficult to find counter-examples.

**Lemma 6.2.11.** *[BV14, Lem. 4.7] Let $(E, A, B, C) \in \Sigma_{n,m,p}$ with transfer function $G(s) \in \mathbb{R}(s)^{p \times m}$ be given and let Assumption 6.2.10 be satisfied. If $\lambda_* \in \mathbb{C}$ is a local maximizer of the optimization problem*

$$\sup \{\mathrm{Re}(\lambda) : \lambda \in \mathfrak{P}_\varepsilon(G)\}, \tag{6.18}$$

*then it holds that*

$$\|G(\lambda_*)\|_2 = \varepsilon^{-1} \quad \text{and} \quad w_*^{\mathsf{H}} C \left(\lambda_* E - A\right)^{-1} E \left(\lambda_* E - A\right)^{-1} B v_* > 0,$$

*where $v_* \in \mathbb{C}^m$ and $w_* \in \mathbb{C}^p$ are the normalized right and left singular vectors of $G(\lambda_*)$.*

*Proof.* Our proof follows similar arguments as in the proof of [GGO13, Lem. 2.21]. The assertion that we have $\|G(\lambda_*)\|_2 = \varepsilon^{-1}$ directly follows from the fact that $\lambda_*$ is on the boundary of $\mathfrak{P}_\varepsilon(G)$. Next, the optimization problem (6.18) is equivalent to

$$\max \left\{ \operatorname{Re}(\lambda) : \|G(\lambda)\|_2 \geq \varepsilon^{-1} \right\}.$$

By identifying $\lambda = \gamma + \mathrm{i}\delta \in \mathbb{C}$ with the vector $\begin{pmatrix} \gamma \\ \delta \end{pmatrix} \in \mathbb{R}^2$, this is furthermore equivalent to

$$\max \left\{ g(\gamma, \delta) : h(\gamma, \delta) \leq 0 \right\}$$

with $g(\gamma, \delta) = \gamma$ and $h(\gamma, \delta) = \varepsilon^{-1} - \|G(\gamma + \mathrm{i}\delta)\|_2$. At the optimum $\begin{pmatrix} \gamma_* \\ \delta_* \end{pmatrix}$ we must now either have

a) $\nabla h(\gamma_*, \delta_*) = 0$; or

b) $\nabla g(\gamma_*, \delta_*) = \mu \nabla h(\gamma_*, \delta_*)$ with a Lagrange multiplier $\mu > 0$.

From Lemma 6.2.9 it follows that in a neighborhood of $\lambda_* = \gamma_* + \mathrm{i}\delta_*$ we have

$$\begin{aligned} G(\lambda) = C(\lambda_* E - A)^{-1} B + (\lambda_* - \lambda) C (\lambda_* E - A)^{-1} E (\lambda_* E - A)^{-1} B& \\ + \mathcal{O}\left((\lambda_* - \lambda)^2\right)&. \end{aligned}$$

By applying Lemma 6.2.8 to $G(\lambda)$, we obtain

$$\nabla h^{\mathsf{T}}(\gamma_*, \delta_*) = \begin{pmatrix} \operatorname{Re}\left( w_*^{\mathsf{H}} C \left(\lambda_* E - A\right)^{-1} E \left(\lambda_* E - A\right)^{-1} B v_* \right) \\ \operatorname{Im}\left( w_*^{\mathsf{H}} C \left(\lambda_* E - A\right)^{-1} E \left(\lambda_* E - A\right)^{-1} B v_* \right) \end{pmatrix}.$$

Since $\lambda_*$ is an eigenvalue of the pencil $sE - \left(A + \varepsilon B v_* w_*^{\mathsf{H}} C\right) \in \mathbb{C}[s]^{n \times n}$ with right and left eigenvectors $x_*$, $y_* \in \mathbb{C}^n$ we obtain

$$\begin{aligned} x_* &= \varepsilon (\lambda_* E - A)^{-1} B v_* w_*^{\mathsf{H}} C x_*, \\ y_*^{\mathsf{H}} &= \varepsilon y_*^{\mathsf{H}} B v_* w_*^{\mathsf{H}} C (\lambda_* E - A)^{-1}, \end{aligned}$$

and hence

$$0 \neq y_*^{\mathsf{H}} E x_* = \varepsilon^2 y_*^{\mathsf{H}} B v_* w_*^{\mathsf{H}} C (\lambda_* E - A)^{-1} E (\lambda_* E - A)^{-1} B v_* w_*^{\mathsf{H}} C x_*.$$

This means that $w_*^{\mathsf{H}} C(\lambda_* E - A)^{-1} E(\lambda_* E - A)^{-1} B v_* \neq 0$ and as a consequence we can exclude case a) above. Since $\nabla g(\gamma_*, \delta_*) = \begin{pmatrix} 1 & 0 \end{pmatrix}$ and b) holds true, we directly obtain

$$w_*^{\mathsf{H}} C \left(\lambda_* E - A\right)^{-1} E \left(\lambda_* E - A\right)^{-1} B v_* = 1/\mu > 0$$

and the proof is complete. □

Lemma 6.2.11 gives necessary first-order optimality conditions for $\lambda_* \in \mathbb{C}$ to be a locally rightmost point in the $\varepsilon$-pseudopole set. Now we can start analyzing the fixed points of the iteration presented in Algorithm 6.1. First we introduce the notion of a fixed point similar as in [GGO13, Def. 3.1]. Here, the term "dominant pole" is either understood with respect to (6.15) or (6.16).

**Definition 6.2.12** (Fixed point)**.** [BV14, Def. 4.8] Let $\varepsilon > 0$ be given such that the associated $\varepsilon$-pseudopole set is regular. Furthermore, assume that $((v_j, w_j))_{j \in \mathbb{N}_0} = \left(\left(B^{\mathsf{T}} y_j / \left\|B^{\mathsf{T}} y_j\right\|_2, C x_j / \left\|C x_j\right\|_2\right)\right)_{j \in \mathbb{N}_0}$ is a sequence of perturbations constructed by Algorithm 6.1. Assume that $\lambda_j$ is the unique dominant pole of the perturbed transfer function $G_{\varepsilon v_j w_j^{\mathsf{H}}}(s)$. A vector pair $(v_j, w_j)$ is a *fixed point* of this iteration if $\lambda_j$ is simple, $v_j w_j^{\mathsf{H}} = v_{j+1} w_{j+1}^{\mathsf{H}}$, and consequently $\lambda_j = \lambda_{j+1}$.

Next, we get a similar theorem as [GGO13, Thm. 3.2] which we will prove in an analogous fashion.

**Theorem 6.2.13.** *[BV14, Thm. 4.9]*

*a) Let $\varepsilon > 0$ be chosen such that the corresponding $\varepsilon$-pseudopole set is regular. Let $(v_*, w_*)$ be a fixed point of the iteration induced by Algorithm 6.1 corresponding to the dominant pole $\lambda_* \in \mathbb{C}$ of $G_{\varepsilon v_* w_*^{\mathsf{H}}}(s)$ which we assume to be unique. Then $G(\lambda_*)$ has a singular value equal to $\varepsilon^{-1}$ and if this is the largest one, the first-order optimality conditions presented in Lemma 6.2.11 hold true.*

*b) Conversely, assume that $\varepsilon > 0$ is chosen such that the corresponding $\varepsilon$-pseudopole set is regular and that the first-order optimality conditions in Lemma 6.2.11 hold true for some $\lambda_* \in \mathbb{C}$ and $(v_*, w_*)$. Then $\lambda_*$ is a pole of $G_{\varepsilon v_* w_*^{\mathsf{H}}}(s)$ and if it is the unique dominant one and simple, then $(v_*, w_*)$ is a fixed point of the iteration induced by Algorithm 6.1.*

*Proof.* First we show a): Let $(v_*, w_*)$ be a fixed point of the iteration induced by Algorithm 6.1 corresponding the dominant pole $\lambda_* \in \mathbb{C}$ of $G_{\varepsilon v_* w_*^{\mathsf{H}}}(s)$. Then we have $v_* = B^{\mathsf{T}} y_* / \left\|B^{\mathsf{T}} y_*\right\|_2$ and $w_* = C x_* / \left\|C x_*\right\|_2$, where $x_*$ and $y_*$ are the right and left eigenvectors of the perturbed matrix pencil

$$\lambda E - \left(A + \varepsilon \frac{B B^{\mathsf{T}} y_* x_*^{\mathsf{H}} C^{\mathsf{T}} C}{\left\|B^{\mathsf{T}} y_*\right\|_2 \left\|C x_*\right\|_2}\right) \tag{6.19}$$

to the eigenvalue $\lambda_*$. That $\varepsilon^{-1}$ is indeed a singular value of $G(\lambda_*)$ follows by Corollary 6.2.7. Let $\varepsilon^{-1}$ now be the largest singular value of $G(\lambda_*)$. Then we have

$$\begin{aligned} x_* &= \varepsilon \frac{(\lambda_* E - A)^{-1} B B^\mathsf{T} y_* x_*^\mathsf{H} C^\mathsf{T} C x_*}{\|B^\mathsf{T} y_*\|_2 \|C x_*\|_2}, \\ y_*^\mathsf{H} &= \varepsilon \frac{y_*^\mathsf{H} B B^\mathsf{T} y_* x_*^\mathsf{H} C^\mathsf{T} C (\lambda_* E - A)^{-1}}{\|B^\mathsf{T} y_*\|_2 \|C x_*\|_2}. \end{aligned}$$

Due to the normalization $y_*^\mathsf{H} E x_* = 1$ we now obtain

$$\begin{aligned} y_*^\mathsf{H} E x_* &= \varepsilon^2 \|B^\mathsf{T} y_*\|_2 \|C x_*\|_2 \frac{x_*^\mathsf{H} C^\mathsf{T}}{\|C x_*\|_2} C \left(\lambda_* E - A\right)^{-1} E \left(\lambda_* E - A\right)^{-1} B \frac{B^\mathsf{T} y_*}{\|B^\mathsf{T} y_*\|_2} \\ &= 1 > 0, \end{aligned} \tag{6.20}$$

i. e., the first-order optimality conditions hold true.

Now we prove statement b): Assume that the first-order optimality conditions in Lemma 6.2.11 are satisfied for some $\lambda_* \in \mathbb{C}$ and $(v_*, w_*)$. Then $\sigma_{\max}(G(\lambda_*)) = \varepsilon^{-1}$ with right and left normalized singular vectors $v_* = B^\mathsf{T} y_* / \|B^\mathsf{T} y_*\|_2$ and $w_* = C x_* / \|C x_*\|_2$. From the second optimality condition in Lemma 6.2.11 and with (6.9) we obtain

$$\frac{x_*^\mathsf{H} C^\mathsf{T} C (\lambda_* E - A)^{-1} E (\lambda_* E - A)^{-1} B B^\mathsf{T} y_*}{\|C x_*\|_2 \|B^\mathsf{T} y_*\|_2} > 0,$$

where $x_*$ and $y_*$ are now again the right and left eigenvectors of the perturbed pencil (6.19) to the eigenvalue $\lambda_*$. So, if $\lambda_*$ is dominant and simple, then the pair $(v_*, w_*)$ is a fixed point of the iteration. □

As in [GO11, p. 1176] we argue that the only possible *attractive* fixed points of the iteration given by Algorithm 6.1 are the local maximizers of the optimization problem (6.18).

**Local Convergence and Error Analysis** Similarly as in [GO11] it is possible to show that for sufficiently small values of $\varepsilon$ we have local convergence to a fixed point with linear rate. The generalization of the proof is analogous to the one for the algorithm in [GGO13], as discussed by [GO13]. Since this analysis is rather lengthy we omit the details. However, as shown by our numerical examples we always have convergence to a fixed point independently of $\varepsilon$. However, similarly as in [GO11], the linear convergence factor might get higher for larger values of $\varepsilon$.

### 6.2.3 Newton's Method for Computing the Complex $\mathcal{H}_\infty$-Radius

In this subsection we derive a Newton-like method for computing the root of $\alpha(\cdot)$. This is a generalization of the method presented in [GGO13] and we will later show that it is slightly faster than the secant method used in [BV12b] or any other superlinearly converging root-finding scheme [BV12a] applied to this problem. The following theorem deals with the derivative of the pseudopole set abscissa with respect to $\varepsilon$, similarly as in [GGO13, Thm. 4.1].

**Theorem 6.2.14.** *[BV14, Thm. 5.1] Let $(E, A, B, C) \in \Sigma_{n,m,p}$ with transfer function $G(s) \in \mathbb{R}(s)^{p\times m}$ be given. Let $\varepsilon > 0$ be such that the associated $\varepsilon$-pseudopole set is regular. Let $\lambda(\varepsilon)$ be the rightmost point of $\mathfrak{P}_\varepsilon(G)$. Let Assumption 6.2.10 be satisfied for all regular $\varepsilon$-pseudopole sets and let $v(\varepsilon) = B^\mathsf{T} y(\varepsilon)/\left\|B^\mathsf{T} y(\varepsilon)\right\|_2$ and $w(\varepsilon) = Cx(\varepsilon)/\left\|Cx(\varepsilon)\right\|_2$ be the normalized singular vectors of $G(\lambda(\varepsilon))$ corresponding to the largest singular value $\varepsilon^{-1}$, where $x(\varepsilon), y(\varepsilon) \in \mathbb{C}^n$ are the right and left eigenvectors of the perturbed pencil*

$$sE - \left(A + \varepsilon \frac{BB^\mathsf{T} y(\varepsilon)x(\varepsilon)^\mathsf{H} C^\mathsf{T} C}{\left\|B^\mathsf{T} y(\varepsilon)\right\|_2 \left\|Cx(\varepsilon)\right\|_2}\right) \in \mathbb{C}[s]^{n\times n},$$

*with $y(\varepsilon)^\mathsf{H} Ex(\varepsilon) = 1$. Furthermore, let $\varepsilon_0 > 0$ be given such that the structured $\varepsilon_0$-pseudopole set is regular and such that the rightmost point $\lambda(\varepsilon_0)$ of $\mathfrak{P}_{\varepsilon_0}(G)$ is uniquely determined. Then $\lambda(\cdot)$ is continuously differentiable at $\varepsilon_0$ and it holds that*

$$\left.\frac{\mathrm{d}\alpha(\varepsilon)}{\mathrm{d}\varepsilon}\right|_{\varepsilon=\varepsilon_0} = \left.\frac{\mathrm{d}\lambda(\varepsilon)}{\mathrm{d}\varepsilon}\right|_{\varepsilon=\varepsilon_0} = \left\|B^\mathsf{T} y(\varepsilon_0)\right\|_2 \left\|Cx(\varepsilon_0)\right\|_2. \tag{6.21}$$

*Proof.* The proof is similar as for [GGO13, Thm. 4.1]. Due to the first part of Assumption 6.2.10, the singular vectors $v(\varepsilon)$ and $w(\varepsilon)$ are unique up to multiplication with a unitary scalar. Therefore, the largest singular value of $G(\lambda(\cdot))$ is differentiable with respect to $\varepsilon$. The second part of Assumption 6.2.10 ensures that $y(\varepsilon)^\mathsf{H} Ex(\varepsilon) \neq 0$ while the uniqueness of the rightmost point $\lambda(\varepsilon_0)$ guarantees the continuity of $\lambda(\cdot)$ and its derivative in a neighborhood of $\varepsilon_0$.

Now we prove (6.21). Differentiating the constraint $G(\lambda(\varepsilon)) = \varepsilon^{-1}$ with respect to $\varepsilon$ yields

$$\begin{aligned} 0 &= \frac{\mathrm{d}}{\mathrm{d}\varepsilon}\left(\varepsilon^{-1} - \left\|C(\lambda(\varepsilon)E - A)^{-1}B\right\|_2\right) \\ &= -\varepsilon^{-2} + w(\varepsilon)^\mathsf{H} C\left(\lambda(\varepsilon)E - A\right)^{-1} E\left(\lambda(\varepsilon)E - A\right)^{-1} Bv(\varepsilon)\cdot\frac{\mathrm{d}\lambda(\varepsilon)}{\mathrm{d}\varepsilon}. \end{aligned}$$

Plugging in $v(\varepsilon) = B^\mathsf{T} y(\varepsilon)/\left\|B^\mathsf{T} y(\varepsilon)\right\|_2$ and $w(\varepsilon) = Cx(\varepsilon)/\left\|Cx(\varepsilon)\right\|_2$ and comparing this with (6.20) gives the desired result and finalizes the proof. □

Now, since we know how to differentiate $\alpha(\cdot)$, we can make use of Newton's method to compute the root of $\alpha(\cdot)$. This method has a local quadratic convergence since $\mathrm{d}\alpha(\varepsilon)/\mathrm{d}\varepsilon > 0$ and $\mathrm{d}^2\alpha(\varepsilon)/\mathrm{d}\varepsilon^2$ is finite. The complete procedure is summarized in Algorithm 6.2. First, we check whether the $\mathcal{H}_\infty$-norm is attained at $\omega = \infty$. This is done by evaluating $\sigma_{\max}(G(\cdot))$ at the imaginary parts of the dominant poles. If all these values are below $g_\infty := \lim_{\omega\to\infty} \sigma_{\max}(G(\mathrm{i}\omega))$, we assume that the norm is attained at infinity and we return $g_\infty$. This step is necessary to avoid possible computations with nonregular pseudopole sets in subsequent steps. To estimate $\lim_{\omega\to\infty} G(\mathrm{i}\omega)$, we evaluate $G(\mathrm{i}\omega)$ for a sufficiently large $\omega$. The largest singular value of $G(\mathrm{i}\omega)$ will converge quickly due to the fact that for large $\omega$ there are no close poles which can introduce peaks. We can give the following upper bound using the residue representation of the transfer function (6.13) with $\lambda_j = \nu_j + \mathrm{i}\omega_j$, $j = 1, \ldots, n$:

$$\begin{aligned} \|G(\mathrm{i}\omega)\|_2 &= \left\| \sum_{j=1}^{n} \frac{R_j}{\mathrm{i}\omega - \mathrm{i}\omega_j - \nu_j} + R_\infty \right\|_2 \\ &\leq \sum_{j=1}^{n} \frac{\|R_j\|_2}{|\mathrm{i}\omega - \mathrm{i}\omega_j - \nu_j|} + \|R_\infty\|_2 \,. \end{aligned} \tag{6.22}$$

For $\omega \gg \max_{1\leq j\leq n} \omega_j$ we can neglect the real parts of the denominators $\nu_j$, $j = 1, \ldots, n$. Since usually there are only very few dominant poles, we can control the desired accuracy by, e. g., choosing $\omega$ such that for the most dominant poles $\lambda_j$, $j = 1, \ldots, \ell$, we have

$$\sum_{j=1}^{\ell} \frac{\|R_j\|_2}{|\omega - \omega_j|} \leq \eta \tag{6.23}$$

for some small $\eta > 0$. Note that in the algorithm, the dominant poles have to be computed anyway, so we can evaluate the left-hand side of (6.23) at no additional cost.

When we assume that the $\mathcal{H}_\infty$-norm is attained at a finite frequency, we compute the root of $\alpha(\cdot)$ as described above. As the initial value we take $\varepsilon_1 = r_{\mathbb{C}}^{\mathrm{est}}(G)$ as in (6.17) which is already very close to the exact value of the complex $\mathcal{H}_\infty$-radius for most of our examples. In our actual implementation of the algorithm we also check if the most dominant poles are purely real. In this case we assume that $\|G\|_{\mathcal{H}_\infty} = \|G(0)\|_2$ and return this value. This is done to improve the performance of the algorithm since there are many examples with this property.

### 6.2.4 Comparison with the Method of Guglielmi, Gürbüzbalaban, and Overton

As already pointed out in the introduction, the work [GGO13] uses a similar idea to compute the $\mathcal{H}_\infty$-norm for standard state-space system $\Sigma := (I_n, A, B, C, D) \in \Sigma_{n,m,p}$

**Algorithm 6.2** Computation of the $\mathcal{H}_\infty$-norm using the pseudopole set approach

**Input:** System $(E, A, B, C) \in \Sigma_{n,m,p}$ with transfer function $G \in \mathcal{RH}_\infty^{p\times m}$.
**Output:** $\|G\|_{\mathcal{H}_\infty}$, optimal frequency $\omega_*$.

1: Compute some dominant poles $\lambda_j = \nu_j + \mathrm{i}\omega_j$, $j = 1, \ldots, \ell$.
2: Compute $g_\infty := \lim_{\omega\to\infty} \sigma_{\max}(G(\mathrm{i}\omega))$.
3: **if** $g_\infty > \sigma_{\max}(G(\mathrm{i}\omega_j))$, $j = 1, \ldots, \ell$ **then**
4: Set $\|G\|_{\mathcal{H}_\infty} = g_\infty$ and $\omega_* = \infty$.
5: Return.
6: **else**
7: Set $\varepsilon_1 = r_{\mathbb{C}}^{\mathrm{est}}(G)$ as in (6.17).
8: **for** $j = 1, 2, \ldots, k$ **do**
9: Compute $\alpha(\varepsilon_j)$, and right and left eigenvectors $x_j$, $y_j \in \mathbb{C}^n$ associated to the optimal pseudopole $\lambda_j$.
10: Perform a Newton step: set $\varepsilon_{j+1} = \varepsilon_j - \frac{\alpha(\varepsilon_j)}{\left\|B^{\mathsf{T}} y(\varepsilon_j)\right\|_2 \left\|Cx(\varepsilon_j)\right\|_2}$.
11: **end for**
12: **end if**
13: Set $\|G\|_{\mathcal{H}_\infty} = \varepsilon_{k+1}^{-1}$ and $\omega_* = \mathrm{Im}\,(\lambda_k)$.

with transfer function

$$G(s) = C(sI_n - A)^{-1}B + D \in \mathcal{RH}_\infty^{p\times m}. \tag{6.24}$$

The presence of a nonzero matrix $D$ makes the derivation and analysis of the algorithm from [GGO13] particularly cumbersome and difficult. In this case the role of the $\varepsilon$-pseudopole sets of this work is attained by the so-called *$\varepsilon$-spectral value sets* which are defined by

$$\Xi_\varepsilon(\Sigma) := \big\{\lambda \in \mathbb{C} : \lambda \in \Lambda\left(A + B\Delta(I_p - D\Delta)^{-1}C\right) \\ \text{for some } \Delta \in \mathbb{C}^{m\times p} \text{ with } \|\Delta\|_2 < \varepsilon\big\}.$$

Then it is shown that $\Xi_\varepsilon(\Sigma)$ can be constructed by only using rank-1 perturbations $\Delta$, and therefore, also the rightmost point in the spectral value set can be realized by a rank-1 perturbation $\varepsilon v_* w_*^{\mathsf{H}}$ with $\|v_*\|_2 = \|w_*\|_2 = 1$. The construction of this optimizing perturbation is done by an iteration that yields a sequence of normalized vector pairs $((v_j, w_j))_{j\in\mathbb{N}_0}$. For each $j > 0$, $v_j$ and $w_j$ are determined as solutions of the optimization problem

$$\max_{\|u\|_2=\|v\|_2=1} \mathrm{Re}\left(y_j^{\mathsf{H}} B \left(\frac{vw^{\mathsf{H}}}{1 - \varepsilon w^{\mathsf{H}} Dv}\right) Cx_j\right), \tag{6.25}$$

where $x_j$ and $y_j$ are the right and left eigenvectors to the rightmost eigenvalue of the perturbed matrix $A + \varepsilon Bv_{j-1}w_{j-1}^{\mathsf{H}}\left(I_p - \varepsilon Dv_{j-1}w_{j-1}^{\mathsf{H}}\right)^{-1}C$, respectively. An explicit solution of this optimization problem is derived in [GGO13]. In our method we circumvent the solution of such an optimization problem, since we always eliminate $D$. Then we can directly construct the next perturbation matrix by using (6.12). Of course, then we have to deal with a descriptor system, even if the original problem comes from a standard state-space system.

Note that both methods are also *not* equivalent in the context of standard state-space systems. To see this we rewrite (6.24) as

$$G(s) = \begin{bmatrix} C & I_p \end{bmatrix} \left( s \begin{bmatrix} I_n & 0 \\ 0 & 0 \end{bmatrix} - \begin{bmatrix} A & 0 \\ 0 & -I_p \end{bmatrix} \right)^{-1} \begin{bmatrix} B \\ D \end{bmatrix}.$$

Let $\lambda_0$ be the rightmost finite eigenvalue of the pencil $s \begin{bmatrix} I_n & 0 \\ 0 & 0 \end{bmatrix} - \begin{bmatrix} A & 0 \\ 0 & -I_p \end{bmatrix} \in \mathbb{R}[s]^{n+p \times n+p}$ with right and left eigenvectors $x_0 = \begin{pmatrix} x_{01}^\mathsf{T} & x_{02}^\mathsf{T} \end{pmatrix}^\mathsf{T} \in \mathbb{C}^{n+p}$ and $y_0 = \begin{pmatrix} y_{01}^\mathsf{T} & y_{02}^\mathsf{T} \end{pmatrix}^\mathsf{T} \in \mathbb{C}^{n+p}$. Then we have $x_{02} = 0$ and $y_{02} = 0$. Therefore, the first perturbation in the descriptor system case is constructed by the vectors

$$v_0 := B^\mathsf{T} y_{01} / \left\| B^\mathsf{T} y_{01} \right\|_2, \quad w_0 := C x_{01} / \left\| C x_{01} \right\|_2. \tag{6.26}$$

Furthermore, since $x_{01}$ and $y_{01}$ are the right and left eigenvectors of the matrix $A$ with respect to the eigenvalue $\lambda_0$, the first perturbation in the standard system case is computed via the solution of the optimization problem (6.25) with $x_j = x_{01}$ and $y_j = y_{01}$. However, its solution only corresponds to (6.26) if $D = 0$. In our algorithm, $D$ is completely ignored, whereas in [GGO13] both $D$ and $\varepsilon$ play a certain role in the optimization process. So, even if both methods converge to the same locally rightmost point in the pseudopole set or spectral value set, the path of intermediate iterates might be different. It is far from obvious to see whether one of the approaches works better than the other one, but this is also out of the scope of this thesis. However, this question is particularly interesting in the context of generalizing the approach from [GGO13] to the descriptor system case, since there is a large freedom of choosing $E$ and $D$ to obtain the same transfer function.

The second main difference to [GGO13] is that in our algorithm we explicitly consider poles instead of eigenvalues. This difference seems to be of minor nature but it has some important consequences. Consider for example the case that $E = I_n$ and $\Lambda(A) \subset \mathbb{C}^-$. Now we could inflate $A$ by an uncontrollable or unobservable mode $\lambda_0$ to still obtain the same transfer function. Since we can choose $\lambda_0$ arbitrarily we can also place it far in the right half-plane. The method from [GGO13] would now choose $\lambda_0$ as a starting value and directly return $\mathrm{Re}(\lambda_0)$ as spectral value set abscissa since there is no spectral value set component around $\lambda_0$. This could then lead to a wrong result, since $\mathrm{Re}(\lambda_0)$ can be much larger than the corresponding spectral value set abscissa obtained by using the original data. This problem is avoided by our method since obviously $\lambda_0$ is no pole of the transfer function. In fact, it is only a removable singularity. The further incorporation of only the most dominant poles by our method enables us to find good starting values that can be used to indeed find a global instead of a local maximizer of $\|G(\mathrm{i}\cdot)\|_2$. This is not guaranteed by [GGO13], since controllability and observability do not play a role for choosing the initial eigenvalue.

### 6.2.5 Numerical Results

**Test Setup** In this subsection we present some numerical results of our downloadable implementation[1]. To compute the (exponentially) dominant poles we use a (slightly modified) version of Rommes' MATLAB codes[2]. The data for the numerical examples was taken from [RM06a, MPR07, FRM08, CD02] and it can also be downloaded from Rommes' website or from the NICONET page[3]. As dominance measure we use (6.15). In Paragraph 6.2.5 we also analyze the behavior of the algorithm using the exponential dominance measure. The tolerance on the relative change of the iterates in Algorithm 6.1 is set to $\tau = 10^{-3}$. We also abort the iteration when the iterates start to cycle which typically happens when they are approaching zero. In Algorithm 6.2 we abort this iteration when the relative change of the iterates is below $10^{-6}$, i. e., if $\left|1 - \frac{\varepsilon_k}{\varepsilon_{k+1}}\right| < 10^{-6}$. To obtain $r_{\mathbb{C}}^{\mathrm{est}}(G)$, we compute 20 dominant poles using SAMDP (40 for the `peec` example and 30 for the `bips07_1693` system). For every further outer iteration we compute only 5 dominant poles (10 for the `peec` example). Note that we have to compute more dominant poles for `peec` and `bips07_1693` to ensure that the most dominant poles are really found. Both examples are particularly difficult, so it is necessary to deviate from the default values mentioned above.

**Test Results** In Table 6.1 we summarize the results of 33 numerical tests. The first 13 examples are standard or generalized state space systems whereas the other 20 ones are descriptor systems (with singular $E$). With $n_{\mathrm{out}}$ we denote the number of outer iterations, i. e., the number of steps needed by Algorithm 6.2 to find the root. By $n_{\mathrm{in}}$ we refer to the total number of inner iterations, i. e., the total number of steps needed by Algorithm 6.1.

For all tests, the correct value of $\|G\|_{\mathcal{H}_\infty}$ was found. In 30 tests the first outer iteration returned a positive value. However, for 3 of the tests (`M10PI_n1`, `M10PI_n`, `bips07_1693`), a negative value was returned and therefore, we have to try more dominant poles (one for each `M10PI_n1`, `M10PI_n` and four for `bips07_1693`) to converge to the correct initial value.

[1] http://www.mpi-magdeburg.mpg.de/mpcsc/software/infnorm/
[2] http://sites.google.com/site/rommes/software
[3] http://www.icm.tu-bs.de/NICONET/benchmodred.html

Table 6.1: Numerical results for $\mathcal{H}_\infty$-norm computation for 33 test examples

| # | example | $n$ | $m$ | $p$ | $\|G\|_{\mathcal{H}_\infty}$ | $\omega_*$ | $\alpha_{r^f_{\mathbb{C}}(G)}(G)$ | $n_{\text{out}}$ | $n_{\text{in}}$ | time in s |
|---|---|---|---|---|---|---|---|---|---|---|
| 1 | build | 48 | 1 | 1 | 5.27633e–03 | 5.20608e+00 | 4.3258e–15 | 3 | 13 | 1.54 |
| 2 | pde | 84 | 1 | 1 | 1.08358e+01 | 0 | -1.4872e–15 | 1 | 6 | 2.08 |
| 3 | CDplayer | 120 | 2 | 2 | 2.31982e+06 | 2.25682e+01 | 8.6683e–16 | 2 | 8 | 2.70 |
| 4 | iss | 270 | 3 | 3 | 1.15887e–01 | 7.75093e–01 | 7.1406e–17 | 1 | 3 | 2.69 |
| 5 | beam | 348 | 1 | 1 | 4.55487e+03 | 1.04575e–01 | -2.0176e–13 | 2 | 8 | 50.22 |
| 6 | S10PI_n1 | 528 | 1 | 1 | 3.97454e+00 | 7.53151e+03 | 3.6808e–11 | 2 | 7 | 1.78 |
| 7 | S20PI_n1 | 1028 | 1 | 1 | 3.44317e+00 | 7.61831e+03 | 4.0418e–13 | 3 | 13 | 4.22 |
| 8 | S40PI_n1 | 2028 | 1 | 1 | 3.34732e+00 | 6.95875e+03 | 1.9663e–12 | 2 | 8 | 4.77 |
| 9 | S80PI_n1 | 4028 | 1 | 1 | 3.37016e+00 | 6.96149e+03 | 1.3880e–12 | 2 | 8 | 10.50 |
| 10 | M10PI_n1 | 528 | 3 | 3 | 4.05662e+00 | 7.53181e+03 | 3.3271e–12 | 3 | 11 | 3.08 |
| 11 | M20PI_n1 | 1028 | 3 | 3 | 3.87260e+00 | 5.06412e+03 | 1.0967e–12 | 2 | 8 | 3.63 |
| 12 | M40PI_n1 | 2028 | 3 | 3 | 3.81767e+00 | 5.07107e+03 | 5.9180e–13 | 2 | 8 | 5.83 |
| 13 | M80PI_n1 | 4028 | 3 | 3 | 3.80375e+00 | 5.07279e+03 | 3.7053e–13 | 2 | 8 | 10.88 |
| 14 | peec | 480 | 1 | 1 | 3.52651e–01 | 5.46349e+00 | -1.8202e–10 | 2 | 7 | 20.80 |
| 15 | S10PI_n | 682 | 1 | 1 | 3.97454e+00 | 7.53151e+03 | 3.8015e–11 | 2 | 7 | 2.29 |
| 16 | S20PI_n | 1182 | 1 | 1 | 3.44317e+00 | 7.61831e+03 | -4.3232e–13 | 3 | 13 | 4.67 |
| 17 | S40PI_n | 2182 | 1 | 1 | 3.34732e+00 | 6.95875e+03 | 1.0369e–12 | 2 | 8 | 5.35 |
| 18 | S80PI_n | 4182 | 1 | 1 | 3.37016e+00 | 6.96149e+03 | -6.3127e–14 | 2 | 8 | 10.41 |
| 19 | M10PI_n | 682 | 3 | 3 | 4.05662e+00 | 7.53181e+03 | 1.4553e–12 | 3 | 11 | 3.56 |
| 20 | M20PI_n | 1182 | 3 | 3 | 3.87260e+00 | 5.06412e+03 | -1.8303e–13 | 2 | 8 | 4.03 |
| 21 | M40PI_n | 2182 | 3 | 3 | 3.81767e+00 | 5.07107e+03 | 8.4266e–13 | 2 | 8 | 6.11 |
| 22 | M80PI_n | 4182 | 3 | 3 | 3.80375e+00 | 5.07279e+03 | 1.5073e–12 | 2 | 8 | 11.28 |
| 23 | bips98_606 | 7135 | 4 | 4 | 2.01956e+02 | 3.81763e+00 | 2.6525e–14 | 3 | 13 | 35.43 |
| 24 | bips98_1142 | 9735 | 4 | 4 | 1.60427e+02 | 4.93005e+00 | 1.1867e–13 | 3 | 22 | 69.65 |
| 25 | bips98_1450 | 11305 | 4 | 4 | 1.97389e+02 | 5.64575e+00 | 1.2481e–14 | 3 | 16 | 61.65 |
| 26 | bips07_1693 | 13275 | 4 | 4 | 2.04168e+02 | 5.53766e+00 | 2.1761e–12 | 7 | 45 | 167.10 |
| 27 | bips07_1998 | 15066 | 4 | 4 | 1.97064e+02 | 6.39968e+00 | 4.8834e–13 | 2 | 20 | 102.11 |
| 28 | bips07_2476 | 16861 | 4 | 4 | 1.89579e+02 | 5.88971e+00 | 4.7231e–11 | 2 | 30 | 146.18 |
| 29 | bips07_3078 | 21128 | 4 | 4 | 2.09445e+02 | 5.55792e+00 | 1.5001e–13 | 3 | 13 | 91.05 |
| 30 | xingo_afonso_itaipu | 13250 | 1 | 1 | 4.05605e+00 | 1.09165e+00 | -4.1368e–14 | 2 | 7 | 39.24 |
| 31 | mimo8x8_system | 13309 | 8 | 8 | 5.34292e–02 | 1.03313e+00 | 2.4820e–12 | 2 | 14 | 78.47 |
| 32 | mimo28x28_system | 13251 | 28 | 28 | 1.18618e–01 | 1.07935e+00 | 6.1195e–14 | 2 | 9 | 85.36 |
| 33 | mimo46x46_system | 13250 | 46 | 46 | 2.05631e+02 | 1.07908e+00 | 4.8454e–14 | 2 | 8 | 115.91 |

Table 6.2: Convergence history for the M20PI_n example, $\lambda_{\text{dom}}$ denotes the dominant pole

| | $k$ | |
|---|---|---|
| | 1 | 2 |
| $\text{Re}(\lambda_{\text{dom}})$ | -6.7945e–02 | 1.6145e–08 |
| | 2.3140e–03 | 1.6256e–08 |
| | 3.0285e–03 | 1.6257e–08 |
| | 3.0355e–03 | — |
| | 3.0356e–03 | — |
| $\varepsilon_k$ | 2.58250e–01 | 2.58224e–01 |

For a more detailed impression of the behavior of the algorithm, Tables 6.2 and 6.3 summarize the convergence history for the M20PI_n and the bips07_2476 examples, listing each intermediate iterate for each iteration of the root-finding algorithm. We clearly observe a linear convergence of Algorithm 6.1 with different convergence rates depending on the problem, see also the last paragraph in Subsection 6.2.2.

In Figure 6.2 we also depict a set of pseudopole sets for the M20PI_n example and the iterates of the first iteration of Algorithm 6.1. Blue contours in Figure 6.2(a) correspond to small perturbation levels whereas yellow and red contours indicate areas which need larger perturbations to be reached by the perturbed poles. Therefore, poles that correspond to the blue contours are particularly controllable and observable. Intuitively, it is clear that the $\mathcal{H}_\infty$-norm will most likely be attained at frequencies close to these poles. This is also confirmed by having a look at Figure 6.3 that shows transfer function plots for the M20PI_n and bips07_2476 examples together with the computed $\mathcal{H}_\infty$-norms. For the M20PI_n example we observe that the correct norm value is computed even though there are lots of close-by peaks of similar height. We can also see that lots of peaks are introduced in the transfer function at frequencies that correspond to areas that are covered by the blue contours in Figure 6.2(a).

**Comparison of Dominance Measures** Since our exponential dominance measure is a heuristic method, we provide some results that emphasize that using (6.16) with larger values of $\beta$ as an alternative dominance measure can also be a good choice. In particular, we perform a study of the behavior of our algorithm for different values of $\beta$, but also with the standard measure given by (6.15). To ensure that the most exponentially dominant poles are really found in the inner iterations we increase $n_{\text{in}}$ to 20, but all other options are the same as before. In Table 6.4 we list those examples that fail when trying to compute the $\mathcal{H}_\infty$-norm using different dominance

Table 6.3: Convergence history for the `bips07_2476` example, $\lambda_{\mathrm{dom}}$ denotes the dominant pole

| | $k$ = 1 | $k$ = 2 |
|---|---|---|
| $\mathrm{Re}(\lambda_{\mathrm{dom}})$ | -8.1617e–02 | 2.1202e–08 |
| | -9.6637e–03 | 3.0116e–08 |
| | -2.6717e–03 | 3.4986e–08 |
| | -9.8184e–04 | 3.7652e–08 |
| | -3.9561e–04 | 3.9111e–08 |
| | -1.5362e–04 | 3.9910e–08 |
| | -4.3525e–05 | 4.0348e–08 |
| | 9.6185e–06 | 4.0587e–08 |
| | 3.6273e–05 | 4.0718e–08 |
| | 4.9989e–05 | 4.0790e–08 |
| | 5.7173e–05 | 4.0830e–08 |
| | 6.0984e–05 | — |
| | 6.3022e–05 | — |
| | 6.4120e–05 | — |
| | 6.4714e–05 | — |
| | 6.5036e–05 | — |
| | 6.5211e–05 | — |
| | 6.5307e–05 | — |
| | 6.5359e–05 | — |
| $\varepsilon_k$ | 5.27515e–03 | 5.27485e–03 |

measures. Unfortunately, for all tested values of $\beta$, some of the examples fail which is always due to a wrong selection of the poles with the respective dominance measure. Nevertheless, also all examples could be solved for some choice of $\beta$. For the `iss` example, we obtain $\|G\|_{\mathcal{H}_\infty} = 0.115885$ for $\beta = 10$ and $\beta = 100$ which is slightly smaller than the true value. This inaccuracy is also related to a different pole selection in the iteration for different dominance measures. However, the algorithm converges to the same peak in the transfer function.

In conclusion, it is important to choose a good value of $\beta$ *depending on the actual example* when applying (6.16). The table also makes clear that it is advisable to put a rather high weight on the real part of the poles which corresponds to a high value of $\beta$. However, it is still important to have an eye on the perturbability of the poles which is neglected if $\beta$ becomes too large. The standard measure (6.15) works particularly well for our examples since the computed dominant poles already give

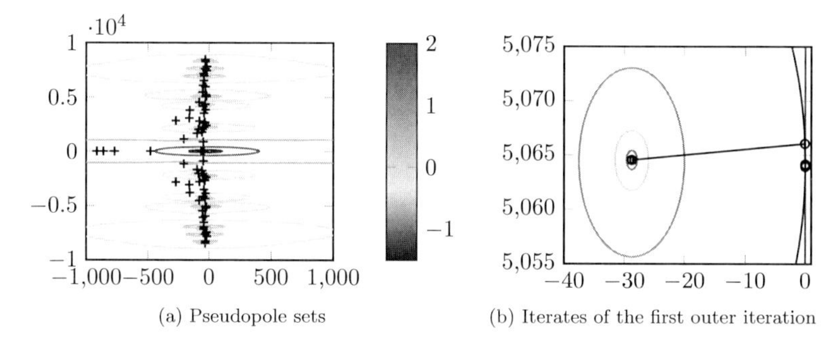


(a) Pseudopole sets

(b) Iterates of the first outer iteration

Figure 6.2: Pseudopole sets with the most dominant poles (black crosses) and first outer iteration for the M20PI_n example

Table 6.4: Comparison of different dominance measures for the 33 test examples from Table 6.1

| dominance measure | failed examples |
|---|---|
| (6.15) | — |
| (6.16) with $\beta = 1$ | iss, peec, bips98_606, bips98_1142, bips98_1450, bips07_1693, xingo_afonso_itaipu, mimo8x8_system, mimo46x46_system |
| (6.16) with $\beta = 10$ | (iss), peec |
| (6.16) with $\beta = 100$ | (iss), peec |
| (6.16) with $\beta = 1000$ | peec |
| (6.16) with $\beta = 10000$ | M10PI_n1, M10PI_n |

very good initial estimates of the actual value of the $\mathcal{H}_\infty$-norm. However, if this is not the case, the rightmost pseudopole in the first iteration might be "far" in the right half-plane which makes the exponential dominance measure more advisable.

**Comparison with Other Methods** This subsection gives a brief comparison with other methods, in particular those based on the solution of Hamiltonian and skew-Hamiltonian/Hamiltonian eigenvalue problems. These are implemented in the MATLAB Control Systems Toolbox as the function norm and in SLICOT[4] as the upcoming subroutine AB13HD. Both implementations rely on dense matrix algebra and therefore, they do not exploit the *sparse structure* of the involved matrices. We illustrate

[4] http://www.slicot.org/

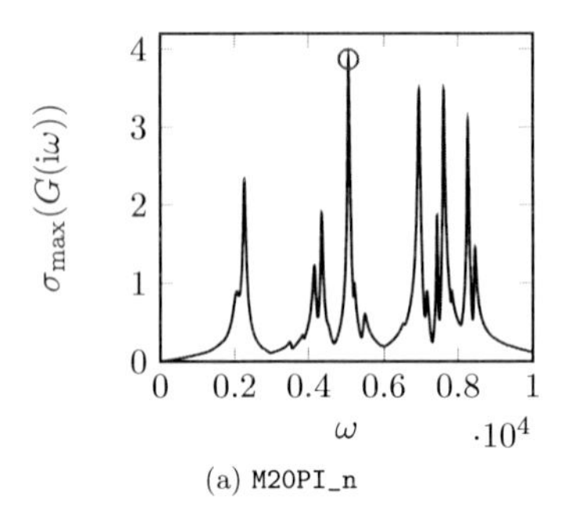

(a) M20PI_n

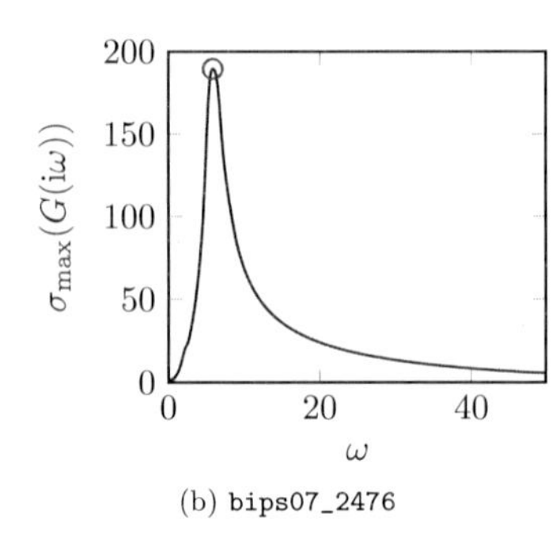

(b) bips07_2476

Figure 6.3: Transfer function plots for the M20PI_n and bips07_2476 test examples with computed $\mathcal{H}_\infty$-norm (red circle)

Table 6.5: Comparison of the pseudopole set method with standard approaches

| | computed $\mathcal{H}_\infty$-norm | | | time in s | | |
|---|---|---|---|---|---|---|
| example | norm | AB13HD | new method | norm | AB13HD | new method |
| M10PI_n | 4.05662 | 4.05662 | 4.05662 | 40.44 | 50.59 | 3.56 |
| M20PI_n | **9.92404** | 3.87260 | 3.87260 | 296.65 | 276.41 | 4.03 |
| M40PI_n | 3.81766 | 3.81767 | 3.81767 | 2322.66 | 1998.26 | 6.11 |

the speed-up of the method compared to the standard approaches in Table 6.5 for some smaller examples that can be solved by all available implementations. Both, the MATLAB function norm and SLICOT's AB13HD are used with a relative tolerance of $10^{-6}$ to obtain a similar accuracy as in Table 6.1. From the table we conclude that the new method is much faster than existing algorithms. A high speed-up can already be observed for small and medium-sized examples. Furthermore, for the M20PI_n example, the new method and the SLICOT solver AB13HD are able to compute the correct result whereas the MATLAB solver norm returns the wrong norm value without printing an error or warning message.

Furthermore, we compare our approach with the method in [GGO13] for standard state-space examples. The MATLAB code for [GGO13] is freely available[5]. As examples we use the state-space examples used in Table 6.1 (#1–5). Moreover, we generated test systems using state matrices $A$ from EigTool [Wri02] or the COMPl$_e$ib package [Lei04, LL04] with randomly generated sparse input and output matrices $B$

[5] http://cims.nyu.edu/~mert/software/hinfinity.html

and $C$ and random feedthrough matrix $D$. For a good comparison we choose similar termination tolerances in both methods. For our algorithm we use the same tolerances as above. For the method from [GGO13] we choose tolerances for the relative and absolute error of $10^{-6}$ and $10^{-12}$, respectively. As tolerance of the absolute error of the spectral value set abscissa computation we take $10^{-3}$. In order to apply our method, we first eliminate the matrix $D$ as in (6.3). Furthermore, we remark that we let the algorithm from [GGO13] run using *dense* arithmetics in case that $n \leq 500$ since its performance is much better in this situation. The computed $\mathcal{H}_\infty$-norms and the runtimes of the algorithms are listed in Table 6.6. First, we see that for both algorithms there are examples that could not be solved which is emphasized by boldface font in the table. For the `iss` system this is caused by convergence to a locally but not globally rightmost pseudopole whereas for `NN18` and `tolosa`, SAMDP or `eigs` from MATLAB fail, respectively. The `convdiff_fd` system is difficult in the sense that we need to increase the relative tolerance of the inner iteration of our algorithm to at least $10^{-4}$ to obtain an accurate result. On the other hand, even decreasing the tolerance of the absolute error of the spectral value set abscissa computation in the algorithm from [GGO13] does not improve the accuracy of the result in Table 6.6 which is obtained by using a tolerance of $10^{-3}$.

Concerning the runtimes there is no pattern observable. There are examples for which the method from [GGO13] performs better, but on the other hand there are many examples for which our method is faster. This is particularly true for those systems whose $\mathcal{H}_\infty$-norm is attained at $\omega = 0$. These are `dwave`, `HF1`, `markov`, and `olmstead`. When all dominant poles are purely real, our algorithm directly returns $\|G\|_{\mathcal{H}_\infty} = \|G(0)\|_2$ as result whereas this property is not explicitly checked in [GGO13] which performs the full iteration. This is a particular advantage of our method since by employing the residues of the poles we can get information about the shape of the transfer function which is not possible by looking only at the location of the eigenvalues as in [GGO13].

**Limitations of the Method** In this paragraph we explain limitations of our method. For illustrate these we use the `peec` example. We plot the transfer function of this example in Figure 6.4, once in the interval $(0, 10)$ and once for the interval $(5.2, 5.5)$, where the maximum peak is located. First of all we see, the transfer function has lots of peaks which is due to the high amount of poles close to the imaginary axis. We plot the eigenvalues of the corresponding pencil $sE - A \in \mathbb{R}[s]^{480 \times 480}$ in Figure 6.5, together with the ten most dominant poles. It is very hard for SAMDP to find the most dominant pole. In fact, if we only compute 20 dominant poles, the actually most dominant one is not found. This is only the case if we increase the number of wanted poles up to 40. Another problem is that the maximum peak is extremely thin and spiky (see Figure 6.4(b)). We do not even see it with the resolution used for

Table 6.6: Comparison of our method with [GGO13] for standard state-space systems, bold values indicate failure of the method

| # | example | $n$ | $m$ | $p$ | computed $\mathcal{H}_\infty$-norm [GGO13] | computed $\mathcal{H}_\infty$-norm new meth. | time in s [GGO13] | time in s new meth. |
|---|---|---|---|---|---|---|---|---|
| 1 | build | 48 | 1 | 1 | 5.27633e–03 | 5.27633e–03 | 0.93 | 1.54 |
| 2 | pde | 84 | 1 | 1 | 1.08358e+01 | 1.08358e+01 | 0.99 | 2.08 |
| 3 | CDplayer | 120 | 2 | 2 | 2.31982e+06 | 2.31982e+06 | 9.68 | 2.70 |
| 4 | iss | 270 | 3 | 3 | **1.20261e–02** | 1.15887e–01 | 5.53 | 2.69 |
| 5 | beam | 348 | 1 | 1 | 4.55487e+03 | 4.55487e+03 | 10.97 | 50.22 |
| 6 | convdiff_fd | 400 | 4 | 6 | **1.46898e+01** | 1.46968e+01 | 1247.09 | 92.21 |
| 7 | dwave | 2048 | 4 | 6 | 1.27874e+01 | 1.27874e+01 | 58.34 | 5.55 |
| 8 | HF1 | 130 | 1 | 2 | 8.42158e–01 | 8.42158e–01 | 1.15 | 0.34 |
| 9 | markov | 5050 | 4 | 6 | 3.37266e+00 | 3.37266e+00 | 98.70 | 14.56 |
| 10 | NN18 | 1006 | 1 | 2 | 9.29638e+00 | **fail** | 63.84 | — |
| 11 | olmstead | 500 | 4 | 6 | 2.79331e+00 | 2.79331e+00 | 23.17 | 1.68 |
| 12 | tolosa | 4000 | 4 | 6 | **fail** | 6.76382e+01 | — | 15.40 |

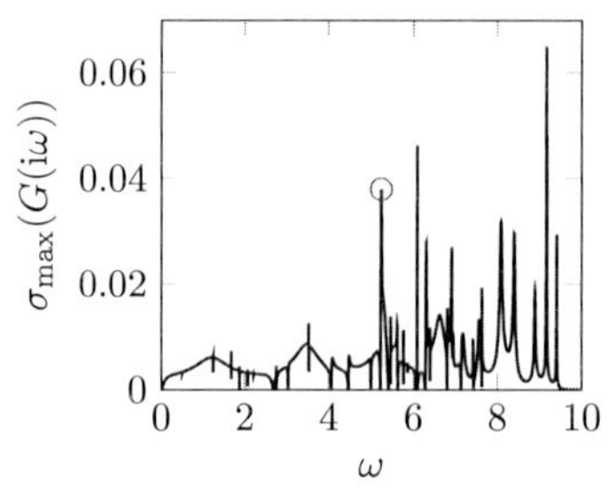


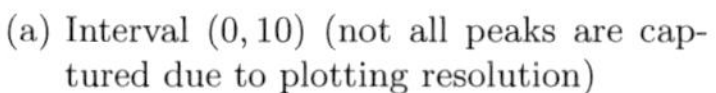

(a) Interval $(0, 10)$ (not all peaks are captured due to plotting resolution)

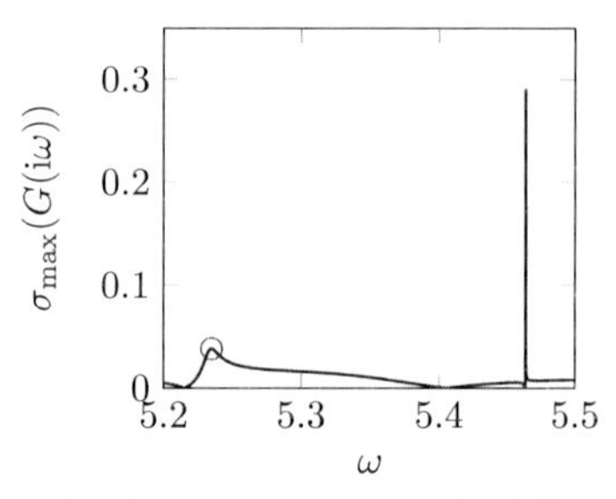


(b) Zoom into the interval $(5.2, 5.5)$

Figure 6.4: Transfer function plots for the peec example with computed (wrong) $\mathcal{H}_\infty$-norm

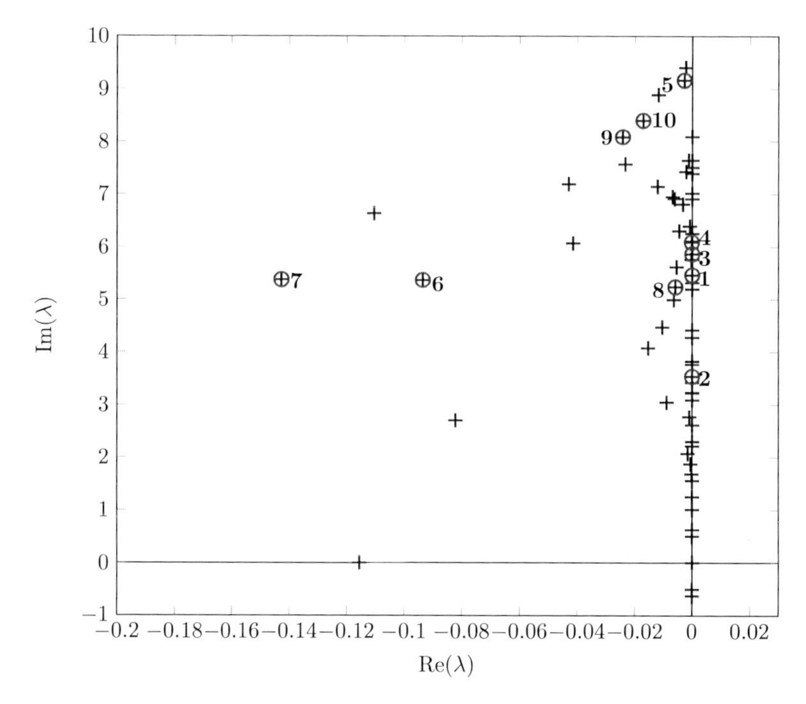


Figure 6.5: Eigenvalues of $sE - A$ for the peec example (blue crosses) and the 10 most dominant poles (red circles)

plotting Figure 6.4(a). To find it we would actually need very good approximations to the eigenvectors which are needed to construct the *optimal* rank-1 perturbation. This is only the case if we initialize the iteration with the most dominant pole and take an appropriate dominance measure to continue with the most dominant poles in each further iteration as done in Table 6.1. On the other hand, even if we start from a pole very close to this optimal pole, e. g., when we only compute 20 dominant poles and take, e. g., $\beta = 100$ in the beginning, our initial eigenvectors (the ones of the initial pencil) are not that good. Therefore, we only find the close-by peak which is much "wider" than the one we actually seek, see also the computed norm value in Figure 6.4. However, we can check if the computed norm value is larger than the 2-norms of the transfer function evaluated at the test frequencies. For this example, the test that we have implemented is not satisfied and therefore, we can at least return an error indicator. The main problem of our method is that the user has to provide certain parameters to the algorithm, for instance the number of dominant poles that should be computed in the beginning and every further iteration, since a high number can be necessary in order to find the most dominant pole.

Unfortunately, there also exist examples for which we observe an extremely slow convergence of SAMDP, in particular for those that have all real poles. Typically, the circuit examples from the `MNA` group of [CD02] are of this kind. However, for these problems, the $\mathcal{H}_\infty$-norm is attained at zero, i. e., $\|G\|_{\mathcal{H}_\infty} = \|G(0)\|_2$, see [RS11]. This behavior is captured by our algorithm by returning this value if the most dominant poles are all real, provided that SAMDP is able to compute the poles.

## 6.3 The Even Pencil Approach

In this section we turn back to the original approach using the even pencil (6.1) and discuss modifications to make this idea applicable to large-scale systems, too. In particular, we will exploit structured iterative eigensolvers to reduce the complexity of the eigenvalue computation (and thus, of the overall algorithm). For more details, especially on the implementation, we refer to [LV13].

### 6.3.1 Theoretical Preliminaries and Algorithm Outline

The following two theorems form the basis for the algorithm to calculate the $\mathcal{H}_\infty$-norm. The first relates the singular values of $G(\mathrm{i}\omega)$ with the purely imaginary eigenvalues of $s\mathcal{E} - \mathcal{A}(\gamma)$ in (6.1), see [BSV12a, BSV12b].

**Theorem 6.3.1.** *Let $(E, A, B, C, D) \in \Sigma_{n,m,p}$ with transfer function $G \in \mathcal{RH}_\infty^{p\times m}$ be given. Assume that $sE - A$ has no purely imaginary eigenvalues and let $\omega_0 \in \mathbb{R}$ be given. Then $\gamma$ is a singular value of $G(\mathrm{i}\omega_0)$ if and only if $\mathrm{i}\omega_0\mathcal{E} - \mathcal{A}(\gamma)$ is singular.*

The following theorem is a direct consequence of Theorem 6.3.1 [BSV12a].

**Theorem 6.3.2.** *Let $(E, A, B, C, D) \in \Sigma_{n,m,p}$ with transfer function $G \in \mathcal{RH}_\infty^{p\times m}$ be given. Assume that $sE - A$ has no purely imaginary eigenvalues and let $\gamma > \inf_{\omega\in\mathbb{R}} \sigma_{\max}(G(\mathrm{i}\omega))$. Then it holds that $\|G\|_{\mathcal{H}_\infty} \geq \gamma$ if and only if $s\mathcal{E} - \mathcal{A}(\gamma)$ has purely imaginary eigenvalues.*

Using the theorems from above we are able to state an iterative method, as presented in [BSV12a, BSV12b], to calculate the $\mathcal{H}_\infty$-norm. First, an initial $\gamma$ is calculated that is less than the $\mathcal{H}_\infty$-norm. In each step, a check is performed to verify whether the matrix pencil $s\mathcal{E} - \mathcal{A}(\gamma)$ has purely imaginary eigenvalues. If such eigenvalues are found, $\gamma$ is incremented and the process is repeated. Finally, when no imaginary eigenvalues are found, $\gamma$ serves as an upper bound for $\|G\|_{\mathcal{H}_\infty}$. By updating $\gamma$ in a certain way, this process converges monotonically at a quadratic rate, with

**Algorithm 6.3** Computation of the $\mathcal{H}_\infty$-norm using the even pencil approach

**Input:** System $(E, A, B, C, D) \in \Sigma_{n,m,p}$ with transfer function $G(s) \in \mathcal{RH}_\infty^{p\times m}$, relative tolerance $\varepsilon$.

**Output:** $\|G\|_{\mathcal{H}_\infty}$, optimal frequency $\omega_*$.

1: Compute an initial value $\gamma_{\mathrm{lb}} \leq \|G\|_{\mathcal{H}_\infty}$.
2: **repeat**
3: Set $\gamma := (1 + 2\varepsilon)\gamma_{\mathrm{lb}}$.
4: Compute some desired eigenvalues of the matrix pencil $s\mathcal{E} - \mathcal{A}(\gamma)$.
5: **if** no purely imaginary eigenvalues **then**
6: Set $\gamma_{\mathrm{ub}} = \gamma$.
7: Break.
8: **else**
9: Set $\{\mathrm{i}\omega_1, \ldots, \mathrm{i}\omega_k\}$ = purely imaginary eigenvalues of $s\mathcal{E} - \mathcal{A}(\gamma)$, with $\omega_j \geq 0$ for $j = 1, \ldots, k$.
10: Set $m_j = \sqrt{\omega_j \omega_{j+1}}$, $j = 1, \ldots, k-1$.
11: Compute the largest singular value of $G(\mathrm{i}m_j)$ for $j = 1, \ldots, k-1$.
12: Set $\gamma_{\mathrm{lb}} = \max_{1\leq j\leq k-1} \sigma_{\max}(G(\mathrm{i}m_j))$.
13: Set $\omega_* = \operatorname{argmax}_{m_1\leq m_j\leq m_{k-1}} \sigma_{\max}(G(\mathrm{i}m_j))$.
14: **end if**
15: **until** break
16: Set $\|G\|_{\mathcal{H}_\infty} = \frac{1}{2}(\gamma_{\mathrm{lb}} + \gamma_{\mathrm{ub}})$.

a relative error of at most the desired tolerance $\varepsilon$, as long as the arithmetic is exact. The complete process is summarized in Algorithm 6.3.

### 6.3.2 Structured Iterative Eigensolvers

In the large-scale setting there are two main problems:

a) How do we efficiently determine a good initial value of $\gamma_{\mathrm{lb}}$ in Step 1 of Algorithm 6.3?

b) How can we compute the desired eigenvalues of $s\mathcal{E} - \mathcal{A}(\gamma)$ in Step 4 of Algorithm 6.3?

First we discuss problem a): In general, the important factors that affect the computational costs of the algorithm are the total number of iterations and the number of purely imaginary eigenvalues of the matrix pencil $s\mathcal{E} - \mathcal{A}(\gamma)$ in each iteration Therefore, the initial choice of $\gamma_{\mathrm{lb}}$ is an important issue, since it can have a significant impact on both the total number of iterations and the number of imaginary eigenvalues. Here we choose

$$\gamma_{\text{lb}} := \max\left\{\sigma_{\max}(G(0)), \sigma_{\max}(G(\mathrm{i}\omega_{\text{p}})), \lim_{\omega\to\infty}\sigma_{\max}(G(\mathrm{i}\omega))\right\},$$

where $\omega_{\text{p}}$ is a test frequency that gives the maximum singular value [BSV12a]. To determine the test frequencies we make use of the dominant poles of $G(s)$. This is reasonable strategy due to the relation (6.14). Therefore, let $\lambda_j$ for $j = 1, \ldots, \ell$ be some computed dominant poles. Then we define the test frequency $\omega_j$ as

$$\omega_j = \operatorname{Im}(\lambda_j),$$

and thus we have

$$\omega_{\text{p}} := \operatorname{argmax}_{\omega_1 \le \omega_j \le \omega_\ell} \sigma_{\max}(G(\mathrm{i}\omega_j)).$$

Several dominant poles are used to potentially increase the value of $\gamma_{\text{lb}}$.

Next we discuss problem b): Since the standard algorithm [BSV12a, BSV12b] uses full structured factorizations of skew-Hamiltonian/Hamiltonian pencils related to $s\mathcal{E} - \mathcal{A}(\gamma)$ and relies on dense matrix algebra, it has a cubic complexity and therefore, it is infeasible in the large-scale setting.

Instead, we propose to use the *even IRA* algorithm [MSS12] which can compute several eigenvalues *close to a prespecified shift* (and, if desired, the associated eigenvectors) of large and sparse even pencils.

In order to solve problems of this type, we limit ourselves to the case where the structure of $\mathcal{E}$ and $\mathcal{A}(\gamma)$ allows for the use of sparse direct LU factorizations of $\sigma\mathcal{E} - \mathcal{A}(\gamma)$ for some shift $\sigma$. In particular, the even IRA algorithm is a structure-preserving method based on Krylov subspaces with implicit restarts [MSS12]. A prespecified number of eigenvalues is calculated in a neighborhood of the shift $\sigma$, which is allowed to be either real or purely imaginary. The algorithm then implicitly solves a related eigenvalue problem of the form $\mathcal{K}(\gamma)x = \theta x$, with

$$\mathcal{K}(\gamma) := (\mathcal{A}(\gamma) + \sigma\mathcal{E})^{-1}\mathcal{E}(\mathcal{A}(\gamma) - \sigma\mathcal{E})^{-1}\mathcal{E}.$$

Then an eigenvalue pair $(\lambda, -\lambda)$ can be easily extracted from $\theta$ by a simple transformation. Since it is desirable for the shifts to be as near as possible to the calculated eigenvalues, the ones used in this chapter are the midpoints of the imaginary parts of the eigenvalues of $s\mathcal{E} - \mathcal{A}(\gamma)$ calculated in the previous iteration, with a slight offset which is explained in Subsection 6.3.3. In the case of the first iteration, when such eigenvalues have yet to be calculated, the imaginary parts of some of the dominant poles are used instead. In order to improve the accuracy of the eigenvalue calculation method, and to ensure that all desired purely imaginary eigenvalues of the matrix pencil are found, often multiple shifts are used in a loop.

Table 6.7: Most important design parameters of Algorithm 6.3

| param. | description | default |
|---|---|---|
| $\varepsilon$ | desired relative accuracy for the $\mathcal{H}_\infty$-norm | 1e–06 |
| $n_{\text{dom}}$ | number of dominant poles computed in initial stage | 20 |
| $\tau_{\text{dom}}$ | relative cutoff value for the dominance of the poles to determine the initial eigenvalues | 0.5 |
| $\tau_{\text{shift}}$ | minimum relative distance between two subsequent shifts | 0.01 |
| $m_{\text{max}}$ | maximum search space dimension even IRA | 8 |
| $r_{\text{max}}$ | maximum number of restarts for even IRA | 30 |
| $n_{\text{eig}}$ | number of eigenvalues calculated per shift in even IRA | 4 |
| $\tau_{\text{eig}}$ | tolerance on the eigenvalue residual for even IRA | 1e–06 |
| $\tau_{\text{sweep}}$ | relative eigenvalue distance for frequency sweep | 1e–05 |
| $\delta$ | relative shift displacement | 5e–04 |

### 6.3.3 Implementation Details

In this subsection, we outline some of the details regarding the implementation of Algorithm 6.3[6].

In our implementation we use a large amount of design parameters which are summarized in Table 6.7.

The parameter $n_{\text{dom}}$ specifies the number of dominant poles returned by SAMDP. Since the most dominant poles do not always correlate to the test frequencies that provides the highest singular value, this parameter is not trivially 1. Furthermore, for some examples this number must be much higher before the pole with the highest dominance is actually found; as such, the default value is set to 20, although this must be increased for some examples, see also the last paragraph in Subsection 6.2.5.

To compute the purely imaginary eigenvalues of $s\mathcal{E} - \mathcal{A}(\gamma)$ in the initial iteration we use the imaginary parts of the dominant poles as shifts. When we take a large number of shifts we observe that some eigenvalues occur repeatedly. Therefore, we only use a limited amount of them, determined by the parameter $\tau_{\text{dom}}$. Let $\lambda_k$, $k = 1, \ldots, \ell$, be the dominant poles of $G(s)$. With $R_k$ as in (6.13) we define the *dominance* by

$$\text{dom}(\lambda_k) := \left\|R_k\right\|_2 / \operatorname{Re}(\lambda_k).$$

Then we determine the shifts only by those dominant poles $\lambda_j$ that satisfy

$$\text{dom}(\lambda_j) > \tau_{\text{dom}} \max_{1\le k\le \ell} \text{dom}(\lambda_k).$$

[6]downloadable from `http://www.mpi-magdeburg.mpg.de/mpcsc/software/infnorm`

Moreover, we reduce the number of shifts, if some of them are close together. Therefore, let the shifts $\mathrm{i}\omega_1, \ldots, \mathrm{i}\omega_\ell \in \mathrm{i}\mathbb{R}$ be given in increasing order. If

$$|\omega_{j+1} - \omega_j| < \tau_{\text{shift}}\omega_\ell,$$

then we replace the shifts $\mathrm{i}\omega_j$ and $\mathrm{i}\omega_{j+1}$ by a single shift. In the code this is also done for multiple close-by shifts.

The even IRA solver, while providing a fast and relatively robust method for the computation of the eigenvalues of even matrix pencils, can return inaccurate eigenvalues under certain circumstances. Additional security measures are put into the code to alleviate these issues.

A problem with the even IRA solver arises in the case of close proximity of the given shift and one of the eigenvalues of the matrix pencil to be calculated. In such a case, the eigenvalue close to the shift is calculated correctly, however all *other* eigenvalues have a high degree of inaccuracy. This is thought to be caused by resulting ill-conditioned matrices used in the even IRA solver [Sch13]. This problem occurs fairly often, when taking the eigenvalues from the previous iteration as shifts. Therefore, we perform a slight relative offset by the parameter $\delta$ – if the shifts are displaced by a small relative amount, all inaccuracies in the eigenvalue calculation associated with this problem are eliminated.

Another possible scenario involves a transfer function where the slope of a singular value plot around the optimal frequency $\omega_*$ is very steep – in other words, the $\mathcal{H}_\infty$-norm is achieved at the tip of a very thin spike on the singular value vs. frequency graph, see for instance Figure 6.4(a). In this case, the purely imaginary eigenvalues of the matrix pencil $s\mathcal{E} - \mathcal{A}(\gamma)$ will be very close together in the later stages of computation. Even IRA is not able to accurately calculate eigenvalues that are extremely close together which can result in a decreasing value of $\gamma$ in the next steps of the iteration. To prevent this, if the previously calculated eigenvalues have a relative distance less than $\tau_{\text{sweep}}$, a frequency sweep is performed around the imaginary parts of the eigenvalues to find the $\mathcal{H}_\infty$-norm. Since this process is relatively computationally inexpensive, many points can be evaluated over a large range (relative to the distance between the eigenvalues), so the final norm is usually very accurate. The drawback to this approach is, of course, that the accuracy of the resulting $\mathcal{H}_\infty$-norm is not *guaranteed*. However, since the eigenvalues close to the location of the frequency sweep are so close to the optimal frequency, this method usually still gives a good approximation.

Finally, in some rare occasions, even IRA does not find any purely imaginary eigenvalues of the matrix pencil, even if the provided $\gamma$ is smaller than the known $\mathcal{H}_\infty$-norm calculated using other methods. Often, the reason for this problem is that the shifts used for the even IRA solver are too far from the imaginary eigenvalues, for instance if the range in which shifts are averaged, is too big. In such a case, adjusting

$\tau_{\text{dom}}$ and $\tau_{\text{shift}}$ such that more shifts are obtained, usually solves the problem. Another possible reason is that the tolerance of the eigenvalue residuals given by $\tau_{\text{eig}}$ is too big. Then the code of even IRA does not return the inaccurate eigenvalues. To obtain a correct result, $\tau_{\text{eig}}$ must be further decreased.

Summarizing, a major disadvantage of our method is the fact that the user must supply good values for the design parameters. Even if the default values work well for most examples, for a few tests we have to change them to get good results.

### 6.3.4 Numerical Results

**Test Setup** For our MATLAB implementation we use the code of SAMDP by Joost Rommes and the MATLAB implementation for even IRA from Christian Schröder[7]. For all examples we use the default values as in Table 6.7, except for `beam` ($\tau_{\text{eig}} = 1\text{e}-5$), `peec` ($n_{\text{dom}} = 40$), `bips07_1693` ($n_{\text{dom}} = 30$), and `bips07_1998` ($\tau_{\text{dom}} = 0.2$) which are particularly difficult.

**Test Results and Comparison with the Pseudopole Set Approach** In this paragraph we analyze computational results of the new approach and compare them with those obtained by the pseudopole set approach from Section 6.2. The results for 33 test examples are summarized in Table 6.8. Using the even pencil approach, the correct value of $\|G\|_{\mathcal{H}_\infty}$ was found for all cases. For a few examples (`build`, `beam`, `M80PI_n1`, `M80PI_n`, `mimo8x8_system`), the result differs by only 1e–6 for the different methods, which is in the range of desired accuracy. Only for the `peec` example, the difference between the calculated $\mathcal{H}_\infty$-norm values is significant. The reason is the very thin spike at which the $\mathcal{H}_\infty$-norm is attained, see also Figure 6.4(a). In the run using the even pencil approach we actually did not find purely imaginary eigenvalues in the first iteration and therefore, we can only take $\|G\|_{\mathcal{H}_\infty} = \sigma_{\max}(G(\mathrm{i}\omega_{\mathrm{p}}))$ with the optimal test frequency $\omega_{\mathrm{p}}$ as defined above. When decreasing $\tau_{\text{eig}}$ we get purely imaginary eigenvalues, which are so close together such that a frequency sweep is performed. This leads to a similar result.

In terms of the runtime, the method based on even matrix pencils performs remarkably better than the pseudopole set method. In fact, every single numerical example resulted in a decreased runtime. For almost all tests the runtime was reduced to 50% or less, especially for some larger examples we can have a speedup by a factor of 5. Only in the `beam` example, where the dominant pole calculation is the limiting factor, the speedup is much smaller. To demonstrate the behavior of the method, we show the intermediate iterates for the `bips07_3078` example in Figure 6.6. This illustration indicates how quickly the algorithm converges – in fact, all 33 examples took three iterations or fewer.

[7] `http://www.math.tu-berlin.de/fachgebiete_ag_modnumdiff/fg_numerische_mathematik/v-menue/mitarbeiter/christianschroeder/software/even_ira/`

Table 6.8: Comparison of the pseudopole set with the even pencil approach for 33 test examples

| # | example | n | m | p | computed $\mathcal{H}_\infty$-norm pseudop. | computed $\mathcal{H}_\infty$-norm even p. | optimal frequency $\omega_{opt}$ pseudop. | optimal frequency $\omega_{opt}$ even p. | time in s pseudop. | time in s even p. |
|---|---|---|---|---|---|---|---|---|---|---|
| 1 | build | 48 | 1 | 1 | 5.27633e–03 | 5.27634e–03 | 5.20608e+00 | 5.20584e+00 | 1.54 | 0.54 |
| 2 | pde | 84 | 1 | 1 | 1.08358e+01 | 1.08358e+01 | 0.00000e+00 | 0.00000e+00 | 2.08 | 0.61 |
| 3 | CDplayer | 120 | 2 | 2 | 2.31982e+06 | 2.31982e+06 | 2.25682e+01 | 2.25682e+01 | 2.70 | 0.54 |
| 4 | iss | 270 | 3 | 3 | 1.15887e–01 | 1.15887e–01 | 7.75093e–01 | 7.75089e–01 | 2.69 | 0.61 |
| 5 | beam | 348 | 1 | 1 | 4.55487e+03 | 4.55488e+03 | 1.04575e–01 | 1.04573e-01 | 50.22 | 38.15 |
| 6 | S10PI_n1 | 528 | 1 | 1 | 3.97454e+00 | 3.97454e+00 | 7.53151e+03 | 7.53152e+03 | 1.78 | 0.79 |
| 7 | S20PI_n1 | 1028 | 1 | 1 | 3.44317e+00 | 3.44317e+00 | 7.61831e+03 | 7.61835e+03 | 4.22 | 1.33 |
| 8 | S40PI_n1 | 2028 | 1 | 1 | 3.34732e+00 | 3.34732e+00 | 6.95875e+03 | 6.95873e+03 | 4.77 | 2.12 |
| 9 | S80PI_n1 | 4028 | 1 | 1 | 3.37016e+00 | 3.37016e+00 | 6.96149e+03 | 6.96147e+03 | 10.50 | 4.80 |
| 10 | M10PI_n1 | 528 | 3 | 3 | 4.05662e+00 | 4.05662e+00 | 7.53181e+03 | 7.53182e+03 | 3.08 | 1.29 |
| 11 | M20PI_n1 | 1028 | 3 | 3 | 3.87260e+00 | 3.87260e+00 | 5.06412e+03 | 5.06412e+03 | 3.63 | 1.62 |
| 12 | M40PI_n1 | 2028 | 3 | 3 | 3.81767e+00 | 3.81767e+00 | 5.07107e+03 | 5.07107e+03 | 5.83 | 2.83 |
| 13 | M80PI_n1 | 4028 | 3 | 3 | 3.80375e+00 | 3.80376e+00 | 5.07279e+03 | 5.07279e+03 | 10.88 | 5.23 |
| 14 | peec | 480 | 1 | 1 | 3.52651e–01 | 3.52541e–01 | 5.46349e+00 | 5.46349e+00 | 20.80 | 9.31 |
| 15 | S10PI_n | 682 | 1 | 1 | 3.97454e+00 | 3.97454e+00 | 7.53151e+03 | 7.53152e+03 | 2.29 | 1.03 |
| 16 | S20PI_n | 1182 | 1 | 1 | 3.44317e+00 | 3.44317e+00 | 7.61831e+03 | 7.61835e+03 | 4.67 | 1.53 |
| 17 | S40PI_n | 2182 | 1 | 1 | 3.34732e+00 | 3.34732e+00 | 6.95875e+03 | 6.95873e+03 | 5.35 | 2.47 |
| 18 | S80PI_n | 4182 | 1 | 1 | 3.37016e+00 | 3.37016e+00 | 6.96149e+03 | 6.96147e+03 | 10.41 | 4.51 |
| 19 | M10PI_n | 682 | 3 | 3 | 4.05662e+00 | 4.05662e+00 | 7.53181e+03 | 7.53182e+03 | 3.56 | 1.36 |
| 20 | M20PI_n | 1182 | 3 | 3 | 3.87260e+00 | 3.87260e+00 | 5.06412e+03 | 5.06412e+03 | 4.03 | 1.73 |
| 21 | M40PI_n | 2182 | 3 | 3 | 3.81767e+00 | 3.81767e+00 | 5.07107e+03 | 5.07107e+03 | 6.11 | 2.85 |
| 22 | M80PI_n | 4182 | 3 | 3 | 3.80375e+00 | 3.80376e+00 | 5.07279e+03 | 5.07279e+03 | 11.28 | 5.19 |
| 23 | bips98_606 | 7135 | 4 | 4 | 2.01956e+02 | 2.01956e+02 | 3.81763e+00 | 3.81763e+00 | 35.43 | 14.43 |
| 24 | bips98_1142 | 9735 | 4 | 4 | 1.60427e+02 | 1.60427e+02 | 4.93005e+00 | 4.93005e+00 | 69.65 | 16.37 |
| 25 | bips98_1450 | 11305 | 4 | 4 | 1.97389e+02 | 1.97389e+02 | 5.64575e+00 | 5.64650e+00 | 61.65 | 17.91 |
| 26 | bips07_1693 | 13275 | 4 | 4 | 2.04168e+02 | 2.04168e+02 | 5.53766e+00 | 5.53767e+00 | 167.10 | 31.98 |
| 27 | bips07_1998 | 15066 | 4 | 4 | 1.97064e+02 | 1.97064e+02 | 6.39968e+00 | 6.39879e+00 | 102.11 | 29.99 |
| 28 | bips07_2476 | 16861 | 4 | 4 | 1.89579e+02 | 1.89579e+02 | 5.88971e+00 | 5.89023e+00 | 146.18 | 31.62 |
| 29 | bips07_3078 | 21128 | 4 | 4 | 2.09445e+02 | 2.09445e+02 | 5.55792e+00 | 5.55839e+00 | 91.05 | 34.73 |
| 30 | xingo_afonso_itaipu | 13250 | 1 | 1 | 4.05605e+00 | 4.05605e+00 | 1.09165e+00 | 1.09165e+00 | 39.24 | 16.80 |
| 31 | mimo8x8_system | 13309 | 8 | 8 | 5.34292e–02 | 5.34293e–02 | 1.03313e+00 | 1.03308e+00 | 78.47 | 23.25 |
| 32 | mimo28x28_system | 13251 | 28 | 28 | 1.18618e–01 | 1.18618e–01 | 1.07935e+00 | 1.07935e+00 | 85.36 | 35.45 |
| 33 | mimo46x46_system | 13250 | 46 | 46 | 2.05631e+02 | 2.05631e+02 | 1.07908e+00 | 1.07908e+00 | 115.91 | 49.13 |

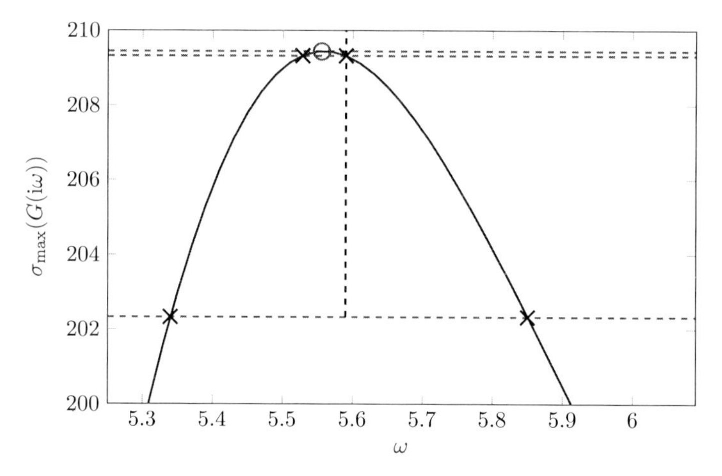


Figure 6.6: Illustration of the convergence of Algorithm 6.3 for the `bips07_3078` example, intermediate values of $\gamma$ are depicted by dotted red lines and the purely imaginary eigenvalues by black crosses

## 6.4 Conclusions and Future Research Perspectives

In this chapter we have introduced and compared two methods to compute the $\mathcal{H}_\infty$-norm of large-scale descriptor systems.

The first method uses the relationship between the $\mathcal{H}_\infty$-norm and the complex $\mathcal{H}_\infty$-radius of a matrix or a pencil. Based on the method introduced in [GO11], the algorithm computes a sequence of pseudopole set abscissae. This is done by computing an optimal rank-1 perturbation of the pencil $sE - A$ such that one of the eigenvalues of the perturbed pencil converges to the rightmost pseudopole of the transfer function. Finally, the complex $\mathcal{H}_\infty$-radius is computed by driving the pseudopole set abscissa to zero.

The second method is based on even matrix pencils similarly as in [BSV12a] and uses structured iterative eigensolvers to make the algorithm applicable to large-scale problems. In particular, it uses the even IRA algorithm to calculate the eigenvalues of the even matrix pencil, and employs the dominant pole algorithm to find appropriate test frequencies and to get good shifts for even IRA. The calculated $\mathcal{H}_\infty$-norm values were identical (up to the desired tolerance) or very close to those calculated by the pseudopole set method, and were accompanied by a significant drop in runtime.

In following we will briefly compare the features, i. e., advantages and disadvantages of the implementations of both methods:

a) The pseudopole set approach

   + needs a smaller number of user-defined parameters and is easier to use;
   - is slower than the even pencil approach;
   - leads to perturbed pencils that are usually complex and thus we need complex arithmetics;
   - does not guarantee global optimization.

b) The even pencil approach

   + is faster than the pseudopole set approach;
   + works under less restrictive conditions than the pseudopole set method; in particular we can drop the requirement of sparsity of $B$ and $C$ and the method can also be used in the more general context of $\mathcal{L}_\infty$-norm computations;
   - needs a larger number of user-defined design parameters;
   - does not guarantee global optimization.

As mentioned above, both methods suffer from the fact, the we cannot guarantee global optimization. For the pseudopole set method this would mean that we have to ensure convergence to a *globally* rightmost point in the pseudopole sets. For the even pencil approach this would mean that we, e. g., have to ensure that we compute *all* purely imaginary eigenvalues of a pencil $s\mathcal{E} - \mathcal{A}(\gamma)$. Both are challenging tasks for future research.

A further idea for the improvement of the pseudopole set method is the *implicit* handling of the perturbed pencils $sE - \widehat{A}$. All we need for the dominant pole algorithm is actually a method to solve shifted linear systems for matrices of the form $\sigma E - \widehat{A}$. This however, can be done by applying the Sherman-Morrison-Woodbury identity which only requires the solution of shifted linear systems of the form $\sigma E - A$ which avoids the explicit construction of the perturbed pencils. This would also allow to deal with *dense* matrices $B$ and $C$. Moreover, we could think of using adaptive tolerances, i. e., we begin with rather large termination tolerances and decrease them throughout the process. This promises to speedup the algorithm drastically.

Another recent approach to compute the $\mathcal{H}_\infty$-norm has been presented in [FSD14]. Furthermore, the method of [GGO13] has been recently improved in [MO14, Mit14]. In the future, these new algorithms should also be compared to our methods.

The pseudopole set algorithm can also be seen as a basis to solve certain related problems. For instance, one could think of a discrete-time version of the algorithm presented here. This has already been done in [GGO13] via optimizing the structured pseudospectral radius. However, the authors of [GGO13] have not used a method that takes controllability or observability of the poles into account. It is an open problem to develop and implement a dominant pole algorithm for the discrete-time case to employ similar ideas as presented here. Another future research direction could be on the analysis of *real* $\mathcal{H}_\infty$-radii. Most likely, this will also lead to an optimization procedure over rank-2 perturbations as presented in [GL13].

Moreover, one could consider the behavior of structured matrices and pencils under structured perturbations. As shown in [GKL14b, GKL14a] there are also low-rank dynamics for Hamiltonian and symplectic perturbations of Hamiltonian and symplectic matrices. In this manner one could consider the more general concepts of even and palindromic matrix pencils.

# 7 Summary and Outlook

In this thesis we have advanced the linear-quadratic control theory for DAEs as well as their robustness analysis. The contributions of this thesis are of both theoretical and numerical nature.

In Chapter 3 we have analyzed the linear-quadratic optimal control problem for DAEs in detail. We have started with a new version of the Kalman-Yakubovich-Popov lemma for DAEs that relates the positive semidefiniteness of a Popov function on the imaginary axis to the solvability of a descriptor KYP inequality. Based on this result we have developed the descriptor Lur'e equation that generalizes both the concepts of algebraic Riccati equations and Lur'e equations, especially in the context of singular control problems. We have shown that the solutions of this new matrix equation define the rank-minimizing solutions of the descriptor KYP inequality. In order to study the solution theory of descriptor Lur'e equations we have taken certain even matrix pencils into account. In particular, we have given solvability conditions in terms of the spectrum of this matrix pencil and constructed its solutions via the deflating subspaces. Moreover, we have shown that under the conditions of strong stabilizability and strong anti-stabilizability, there exist stabilizing and anti-stabilizing solutions of the descriptor Lur'e equation, respectively. These are shown to define also extremal solutions of the descriptor KYP inequality in a certain sense. In this work we have also studied conditions that allow for nonpositive solutions of the descriptor KYP inequality.

In the second part of Chapter 3 we have considered manifold applications of our new theory. A classical application is linear-quadratic optimal control. We have shown that a linear-quadratic optimal control problem is feasible if and only if an associated descriptor Lur'e equation has a stabilizing solution. Moreover, we have clarified the relation to the zero dynamics of the closed-loop system. Furthermore, we have studied linear-quadratic control problems with free terminal condition. For strongly controllable systems, it has been shown that the existence of a nonpositive solution of the descriptor KYP inequality is equivalent to the feasibility of such an optimal control problem. Resulting from these observations we have extended the theory for dissipative and cyclo-dissipative systems. As special cases we have obtained new versions of the positive real and bounded real lemma, respectively. Moreover, we have shown how we can obtain normalized coprime factorizations and inner-outer factorizations of rational matrices by the solution of descriptor Lur'e equations.

In Chapter 4 we have turned to the concepts of systems with ccw I/O dynamics and negative imaginary transfer functions which are strongly related to passivity and positive realness. In the flavor of the results of Chapter 3 we have given equivalent conditions for negative imaginariness in terms of the spectrum of an even matrix pencil. Moreover, we have developed an algorithm for the enforcement of negative imaginariness by perturbations of the model in case this structure has been lost during the modeling process. The algorithm is based on the perturbation of the purely imaginary eigenvalues of an associated skew-Hamiltonian/Hamiltonian pencil off the imaginary axis. This process requires the corresponding eigenvectors that we have computed in a new structure-exploiting manner.

In Chapter 5 we have considered the influence of perturbations on the cyclo-dissipativity of a system. In particular, we have developed an algorithm to compute the cyclo-dissipativity radius, i. e., the distance of a cyclo-dissipative system to the set of non-cyclo-dissipative systems. Moreover, we have given conditions under which the cyclo-dissipativity radius is equivalent to the dissipativity radius. The whole analysis is based on the structured perturbation theory of even matrix pencils. A big challenge in this context has been the analysis of the influence of perturbations on the singular part and the defective infinite eigenvalues which has needed special treatment. Finally, we have solved a sequence of eigenvalue optimization problems to compute the cyclo-dissipativity radius.

Finally, in Chapter 6 we have developed two algorithm to calculate the $\mathcal{H}_\infty$-norm of large-scale descriptor systems. The first approach is based on a relationship between the complex $\mathcal{H}_\infty$-radius of a transfer function $G \in \mathcal{RH}_\infty^{p\times m}$ and its $\mathcal{H}_\infty$-norm. We have developed a fast iteration that computes the rightmost point of a given $\varepsilon$-pseudopole set for $G(s)$. Then we have applied Newton's method to find the value of $\varepsilon$, for which this rightmost point is on the imaginary axis in order to obtain the complex $\mathcal{H}_\infty$-radius. To improve the behavior with respect to global optimization we have used the dominant poles of $G(s)$. The second approach goes back to the original method for $\mathcal{H}_\infty$-norm computation which is based on an optimization process over even matrix pencils. To make this method applicable to large-scale problems, we have used structured iterative eigensolvers such as the dominant pole algorithm and the even IRA algorithm. Numerical examples have demonstrated that both methods work well, even for rather difficult examples.

Future possible research topics are manifold. For instance, in this thesis we have only considered systems where $sE-A \in \mathbb{R}[s]^{n\times n}$ is a square and regular matrix pencil. However, many of the concepts considered here, are also extendable to systems with nonsquare or singular pencils $sE-A$ or even to higher-order systems. In this context the behavior approach of systems theory comes into play. Some advances into this direction have already been obtained in [Brü11b]. Furthermore, in Chapter 3 we have constantly made the assumption of impulse controllability. In [RRV14] it is shown that the descriptor KYP inequality and the descriptor Lur'e equation can also be extended to the non-impulse controllable case. We strongly believe that also the concepts of linear-quadratic optimal control, (cyclo-)dissipativity, and rational matrix factorizations can be considered in the non-impulse controllable context. Moreover, numerical methods for the solution of descriptor Lur'e equations should be developed.

Concerning the results of Chapters 5 and 6 we have only considered *complex* perturbations, even if the problem is real. A possible future research topic is the extension to real perturbations. In the context of the computation of $\mathcal{H}_\infty$-radii this seems to be rather straightforward, since a related method for unstructured real stability radii of matrices already exists [GL13]. However, in the context of the cyclo-dissipativity radii, this is rather involved. A possible way out is the application of structured pseudospectra for even matrix pencils. However, the block structure of the perturbation makes this a nontrivial issue. Luckily, the solution of this problem would also potentially allow for an application to large-scale systems.

# Bibliography


[And66] B. D. O. Anderson. Algebraic description of bounded real matrices. *Electronics Lett.*, 2(12):464–465, 1966.

[And67] B. D. O. Anderson. A system theory criterion for positive real matrices. *SIAM J. Control*, 5:171–182, 1967.

[And69] B. D. O. Anderson. The inverse problem of stationary covariance generation. *J. Stat. Phys.*, 1(1):133–147, 1969.

[AV73] B. D. O. Anderson and S. Vongpanitlerd. *Network Analysis and Synthesis – A Modern Systems Theory Approach.* Prentice-Hall, Englewood Cliffs, NJ, 1973.

[Ang06] D. Angeli. Systems with counterclockwise input-output dynamics. *IEEE Trans. Automat. Control*, 51(7):1130–1143, 2006.

[Ant05] A. C. Antoulas. *Approximation of Large-Scale Dynamical Systems.* SIAM Publications, Philadelphia, PA, 2005.

[Bac03] A. Backes. Optimale Steuerung der linearen DAE im Fall Index 2. Preprint, Humboldt-Universität zu Berlin, Institut für Mathematik, 2003. Available from `http://webdoc.sub.gwdg.de/ebook/serien/e/preprint_HUB/P-03-04.ps`.

[Bac06] A. Backes. *Extremalbedingungen für Optimierungs-Probleme mit Algebro-Differentialgleichungen.* Logos-Verlag, Berlin, 2006.

[BBSW13] E. Bänsch, P. Benner, J. Saak, and H. K. Weichelt. Riccati-based boundary feedback stabilization of incompressible Navier-Stokes flow. Preprint SPP1253-154, DFG-Schwerpunktprogramm 1253, 2013. Available from `http://www.am.uni-erlangen.de/home/spp1253/wiki/images/b/ba/Preprint-SPP1253-154.pdf`.

[BL87] D. J. Bender and A. J. Laub. The linear-quadratic optimal regulator for descriptor systems. *IEEE Trans. Automat. Control*, AC-32(8):672–688, 1987.

[BB97] P. Benner and R. Byers. Disk functions and their relationship to the matrix sign function. In *Proc. European Control Conference*, Waterloo, Belgium, 1997. BELWARE Information Technology. Paper 936.

[BBMX99] P. Benner, R. Byers, V. Mehrmann, and H. Xu. Numerical computation of deflating subspaces of embedded Hamiltonian pencils, 1999. Unpublished report.

[BBMX02] P. Benner, R. Byers, V. Mehrmann, and H. Xu. Numerical computation of deflating subspaces of skew-Hamiltonian/Hamiltonian pencils. *SIAM J. Matrix Anal. Appl.*, 24(1):165–190, 2002.

[BLMV14] P. Benner, P. Losse, V. Mehrmann, and M. Voigt. Numerical linear algebra methods for linear differential-algebraic equations. MATHEON-Preprint 1074, DFG-Forschungszentrum MATHEON, 2014.

[BMS05] P. Benner, V. Mehrmann, and D. Sorensen, editors. *Dimension Reduction of Large-Scale Systems*, volume 45 of *Lect. Notes Comput. Sci. Eng.* Springer-Verlag, Berlin, Heidelberg, New York, 2005.

[BQO99] P. Benner and E. S. Quintana-Ortí. Solving stable generalized Lyapunov equations with the matrix sign function. *Numer. Algorithms*, 20(1):75–100, 1999.

[BSV12a] P. Benner, V. Sima, and M. Voigt. $\mathcal{L}_\infty$-norm computation for continuous-time descriptor systems using structured matrix pencils. *IEEE Trans. Automat. Control*, 57(1):233–238, 2012.

[BSV12b] P. Benner, V. Sima, and M. Voigt. Robust and efficient algorithms for $\mathcal{L}_\infty$-norm computation for descriptor systems. In *Proc. 7th IFAC Symposium on Robust Control Design*, pages 195–200, Aalborg, Denmark, 2012.

[BSV13a] P. Benner, V. Sima, and M. Voigt. FORTRAN 77 subroutines for the solution of skew-Hamiltonian/Hamiltonian eigenproblems – Part I: Algorithms and applications. Preprint MPIMD/13-11, Max Planck Institute Magdeburg, 2013. Available from `http://www.mpi-magdeburg.mpg.de/preprints/2013/11/`.

[BSV13b] P. Benner, V. Sima, and M. Voigt. FORTRAN 77 subroutines for the solution of skew-Hamiltonian/Hamiltonian eigenproblems – Part II: Implementation and numerical results. Preprint MPIMD/13-12, Max Planck Institute Magdeburg, 2013. Available from `http://www.mpi-magdeburg.mpg.de/preprints/2013/12/`.

[BV11] P. Benner and M. Voigt. On the computation of particular eigenvectors of Hamiltonian matrix pencils. *Proc. Appl. Math. Mech.*, 11(1):753–754, 2011.

[BV12a] P. Benner and M. Voigt. $\mathcal{H}_\infty$-norm computation for large and sparse descriptor systems. *Proc. Appl. Math. Mech.*, 12(1):797–800, 2012.

[BV12b] P. Benner and M. Voigt. Numerical computation of structured complex stability radii of large-scale matrices and pencils. In *Proc. 51st IEEE Conference on Decision and Control*, pages 6560–6565, Maui, HI, USA, 2012.

[BV13] P. Benner and M. Voigt. Spectral characterization and enforcement of negative imaginariness for descriptor systems. *Linear Algebra Appl.*, 439(4):1104–1129, 2013.

[BV14] P. Benner and M. Voigt. A structured pseudospectral method for $\mathcal{H}_\infty$-norm computation of large-scale descriptor systems. *Math. Control Signals Systems*, 26(2):303–338, 2014.

[Ber13] T. Berger. *On Differential-Algebraic Control Systems.* Dissertation, Fakultät für Mathematik und Naturwissenschaften, TU Ilmenau, July 2013.

[BIR12] T. Berger, A. Ilchmann, and T. Reis. Zero dynamics and funnel control of linear differential-algebraic systems. *Math. Control Signals Systems*, 24(3):219–263, 2012.

[BIT12] T. Berger, A. Ilchmann, and S. Trenn. The quasi-Weierstraß form for regular matrix pencils. *Linear Algebra Appl.*, 436:4052–4069, 2012.

[BR13] T. Berger and T. Reis. Controllability of linear differential-algebraic equations – a survey. In A. Ilchmann and T. Reis, editors, *Surveys in Differential-Algebraic Equations I*, Differential-Algebraic Equations Forum, pages 1–61. Springer-Verlag, Berlin, Heidelberg, 2013.

[BT12] T. Berger and S. Trenn. The quasi-Kronecker form for matrix pencils. *SIAM J. Matrix Anal. Appl.*, 33(2):336–368, 2012.

[BT13] T. Berger and S. Trenn. Addition to "The quasi-Kronecker form for matrix pencils". *SIAM J. Matrix Anal. Appl.*, 34(1):94–101, 2013.

[BT14] T. Berger and S. Trenn. Kalman controllability decompositions for differential-algebraic systems. *Systems Control Lett.*, 71:54–61, 2014.

[BCM12] L. T. Biegler, S. L. Campbell, and V. Mehrmann, editors. *Control and Optimization with Differential-Algebraic Constraints.* Adv. Des. Control. SIAM, Philadelphia, PA, 2012.

[BGD92] A. I. Bojanczyk, G. H. Golub, and P. Van Dooren. The periodic Schur decomposition. Algorithms and applications. In F. T. Luk, editor, *Advanced Signal Processing Algorithms, Architectures, and Implementations III*, volume 1770 of *Proc. SPIE*, pages 31–42, 1992.

[BB90a] P. M. M. Bongers and O. H. Bosgra. Low order robust $H\infty$ controller synthesis. In *Proc. 29th Conference on Decision and Control*, pages 194–199, Honolulu, HI, USA, 1990.

[BB90b] S. Boyd and V. Balakrishnan. A regularity result for the singular values of a transfer matrix and a quadratically convergent algorithm for computing its $L_\infty$-norm. *Systems Control Lett.*, 15(1):1–7, 1990.

[BBK89] S. Boyd, V. Balakrishnan, and P. Kabamba. A bisection method for computing the $H_\infty$ norm of a transfer matrix and related problems. *Math. Control Signals Systems*, 2(3):207–219, 1989.

[BS90] N. A. Bruinsma and M. Steinbuch. A fast algorithm to compute the $H_\infty$-norm of a transfer function matrix. *Systems Control Lett.*, 14(4):287–293, 1990.

[Brü11a] T. Brüll. Checking dissipativity of linear behavior systems given in kernel representation. *Math. Control Signals Systems*, 23(1–3):159–175, 2011.

[Brü11b] T. Brüll. *Dissipativity of linear quadratic systems.* Dissertation, Institut für Mathematik, Technische Universität Berlin, March 2011.

[BM07] T. Brüll and V. Mehrmann. STCSSP: A FORTRAN 77 routine to compute a structured staircase form for a (skew-)symmetric/(skew-)-symmetric matrix pencil. Preprint 31-2007, Institut für Mathematik, TU Berlin, 2007. Available from `http://www3.math.tu-berlin.de/preprints/files/Preprint-31-2007.pdf`.

[BS13] T. Brüll and C. Schröder. Dissipativity enforcement via perturbation of para-Hermitian pencils. *IEEE Trans. Circuits Syst. I. Regul. Pap.*, 60(1):164–177, 2013.

[BGMN92] A. Bunse-Gerstner, V. Mehrmann, and N. K. Nichols. Regularization of descriptor systems by derivative and proportional state feedback. *SIAM J. Matrix Anal. Appl.*, 13(1):46–67, 1992.

[BGMN94] A. Bunse-Gerstner, V. Mehrmann, and N. K. Nichols. Regularization of descriptor systems by output feedback. *IEEE Trans. Automat. Control*, 39(8):1742–1748, 1994.

[BLO03] J. V. Burke, A. S. Lewis, and M. L. Overton. Robust stability and a criss-cross algorithm for pseudospectra. *IMA J. Numer. Anal.*, 23(3):359–375, 2003.

[Bye88] R. Byers. A bisection method for measuring the distance of a stable matrix to the unstable matrices. *SIAM J. Sci. Stat. Comput.*, 9(5):875–881, 1988.

[BMX07] R. Byers, V. Mehrmann, and H. Xu. A structured staircase algorithm for skew-symmetric/symmetric pencils. *Electron. Trans. Numer. Anal.*, 26:1–33, 2007.

[BN93] R. Byers and N. K. Nichols. On the stability radius of generalized state-space systems. *Linear Algebra Appl.*, 188–189:113–134, 1993.

[BI84] C. I. Byrnes and A. Isidori. A frequency domain philosophy for nonlinear systems, with applications to stabilization and to adaptive control. In *Proc. 23rd IEEE Conference on Decision and Control*, pages 1569–1573, Las Vegas, NV, USA, 1984.

[CH10] C. Cai and G. Hagen. Coupling of stable subsystems with counterclockwise input-output dynamics. In *Proc. 2010 American Control Conference*, pages 3458–3463, Baltimore, MD, USA, 2010.

[CF07] M. K. Camlibel and R. Frasca. Extension of the Kalman-Yakubovich-Popov lemma to descriptor systems. In *Proc. 46th IEEE Conference on Decision and Control*, pages 1094–1099, New Orleans, LA, USA, 2007.

[Cam95] S. L. Campbell. Linearization of DAEs along trajectories. *Z. Angew. Math. Phys.*, 46(1):70–84, 1995.

[CKM12] S. L. Campbell, P. Kunkel, and V. Mehrmann. Regularization of linear and nonlinear descriptor systems. In Biegler et al. [BCM12], chapter 2, pages 17–36.

[CD02] Y. Chahlaoui and P. Van Dooren. A collection of benchmark examples for model reduction of linear time invariant dynamical systems. SLICOT Working Note 2002-2, NICONET e. V., 2002. Available from `http://slicot.org/objects/software/reports/SLWN2002-2.ps.gz`.

[CGD04] Y. Chahlaoui, K. Gallivan, and P. Van Dooren. $\mathcal{H}_\infty$-norm calculations of large sparse systems. In *Proc. 16th International Symposium of Mathematical Theory of Networks and Systems*, Leuven, Belgium, 2004.

[CGVD07] Y. Chahlaoui, K. Gallivan, and P. Van Dooren. Calculating the $\mathcal{H}_\infty$ norm of a large sparse system via Chandrasekhar iterations and extrapolation. *ESAIM Proc.*, 20:83–92, 2007.

[Cle00] D. J. Clements. A state-space approach to indefinite spectral factorization. *SIAM J. Matrix. Anal. Appl.*, 21(3):743–767, 2000.

[CALM97] D. J. Clements, B. D. O. Anderson, A. J. Laub, and J. B. Matson. Spectral factorization with imaginary-axis zeros. *Linear Algebra Appl.*, 250:225–252, 1997.

[CG89] D. J. Clements and K. Glover. Spectral factorization via Hermitian pencils. *Linear Algebra Appl.*, 122–124:797–846, 1989.

[Dai89] L. Dai. *Singular Control Systems*, volume 118 of *Lecture Notes in Control and Inform. Sci.* Springer-Verlag, Heidelberg, 1989.

[DK87] J. Demmel and B. Kågström. Computing stable eigendecompositions of matrix pencils. *Linear Algebra Appl.*, 88–89:139–186, 1987.

[DK93a] J. Demmel and B. Kågström. The generalized Schur decomposition of an arbitrary pencil $A - \lambda B$: robust software with error bounds and applications. Part I: Theory and algorithms. *ACM Trans. Math. Softw.*, 19(2):160–174, 1993.

[DK93b] J. Demmel and B. Kågström. The generalized Schur decomposition of an arbitrary pencil $A - \lambda B$: robust software with error bounds and applications. Part II: Software and applications. *ACM Trans. Math. Softw.*, 19(2):175–201, 1993.

[dHSdCB95] P. M. J. Van den Hof, R. J. P. Schrama, R. A. de Callefon, and O. H. Bosgra. Identification of normalized coprime plant factors from closed loop experimental data. *Eur. J. Control*, 1(1):62–74, 1995.

[Doo83] P. Van Dooren. Reducing subspaces: Definitions, properties and algorithms. In B. Kågström and A. Ruhe, editors, *Matrix Pencils*, Lecture Notes in Control and Inform. Sci., pages 58–73. Springer-Verlag, Heidelberg, 1983.

[Du08] N. H. Du. Stability radii of differential algebraic equations with structured perturbations. *Systems Control Lett.*, 57(7):546–553, 2008.

[DLM13] N. H. Du, V. H. Linh, and V. Mehrmann. Robust stability of differential-algebraic equations. In A. Ilchmann and T. Reis, editors, *Surveys in Differential-Algebraic Equations I*, Differential-Algebraic Equations Forum, chapter 2, pages 63–95. Springer-Verlag, Berlin, Heidelberg, 2013.

[DTL11] N. H. Du, D. D. Thuan, and N. C. Liem. Stability radius of implicit dynamic equations with constant coefficients on time scales. *Systems Control Lett.*, 60(8):596–603, 2011.

[EEK97] A. Edelmann, E. Elmroth, and B. Kågström. A geometric approach to perturbation theory of matrices and matrix pencils. Part I: versal deformations. *SIAM J. Matrix Anal. Appl.*, 18(3):653–692, 1997.

[EEK99] A. Edelmann, E. Elmroth, and B. Kågström. A geometric approach to perturbation theory of matrices and matrix pencils. Part II: a stratifcation-enhanced staircase algorithm. *SIAM J. Matrix Anal. Appl.*, 20(3):667–699, 1999.

[ESF98] E. Eich-Soellner and C. Führer. *Numerical Methods in Multibody Dynamics.* B. G. Teubner, Stuttgart, 1998.

[FBK13] L. Feng, P. Benner, and J. G. Korvink. System-level modeling of MEMS by means of model order reduction (mathematical approximations) – mathematical background. In T. Bechthold, G. Schrag, and L. Feng, editors, *System-Level Modeling of MEMS*, chapter 3, pages 53–94. Wiley-VCH, 2013.

[Fra87] B. A. Francis. *A Course in $H_\infty$ Control Theory*, volume 88 of *Lecture Notes in Control and Inform. Sci.* Springer-Verlag, Heidelberg, 1987.

[FSD14] M. A. Freitag, A. Spence, and P. Van Dooren. Calculating the $H_\infty$-norm using the implicit determinant method. *SIAM J. Matrix Anal. Appl.*, 35(2):619–634, 2014.

[FRM08] F. Freitas, J. Rommes, and N. Martins. Gramian-based reduction method applied to large sparse power system descriptor models. *IEEE Trans. Power Syst.*, 23(3):1258–1270, 2008.

[FJ04] R. W. Freund and F. Jarre. An extension of the positive real lemma to descriptor systems. *Optim. Methods Softw.*, 19(1):69–87, 2004.

[Gan59] F. R. Gantmacher. *The Theory of Matrices II.* Chelsea Publishing Company, New York, 1959.

[Gee89] T. Geerts. *Structure of Linear-Quadratic Control.* Proefschrift, Technische Universiteit Eindhoven, 1989.

[Gee93] T. Geerts. Regularity and singularity in linear-quadratic control subject to implicit continuous-time systems. *Circuits Systems Signal Process.*, 13(1):19–30, 1993.

[Gee94] T. Geerts. Linear-quadratic control with and without stability subject to general implicit continuous-time systems: Coordinate-free interpretations of the optimal costs in terms of dissipation inequality and linear matrix inequality; existence and uniqueness of optimal controls and state trajectories. *Linear Algebra Appl.*, 203–204:607–658, 1994.

[GDV98] Y. Genin, P. Van Dooren, and V. Vermaut. Convergence of the calculation of $\mathcal{H}_\infty$ norms and related questions. In *Proc. 13th Symposium on Mathematical Theory of Networks and Systems*, pages 429–432, Padova, Italy, 1998.

[Geo88] T. T. Georgiou. On the computation of the gap metric. *Systems Control Lett.*, 11(4):253–257, 1988.

[GL96] G. H. Golub and C. F. Van Loan. *Matrix Computations.* The John Hopkins University Press, Baltimore, MD, 3rd edition, 1996.

[GKK03] R. Granat, B. Kågström, and D. Kressner. Computing periodic deflating subspaces associated with a specified set of eigenvalues. *BIT*, 43(1):1–18, 2003.

[Gre88] M. Green. On inner-outer factorization. *Systems Control Lett.*, 11(2):93–97, 1988.

[GM86] E. Griepentrog and R. März. *Differential-Algebraic Equations and Their Numerical Treatment*, volume 88 of *Teubner-Texte Math.* B. G. Teubner, Leipzig, 1986.

[GT04] S. Grivet-Talocia. Passivity enforcement via perturbation of Hamiltonian matrices. *IEEE Trans. Circuits Syst. I. Regul. Pap.*, 51(9):1755–1769, 2004.

[GTU06] S. Grivet-Talocia and A. Ubolli. On the generation of large passive macromodels for complex interconnect structures. *IEEE Trans. Adv. Packaging*, 29(1):39–54, 2006.

[GJH+13] S. Grundel, L. Jansen, N. Hornung, P. Benner, T. Clees, and C. Tischendorf. Model order reduction of differential algebraic equations arising from the simulation of gas transport networks. Preprint MPIMD/13-09, Max Planck Institute Magdeburg, 2013. Available from http://www2.mpi-magdeburg.mpg.de/preprints/2013/09/.

[GGO13] N. Guglielmi, M. Gürbüzbalaban, and M. L. Overton. Fast approximation of the $H_\infty$ norm via optimization of spectral value sets. *SIAM J. Matrix Anal. Appl.*, 34(2):709–737, 2013.

[GKL14a] N. Guglielmi, D. Kressner, and C. Lubich. Computing extremal points of symplectic pseudospectra and solving symplectic matrix nearness problems. *SIAM J. Matrix Anal. Appl.*, 35(2):1407–1428, 2014.

[GKL14b] N. Guglielmi, D. Kressner, and C. Lubich. Low rank differential equations for Hamiltonian matrix nearness problems. *Numer. Math.*, 2014. Also available from http://link.springer.com/article/10.1007%2Fs00211-014-0637-x.

[GL13] N. Guglielmi and C. Lubich. Low-rank dynamics for computing extremal points of real pseudospectra. *SIAM J. Matrix Anal. Appl.*, 34(1):40–66, 2013.

[GO11] N. Guglielmi and M. L. Overton. Fast algorithms for the approximation of the pseudospectral abscissa and pseudospectral radius of a matrix. *SIAM J. Matrix Anal. Appl.*, 32(4):1166–1192, 2011.

[GO13] N. Guglielmi and M. L. Overton. Local convergence analysis of [GGO13] – private communication with M. L. Overton, 2013.

[Ham82] S. J. Hammarling. Numerical solution of the stable, non-negative definite Lyapunov equation. *IMA J. Numer. Anal.*, 2(3):303–323, 1982.

[HS83] M. L. J. Hautus and L. M. Silverman. System structure and singular control. *Linear Algebra Appl.*, 50:369–402, 1983.

[Hay68] E. V. Haynsworth. Determination of the inertia of a partitioned Hermitian matrix. *Linear Algebra Appl.*, 1(1):73–81, 1968.

[HL94] J. J. Hench and A. J. Laub. Numerical solution of the discrete-time periodic Riccati equation. *IEEE Trans. Automat. Control*, 39(6):1197–1210, 1994.

[HM80] D. J. Hill and P. J. Moylan. Dissipative dynamical systems: basic input-output and state properties. *J. Franklin Inst.*, 309(5):327–357, 1980.

[HP86a] D. Hinrichsen and A. J. Pritchard. Stability radii of linear systems. *Systems Control Lett.*, 7(1):1–10, 1986.

[HP86b] D. Hinrichsen and A. J. Pritchard. Stability radius for structured perturbations and the algebraic Riccati equation. *Systems Control Lett.*, 8(2):105–113, 1986.

[HP90] D. Hinrichsen and A. J. Pritchard. Real and complex stability radii: A survey. In D. Hinrichsen and W. Bengt, editors, *Control of Uncertain Systems*, volume 6 of *Progress in Systems and Control Theory*, pages 119–162. Birkhäuser, Boston, 1990.

[HQ97] T. Hu and L. Qiu. Complex and real performance radii and their computation. *Internat. J. Robust Nonlinear Control*, 7(2):187–209, 1997.

[HQ98] T. Hu and L. Qiu. On structured perturbation of Hermitian matrices. *Linear Algebra Appl.*, 275–276:287–314, 1998.

[Ilc93] A. Ilchmann. *Non-Identifier-Based High-Gain Adaptive Control*, volume 189 of *Lecture Notes in Control and Inform. Sci.* Springer-Verlag, London, 1993.

[IR14] A. Ilchmann and T. Reis. Outer transfer functions of differential-algebraic systems. Hamburger Beiträge zur angewandten Mathematik 2014-19, Universität Hamburg, Fachbereich Mathematik, 2014. Submitted, available from `http://preprint.math.uni-hamburg.de/public/papers/hbam/hbam2014-19.pdf`.

[IW13] A. Ilchmann and F. Wirth. On minimum phase. *at-Automatisierungstechnik*, 61(12):805–817, 2013.

[IOW99] V. Ionescu, C. Oară, and M. Weiss. *Generalized Riccati Theory and Robust Control: A Popov Function Approach.* John Wiley & Sons Ltd., Chichester, 1999.

[JV13] P. Jiang and M. Voigt. MB04BV – A FORTRAN 77 subroutine to compute the eigenvectors associated to the purely imaginary eigenvalues of skew-Hamiltonian/Hamiltonian matrix pencils. SLICOT Working Note 2013-3, NICONET e. V., 2013. Available from `http://slicot.org/objects/software/reports/SLWN2013_3.pdf`.

[KD92] B. Kågström and P. Van Dooren. A generalized state-space approach for the additive decomposition of a transfer matrix. *J. Numer. Linear Algebra Appl.*, 1(2):165–181, 1992.

[KW89] B. Kågström and L. Westin. Generalized Schur methods with condition estimators for solving the generalized Sylvester equation. *IEEE Trans. Automat. Control*, 34(7):745–751, 1989.

[Kai80] T. Kailath. *Linear Systems.* Prentice Hall, Englewood Cliffs, NJ, 1980.

[Kal63] R. E. Kalman. Lyapunov functions for the problem of Lur'e in automatic control. *Proc. Natl. Acad. Sci. USA*, 49:201–205, 1963.

[KK02] A. Kawamoto and T. Katayama. The semi-stabilizing solution of generalized algebraic Riccati equation for descriptor systems. *Automatica*, 38(10):1651–1662, 2002.

[KTK99] A. Kawamoto, K. Takaba, and T. Katayama. On the generalized algebraic Riccati equation for continuous-time descriptor systems, 1999.

[Kre01] D. Kressner. An efficient and reliable implementation of the periodic QZ algorithm. In *Proc. IFAC Workshop on Periodic Control Systems*, 2001.

[KM01] P. Kunkel and V. Mehrmann. Analysis of over- and underdetermined nonlinear differential-algebraic systems with application to nonlinear control problems. *Math. Control Signals Systems*, 14(3):233–256, 2001.

[KM06] P. Kunkel and V. Mehrmann. *Differential-Algebraic Equations. Analysis and Numerical Solution.* EMS Publishing House, Zürich, Switzerland, 2006.

[KM08] P. Kunkel and V. Mehrmann. Optimal control for unstructured nonlinear differential-algebraic equations of arbitrary index. *Math. Control Signals Systems*, 20(3):227–269, 2008.

[KM11] P. Kunkel and V. Mehrmann. Formal adjoints of linear DAE operators and their role in optimal control. *Electron. J. Linear Algebra*, 22:672–693, 2011.

[KMR01] P. Kunkel, V. Mehrmann, and W. Rath. Analysis and numerical solution of control problems in descriptor form. *Math. Control Signals Systems*, 14(1):29–61, 2001.

[KMS14] P. Kunkel, V. Mehrmann, and L. Scholz. Self-adjoint differential-algebraic equations. *Math. Control Signals Systems*, 26(1):47–76, 2014.

[KM04] G. A. Kurina and R. März. On linear-quadratic optimal control problems for time-varying descriptor systems. *SIAM J. Control Optim.*, 42(6):2062–2077, 2004.

[LMT13] R. Lamour, R. März, and C. Tischendorf. *Differential-Algebraic Equations: A Projector Based Approach.* Differential-Algebraic Equations Forum. Springer-Verlag, Berlin, Heidelberg, 2013.

[LR95] P. Lancester and L. Rodman. *Algebraic Riccati Equations.* Oxford University Press, New York, 1995.

[LP08] A. Lanzon and I. R. Petersen. Stability robustness of a feedback interconnection of systems with negative imaginary frequency response. *IEEE Trans. Automat. Control*, 53(4):1042–1046, 2008.

[LTD00] C. T. Lawrence, A. L. Tits, and P. Van Dooren. A fast algorithm for the computation of an upper bound on the $\mu$-norm. *Automatica*, 36(3):449–456, 2000.

[LA10] S. Lefteriu and A. C. Antoulas. A new approach to modeling multiport systems from frequency-domain data. *IEEE Trans. Comput. Aided Des. Integr. Circuits Syst.*, 29(1):14–27, 2010.

[Lei04] F. Leibfritz. COMPl$_e$ib: COnstraint Matrix-optimization Problem library – a collection of test examples for nonlinear semidefinite programs, control system design and related problems. Technical report, 2004. Available from http://www.friedemann-leibfritz.de/COMPlib_Data/COMPlib_Main_Paper.pdf.

[LL04] F. Leibfritz and W. Lipinski. COMPl$_e$ib 1.0 – user manual and quick reference. Technical report, 2004. Available from http://www.friedemann-leibfritz.de/COMPlib_Data/COMPlib_User_Guide.pdf.

[LV13] R. Lowe and M. Voigt. $\mathcal{L}_\infty$-norm computation for large-scale descriptor systems using structured iterative eigensolvers. Preprint MPIMD/13-20, Max Planck Institute Magdeburg, 2013. Available from http://www.mpi-magdeburg.mpg.de/preprints/2013/20/.

[MKPL11a] M. Mabrok, A. G. Kallapur, I. R. Petersen, and A. Lanzon. Enforcing a system model to be negative imaginary via perturbation of Hamiltonian matrices. In *Proc. 50th IEEE Conference on Decision and Control and European Control Conference*, pages 3748–3752, Orlando, FL, USA, 2011.

[MKPL11b] M. Mabrok, A. G. Kallapur, I. R. Petersen, and A. Lanzon. Spectral conditions for the negative imaginary property of transfer function matrices. In *Proc. 18th IFAC World Congress*, pages 1302–1306, Milan, Italy, 2011.

[MKPL12] M. Mabrok, A. G. Kallapur, I. R. Petersen, and A. Lanzon. Generalized negative imaginary lemma for descriptor systems. *J. Mechanics Engineering and Automation*, 2(1):17–22, 2012.

[MPR07] N. Martins, P. C. Pellanda, and J. Rommes. Computation of transfer function dominant zeros with applications to oscillation damping control of large power systems. *IEEE Trans. Power Syst.*, 22(4):1657–1664, 2007.

[Mas06] I. Masubuchi. Dissipativity inequalities for continuous-time descriptor systems with applications to synthesis of control gains. *Systems Control Lett.*, 55(2):158–164, 2006.

[MG89] D. C. McFarlane and K. Glover. *Robust Controller Design Using Normalized Coprime Factor Descriptions*, volume 138 of *Lecture Notes in Control and Inform. Sci.* Springer-Verlag, Berlin, 1989.

[MS14] V. Mehrmann and L. Scholz. Self-conjugate differential and difference operators arising in the optimal control of descriptor systems. *Oper. Matrices*, 8(3):659–682, 2014.

[MSS12] V. Mehrmann, C. Schröder, and V. Simoncini. An implicitly-restarted Krylov subspace method for real symmetric/skew-symmetric eigenproblems. *Linear Algebra Appl.*, 436:4070–4087, 2012.

[MS05] V. Mehrmann and T. Stykel. Balanced truncation model reduction for large-scale systems in descriptor form. In Benner et al. [BMS05], chapter 3, pages 89–116.

[Meh91] V. L. Mehrmann. *The Autonomous Linear Quadratic Control Problem*, volume 163 of *Lecture Notes in Control and Inform. Sci.* Springer-Verlag, Heidelberg, 1991.

[Mey88] D. G. Meyer. A fractional approach to model reduction. In *Proc. 1988 American Control Conference*, pages 1041–1047, Atlanta, GA, USA, 1988.

[Mit14] T. Mitchell. *Robust and efficient methods for approximation and optimization of stability measures.* PhD thesis, Courant Institute of Mathematical Sciences, New York University, September 2014.

[MO14] T. Mitchell and M. L. Overton. Fast approximation of the $H_\infty$-norm via hybrid expansion-contraction using spectral value sets, 2014. Submitted, also available from `http://cims.nyu.edu/~tmitchell/papers/submitted/hinf_hec_submitted.pdf`.

[MRS11] J. Möckel, T. Reis, and T. Stykel. Linear-quadratic Gaussian balancing for model reduction of differential-algebraic systems. *Internat. J. Control*, 84(10):1627–1643, 2011.

[OV00] C. Oară and A. Varga. Computation of general inner-outer and spectral factorizations. *IEEE Trans. Automat. Control*, 45(12):2307–2325, 2000.

[OD05] M. L. Overton and P. Van Dooren. On computing the complex passivity radius. In *Proc. 44th IEEE Conference Decision and Control and European Control Conference*, pages 7960–7964, Seville, Spain, 2005.

[PL10] I. R. Petersen and A. Lanzon. Feedback control of negative-imaginary systems. *IEEE Control Syst. Mag.*, 30(5):54–72, 2010.

[PR11] F. Poloni and T. Reis. A structured doubling algorithm for the numerical solution of Lur'e equations. MATHEON-Preprint 758, DFG-Forschungszentrum MATHEON, 2011. Available from `https://opus4.kobv.de/opus4-matheon/frontdoor/index/index/docId/748`.

[PR12] F. Poloni and T. Reis. A deflation approach for large-scale Lur'e equations. *SIAM J. Matrix Anal. Appl.*, 33(4):1339–1368, 2012.

[Pop62] V. M. Popov. Absolute stability of nonlinear systems of automatic control. *Autom. Remote Control*, 22:857–875, 1962. Russian original in August 1961.

[QBR+95] L. Qiu, B. Bernhardsson, A. Rantzer, E. J. Davison, P. M. Young, and J. C. Doyle. A formula for computation of the real stability radius. *Automatica*, 31(6):879–890, 1995.

[Ran96] A. Rantzer. On the Kalman-Yakubovich-Popov lemma. *Systems Control Lett.*, 28(1):7–10, 1996.

[Rei10] T. Reis. Circuit synthesis of passive descriptor systems – a modified nodal approach. *Int. J. Circ. Theor. Appl.*, 38(1):44–68, 2010.

[Rei11] T. Reis. Lur'e equations and even matrix pencils. *Linear Algebra Appl.*, 434:152–173, 2011.

[Rei14] T. Reis. Mathematical modeling and analysis of nonlinear time-invariant RLC circuits. In P. Benner, R. Findeisen, D. Flockerzi, U. Reichl, and K. Sundmacher, editors, *Large-Scale Networks in Engineering and Life Sciences*, Modeling and Simulation in Science, Engineering and Technology, chapter 2, pages 125–198. 2014.

[RRV14] T. Reis, O. Rendel, and M. Voigt. The Kalman-Yakubovich-Popov inequality for differential-algebraic systems. Hamburger Beiträge zur Angewandten Mathematik 2014-27, Fachbereich Mathematik, Universität Hamburg, 2014. Submitted, available from `http://preprint.math.uni-hamburg.de/public/papers/hbam/hbam2014-27.pdf`.

[RS10a] T. Reis and T. Stykel. PABTEC: Passivity-preserving balanced truncation for electrical circuits. *IEEE Trans. Computer-Aided Design Integr. Circuits Syst.*, 29(9), 2010.

[RS10b] T. Reis and T. Stykel. Positive real and bounded real balancing for model reduction of descriptor systems. *Internat. J. Control*, 83(1):74–88, 2010.

[RS11] T. Reis and T. Stykel. Lyapunov balancing for passivity-preserving model reduction of RC circuits. *SIAM J. Appl. Dyn. Syst.*, 10(1):1–34, 2011.

[Ria08] R. Riaza. *Differential-Algebraic Systems. Analytical Aspects and Circuit Applications.* World Scientific Publishing Co. Pte. Ltd., Singapore, 2008.

[Rie94] K. S. Riedel. Generalized epsilon-pseudospectra. *SIAM J. Numer. Anal.*, 31(4):1219–1225, 1994.

[RS88] R. E. Roberson and R. Schwertassek. *Dynamics of Multibody Systems.* Springer-Verlag, Heidelberg, 1988.

[Rom08] J. Rommes. Arnoldi and Jacobi-Davidson methods for generalized eigenvalue problems $Ax = \lambda Bx$ with singular $B$. *Math. Comp.*, 77:995–1015, 2008.

[RM06a] J. Rommes and N. Martins. Efficient computation of multivariate transfer function dominant poles using subspace acceleration. *IEEE Trans. Power Syst.*, 21(4):1471–1483, 2006.

[RM06b] J. Rommes and N. Martins. Efficient computation of transfer function dominant poles using subspace acceleration. *IEEE Trans. Power Syst.*, 21(3):1218–1226, 2006.

[RS08] J. Rommes and G. L. G. Sleijpen. Convergence of the dominant pole algorithm and Rayleigh quotient iteration. *SIAM J. Matrix Anal. Appl.*, 30(1):346–363, 2008.

[Ros73] H. H. Rosenbrock. The zeros of a system. *Internat. J. Control*, 18(2):297–299, 1973.

[Sch91a] C. Scherer. The solution set of the algebraic Riccati equation and the algebraic Riccati inequality. *Linear Algebra Appl.*, 153:99–122, 1991.

[Sch91b] C. W. Scherer. *The Riccati Inequality and State-Space $H_\infty$-Optimal Control.* Dissertation, Fakultät für Mathematik und Informatik, Bayerische Julius Maximilians-Universität Würzburg, 1991.

[Sch13] C. Schröder. Private communication, 2013.

[SS07] C. Schröder and T. Stykel. Passivation of LTI systems. MATHEON-Preprint 368, DFG-Forschungszentrum MATHEON, 2007. Available from `https://opus4.kobv.de/opus4-matheon/frontdoor/index/index/docId/368`.

[Smi02] M. C. Smith. Synthesis of mechanical networks: The inerter. *IEEE Trans. Automat. Control*, 47(10):1648–1662, 2002.

[SDT96] J. Sreedhar, P. Van Dooren, and A. L. Tits. A fast algorithm to compute the real structured stability radius. volume 121 of *Internat. Ser. Numer. Math.*, pages 219–230. Springer-Verlag, 1996.

[SS90] G. W. Stewart and J.-G. Sun. *Matrix Perturbation Theory.* Academic Press, New York, 1990.

[Sty02] T. Stykel. On criteria for asymptotic stability of differential-algebraic equations. *ZAMM Z. Angew. Math. Mech.*, 82(3):147–158, 2002.

[Sty06] T. Stykel. On some norms for descriptor systems. *IEEE Trans. Automat. Control*, 51(5):842–847, 2006.

[SQO04] X. Sun and E. S. Quintana-Ortí. Spectral division methods for block generalized Schur decompositions. *Math. Comp.*, 73:1827–1847, 2004.

[Tho76] R. C. Thompson. The characteristic polynomial of a principal subpencil of a Hermitian matrix pencil. *Linear Algebra Appl.*, 14(2):135–177, 1976.

[Tho91] R. C. Thompson. Pencils of complex and real symmetric and skew matrices. *Linear Algebra Appl.*, 147:323–371, 1991.

[Tre89] H. L. Trentelman. The regular free-endpoint linear quadratic problem with indefinite cost. *SIAM J. Control Optim.*, 27(1):27–42, 1989.

[Tre99] H. L. Trentelman. When does the algebraic Riccati equation have a negative semi-definite solution? In V. D. Blondel, E. D. Sontag, M. Vidyasagar, and J. C. Willems, editors, *Open Problems in Mathematical Systems and Control Theory*, Comm. Control Engrg. Ser., chapter 44, pages 229–237. Springer-Verlag, London, 1999.

[TSH01] H. L. Trentelman, A. A. Stoorvogel, and M. Hautus. *Control Theory for Linear Systems*. Comm. Control Engrg. Ser. Springer-Verlag, London, 2001.

[Var98] A. Varga. Computation of normalized coprime factorizations of rational matrices. *Systems Control Lett.*, 33(1):37–45, 1998.

[VLK81] G. Verghese, B. Lévy, and T. Kailath. A generalized state-space for singular systems. *IEEE Trans. Automat. Control*, AC-26(4), 1981.

[WB89] F. Y. Wang and M. J. Balas. Doubly coprime fractional representations of generalized dynamical systems. *IEEE Trans. Automat. Control*, 34(7):733–744, 1989.

[Wei96] J. Weickert. Navier-Stokes equations as a differential-algebraic system. Technical Report SFB393/96-08, Technische Universität Chemnitz, Fakultät für Mathematik, 1996.

[Wie49] N. Wiener. *Extrapolation, Interpolation, and Smoothing of Stationary Time Series*. Wiley, New York, 1949.

[Wil71] J. C. Willems. Least squares stationary optimal control and the algebraic Riccati equation. *IEEE Trans. Automat. Control*, AC-16(6):621–634, 1971.

[Wil72a] J. C. Willems. Dissipative dynamical systems part I: General theory. *Arch. Ration. Mech. Anal.*, 45(5):321–351, 1972.

[Wil72b] J. C. Willems. Dissipative dynamical systems part II: Linear systems with quadratic supply rates. *Arch. Ration. Mech. Anal.*, 45(5):352–393, 1972.

[Wil74] J. C. Willems. On the existence of a nonpositive solution to the Riccati equation. *IEEE Trans. Automat. Control*, AC-19(5):592–593, 1974.

[WT98] J. C. Willems and H. L. Trentelman. On quadratic differential forms. *SIAM J. Control Optim.*, 36(5):1703–1749, 1998.

[Wri02] T. G. Wright. Eigtool, 2002. Available from `http://www.comlab.ox.ac.uk/pseudospectra/eigtool/`.

[XPL10] J. Xiong, I. R. Petersen, and A. Lanzon. A negative imaginary lemma and the stability of interconnections of linear negative imaginary systems. *IEEE Trans. Automat. Control*, 55(10):2342–2347, 2010.

[Yak62] V. A. Yakubovich. Solution of certain matrix inequalities in the stability theory of nonlinear control systems. *Dokl. Akad. Nauk. SSSR*, 143:1304–1307, 1962.

[You61] D. C. Youla. On the factorization of rational matrices. *IRE Trans. Inform. Theory*, IT-7:172–189, 1961.

[ZDG96] K. Zhou, J. C. Doyle, and K. Glover. *Robust and Optimal Control*. Prentice-Hall, Englewood Cliffs, NJ, 1996.

# Index

# Theses

1. We have developed a new version of the Kalman-Yakubovich-Popov lemma for DAEs which relates the positive semi-definiteness of a Popov function on the imaginary axis to the solvability of an LMI on a subspace (the descriptor KYP inequality). This new formulation eliminates the restrictions of previous approaches.

2. Resulting from the descriptor KYP inequality we have developed the descriptor Lur'e equation that generalizes the concepts of the algebraic Riccati and the Lur'e equation, in particular in the context of singular control problems for DAEs. We have shown that the solutions of this matrix equation realize the rank-minimizing solutions of the descriptor KYP inequality.

3. We have given criteria for the positive semidefiniteness of a Popov funtion on the imaginary axis in terms of the spectrum of an associated even matrix pencil. For this we have used the even Kronecker canonical form.

4. We have proved how to construct the solutions of the descriptor Lur'e equation using the deflating subspaces of the associated even matrix pencil.

5. We have shown that under the condition of strong stabilizability and strong anti-stabilizability, the descriptor Lur'e equation admits stabilizing and anti-stabilizing solutions. Moreover, these solution simultaneously define extremal solutions of the descriptor KYP inequality.

6. We have shown that each solution of the descriptor KYP inequality realizes a spectral factorization of the Popov function.

7. We have shown that under some conditions on controllability and the structure of the Popov function, the descriptor KYP inequality admits nonpositive solutions. Moreover, we have given a sufficient condition for nonpositivity of all solutions.

8. We have applied the new theory to the linear-quadratic optimal problem with zero terminal conditions. We have shown that the optimal control problem is feasible if and only if the associated descriptor Lur'e equation has a stabilizing solution. Moreover, we have proven conditions on the closed-loop system for the existence and uniqueness of optimal control signals.

9. We have further considered the linear-quadratic optimal control problem with free terminal conditions. For strongly controllable systems it turns out that the

optimal control problem is feasible if and only if the descriptor KYP inequality has a nonpositive solution.

10. We have applied the results for linear-quadratic optimal control for the characterization of dissipativity and cyclo-dissipativity of DAEs. Hereby, dissipativity is related to an optimal control problem with free terminal conditions, whereas cyclo-dissipativity corresponds to an optimal control problem with zero terminal condition.

11. We have considered important special cases of dissipativity, namely contractivity and passivity. These concepts are equivalent to bounded realness and positive realness of the transfer function, respectively. We have formulated and proven new versions of the bounded real lemma and the positive real lemma.

12. We have demonstrated how to construct normalized coprime factorizations by solving descriptor Lur'e equations.

13. We have further shown that inner-outer factorizations can be constructed by employing descriptor Lur'e equations.

14. We have studied systems with counterclockwise I/O dynamics and negative imaginary transfer functions. We have proven a characterization for negative imaginariness in terms of the spectral structure of an even matrix pencil.

15. We have developed an algorithm for enforcing negative imaginariness of a transfer function by perturbations of the system. This algorithm is based on moving all purely imaginary eigenvalues of a related skew-Hamiltonian/Hamiltonian matrix pencil off the imaginary axis. This procedure requires the corresponding eigenvectors that have been computed using a new structure-preserving method.

16. We have studied the influence of perturbations on cyclo-dissipativity of a system. We have developed a computational procedure for computing the distance of a cyclo-dissipative system to the set of non-cyclo-dissipative systems, the so-called cyclo-dissipativity radius. We have given conditions on the system under which the cyclo-dissipativity radius is equivalent to the dissipativity radius.

17. The computation of the cyclo-dissipativity radius is related to the structured perturbation analysis of an even matrix pencil. A particular challenge in this context is the influence of the perturbations on the singular part and the defective infinite eigenvalues. We have presented computable conditions to check whether such perturbations are possible. In order to compute the cyclo-dissipativity radius we have finally solved a sequence of eigenvalue optimization problems.

18. We have developed a method to compute the $\mathcal{H}_\infty$-norm of a large-scale descriptor system. We have generalized a relation between the complex $\mathcal{H}_\infty$-radius of a transfer function and its $\mathcal{H}_\infty$-norm. Particular care has to be taken of perturbations that make the transfer function improper or not well-defined. The algorithm itself is a nested process. In the inner iteration we compute the rightmost point of a certain pseudopole set by constructing a sequence of rank-1 perturbations. The outer iteration is a Newton process to drive this rightmost point to the imaginary axis. The usage of dominant poles improves the behavior of the method with respect to global optimization.

19. We have further modified the standard algorithm to compute the $\mathcal{H}_\infty$-norm based on even matrix pencils to make it applicable to large-scale problems. We have used structured eigensolvers such as the dominant pole algorithm and the even IRA algorithm to solve these issues.

20. Both algorithms for $\mathcal{H}_\infty$-norm computations work well, even for rather difficult examples, however one has to use good design-parameters in the implementation to solve some difficult problems. Moreover, none of the methods is guaranteed to perform a global optimization.

# Statement of Scientific Cooperations

This work is based on articles and reports (published and unpublished) that have been obtained in cooperation with various coauthors. To guarantee a fair assessment of this thesis, this statement clarifies the contributions that each individual coauthor has made. The following people contributed to the content of this work:

- Peter Benner (PB), Max Planck Institute for Dynamics of Complex Technical Systems;
- Timo Reis (TR), University of Hamburg;
- Peihong Jiang (PJ), University of Rochester;
- Ryan Lowe (RL), Queen's University.

## Chapter 3

This Chapter is based on a joint work with TR. The initial result we obtained is the Kalman-Yakubovich-Popov lemma (Theorem 3.3.1), based on an idea of TR. I have formulated a first draft of this theorem and its proof, where I initially only showed statement a) for impulse controllable systems. The final formulation of the theorem was done by TR.

The results of Sections 3.4 and 3.5 were mainly obtained by myself with some corrections and improvements by TR. Similarly as for Theorem 3.3.1, statement a) of Theorem 3.4.9 was initially only formulated and proven for impulse controllable systems, the generalization was done by TR. Finally, TR showed (a more general version of) Lemma 3.4.7 which led to shortened and better understandable proofs of Theorem 3.4.9 (shown by me) and Theorem 3.5.3 (shown by TR in a more general version) which I have included.

Section 3.6 is based on some remarks I wrote in an initial draft of this work. In particular, I made comments about the structure of the spectral factors. This part of the work was strongly improved by TR. In particular, the relation to outer systems was found by TR. Finally, I improved the formulation and the proof of Theorem 3.6.4 to let it also be valid for all solutions of the descriptor Lur'e equation (not only the stabilizing and anti-stabilizing ones).

The results of Section 3.7 were almost solely found by TR. Finally, I did corrections of this section and slightly improved the formulation.

Moreover, large parts of Section 3.8 were obtained by TR. In particular, this includes Theorems 3.8.3, 3.8.7, and the previous auxiliary lemmas, as well as the statements on the existence and uniqueness of the optimal control, i. e., Proposition 3.8.4 and Corollary 3.8.5, as well as Remark 3.8.6. An initial version of Theorem 3.8.9 was proven by me, with strong improvements by TR. Moreover, Lemma 3.8.8 was shown by TR. Furthermore, the results on "lossless" optimal control in Theorem 3.8.11 were shown by TR. On the other hand, the results for optimal control with free terminal condition were shown by me, i. e., those of Subsection 3.8.2.

An initial draft of Section 3.9 was written by TR. However, I performed a large amount of extensions and corrections. All statements about cyclo-dissipativity were added by me. Moreover, I added the remarks about Lyapunov functions and the pole locations of bounded real and positive real transfer functions. Finally, I rearranged the whole structure of the section to improve readability.

A first incomplete draft of Section 3.10 was written by TR. However, I completed the proof of Lemma 3.10.3 by applying the idea of feedback regularization. Furthermore, I proved how to construct left and doubly normalized coprime factorizations, i. e., statements b) and c) in Theorem 3.10.5 and all other statements that are related to these.

Finally, Section 3.11 was mainly obtained by TR with corrections and formulation improvements performed by myself.

## Chapter 4

This chapter is based on an idea by PB. All theoretical results of this chapter were obtained by me but proofread and improved by PB. I received further assistance with the implementation and testing of the algorithm for the computation of the eigenvectors of skew-Hamiltonian/Hamiltonian pencils by PJ during a summer internship under my guidance in 2012. Subsection 4.4.8 is based on her internship report that was written under my guidance. In particular, the computational results of this subsection were obtained by PJ.

## Chapter 5

The idea and results of this chapter were completely obtained by me.

## Chapter 6

The idea of Section 6.2 was inspired by PB by making me aware of [GO11]. All theoretical and computational results were obtained by myself, but proofread and improved by PB.

The idea of Section 6.3 was obtained by me. However, I received assistance by RL during a summer internship under my guidance in 2013. In particular, details of the algorithm like the choice of design parameters and the computational results were worked out by RL. This subsection is based on the internship report that was written under my guidance.

Magdeburg, 27th January 2015

Matthias Voigt

# Declaration of Honor

I hereby declare that I produced this thesis without prohibited assistance and that all sources of information that were used in producing this thesis, including my own publications, have been clearly marked and referenced.

In particular I have not wilfully:

- Fabricated data or ignored or removed undesired results.
- Misused statistical methods with the aim of drawing other conclusions than those warranted by the available data.
- Plagiarised data or publications or presented them in a disorted way.

I know that violations of copyright may lead to injunction and damage claims from the author or prosecution by the law enforcement authorities.

This work has not previously been submitted as a doctoral thesis in the same or a similar form in Germany or in any other country. It hast not previously been published as a whole.

# Schriftliche Ehrenerklärung

Ich versichere hiermit, dass ich die vorliegende Arbeit ohne unzulässige Hilfe Dritter und ohne Benutzung anderer als der angegebenen Hilfsmittel angefertigt habe; verwendete fremde und eigene Quellen sind als solche kenntlich gemacht.

Ich habe insbesondere nicht wissentlich:

- Ergebnisse erfunden oder widersprüchliche Ergebnisse verschwiegen,
- statistische Verfahren absichtlich missbraucht, um Daten in ungerechtfertigter Weise zu interpretieren,
- fremde Ergebnisse oder Veröffentlichungen plagiiert oder verzerrt wiedergegeben.

Mir ist bekannt, dass Verstöße gegen das Urheberrecht Unterlassungs- und Schadenersatzansprüche des Urhebers sowie eine strafrechtliche Ahndung durch die Strafverfolgungsbehörden begründen kann.

Die Arbeit wurde bisher weder im Inland noch im Ausland in gleicher oder ähnlicher Form als Dissertation eingereicht und ist als Ganzes auch noch nicht veröffentlicht.

Magdeburg, den 27. Januar 2015

---

Matthias Voigt